Room B016
Florham Park, NJ 07932-0971
(973) 360-8160

Springer Series in Statistics

Advisors:
D. Brillinger, S. Fienberg, J. Gani, J. Hartigan
J. Kiefer, K. Krickeberg

Pierre Brémaud

Point Processes and Queues
Martingale Dynamics

With 31 Illustrations

Springer-Verlag
New York   Heidelberg   Berlin

PIERRE BRÉMAUD
École Nationale Supérieure
  de Techniques Avancées
Paris
France

AMS Classification (1980): 60G55, 62L99, 62H99, 49D99, 90C99

Library of Congress Cataloging in Publication Data

Brémaud, Pierre.
  Point processes and queues, martingale dynamics.

  (Springer series in statistics)
  Bibliography: p.
  Includes index.
  1. Point processes. 2. Queueing theory. 3. Martin-
gales (Mathematics) I. Title. II. Series.
QA274.42.B73   1981   519.2   81-8791
ISBN 0-387-90536-7   AACR2

© 1981 by Springer-Verlag New York Inc.
All rights reserved. No part of this book may be translated or reproduced in any form without written permission from Springer-Verlag, 175 Fifth Avenue, New York, New York 10010, U.S.A.

Printed in the United States of America.

9 8 7 6 5 4 3 2 1

ISBN 0-387-90536-7   Springer-Verlag New York Heidelberg Berlin
ISBN 3-540-90536-7   Springer-Verlag Berlin Heidelberg New York

To my mother

Contents

Acknowledgments	xi
Introduction	xiii
Special Notations	xix

I Martingales — 1

1. Histories and Stopping Times — 1
2. Martingales — 3
3. Predictability — 8
4. Square-Integrable Martingales — 11
References — 12
Solutions to Exercises, Chapter I — 13

II Point Processes, Queues, and Intensities — 18

1. Counting Processes and Queues — 18
2. Watanabe's Characterization — 23
3. Stochastic Intensity, General Case — 27
4. Predictable Intensities — 30
5. Representation of Queues — 35
6. Random Changes of Time — 40
7. Cryptographic Point Processes — 43
References — 47
Solutions to Exercises, Chapter II — 48

III Integral Representation of Point-Process Martingales — 56

1. The Structure of Internal Histories — 56
2. Regenerative Form of the Intensity — 59
3. The Representation Theorem — 64
4. Hilbert-Space Theory of Poissonian Martingales — 70
5. Useful Extensions — 75
References — 76
Solutions to Exercises, Chapter III — 77

IV	Filtering	83
	1. The Theory of Innovations	83
	2. State Estimates for Queues and Markov Chains	100
	3. Continuous States and Nontrivial Prehistory	107
	References	115
	Solutions to Exercises, Chapter IV	115
V	Flows in Markovian Networks of Queues	122
	1. Single Station: The Historical Results and the Filtering Method	122
	2. Jackson's Networks	131
	3. Burke's Output Theorem for Networks	138
	4. Cascades and Loops in Jackson's Networks	143
	5. Independence and Poissonian Flows in Markov Chains	151
	References	154
	Solutions to Exercises, Chapter V	155
VI	Likelihood Ratios	158
	1. Radon–Nikodym Derivatives and Tests of Hypotheses	158
	2. Changes of Intensities "à la Girsanov"	165
	3. Filtering by the Method of the Probability of Reference	170
	4. Applications	174
	5. The Capacity of a Point-Process Channel	180
	6. Detection Formula	187
	References	189
	Solutions to Exercises, Chapter VI	190
VII	Optimal Control	196
	1. Modeling Intensity Controls	196
	2. Dynamic Programming for Intensity Controls: Complete-Observation Case	202
	3. Input Regulation. A Case Study in Impulsive Control	211
	4. Attraction Controls	219
	5. Existence via Likelihood Ratio	225
	References	229
	Solutions to Exercises, Chapter VII	230
VIII	Marked Point Processes	233
	1. Counting Measure and Intensity Kernels	233
	2. Martingale Representation and Filtering	238
	3. Radon–Nikodym Derivatives	241
	4. Towards a General Theory of Intensity	244
	References	250
	Solutions to Exercises, Chapter VIII	250

A1 Background in Probability and Stochastic Processes — 255

1. Introduction — 255
2. Monotone Class Theorem — 256
3. Random Variables — 261
4. Expectations — 266
5. Conditioning and Independence — 275
6. Convergence — 285
7. Stochastic Processes — 287
8. Markov Processes — 290
References — 295

A2 Stopping Times and Point-Process Histories — 296

1. Stopping Times — 296
2. Changes of Time and Meyer–Dellacherie's Integration Formula — 300
3. Point-Process Histories — 303
References — 311

A3 Wiener-Driven Dynamical Systems — 312

1. Ito's Stochastic Integral — 312
2. Square-Integrable Brownian Martingales — 321
3. Girsanov's Theorem — 327
References — 332

A4 Stieltjes–Lebesgue Calculus — 334

1. The Stieltjes–Lebesgue Integral — 334
2. The Product and Exponential Formulas — 336
References — 339

General Bibliography — 341

Index — 351

Acknowledgments

Jean Jacod and Marc Yor have showed great patience in explaining to me some fine technical points concerning the subject matter of this book. They spent a considerable amount of their time correcting the manuscript at every stage. I wish to express my sincere gratitude to them and to Marie Duflo who made useful comments on an early draft of the manuscript.

I am also indebted to Frederic Beutler, René Boel, Claude Dellacherie, Halim Doss, Raouf Jaïbi, and Jan van Schuppen who kindly accepted to check and correct portions of the book at the proof stage, as well as to Jacqueline Chappuis, Claudine Decalf, Raymonde Kurinckx, Chantal Pichon, Danielle Robiolle, Doris Simpson, Ruth Suzuki, and Jacqueline Vaquier who took part in typing and preparing the manuscript.

Introduction

Point-process models are widely used in a variety of fields: in communications theory (laser transmission at low power levels, photon detection), in operations research (systems of queues in computer networks, inventories), in nuclear medicine (radioactive tracers), in neurophysiology (spike trains along nervous fibers), and in any other domain of applications where a sequence of times—each time corresponding to the occurrence of some event—constitutes the observable data.

In applications involving point processes, one can roughly distinguish two schools (not disjoint). The first school, to which belong the neurophysiologists, favors the moment point of view and aims at fitting a model to given moment functions estimated from the collected data. Brillinger {26} gives a survey of the statistical literature on the subject of moment analysis and identification.

The other school describes its models by means of the *stochastic intensity*, which summarizes at a given instant the potential to generate an event in the near future, given some observation of the past including the complete record of all previous times of occurrences. The notion of intensity is a familiar one in the theory of Poisson or conditional Poisson processes. The corresponding models are very popular, mainly because of their analytical tractability. But stochastic intensity can be defined for a larger class of point processes, a class so large that it actually contains almost all the point processes of practical interest. Snyder {140} gives a number of illustrations of the classical intensity point of view, mainly in communication theory. Srinivasan {141} has examples in statistical physics.

Point processes have also received considerable attention from fundamental mathematicians, and a fairly general theory has been developed which views a point process as a discrete random measure. The book by Kerstan, Matthes, and Mecke {90} is the basic reference to date.

The approach in the present book is radically different from the above measure-theoretical approach. It was motivated by the need for a dynamical model which takes the information dynamics into account in a direct and effective manner, and by the necessity of controlling a system on line on the basis of the data collected at the time of implementation of the control. The ideal tool in this respect is *martingale theory*, which indeed fully acknowledges the existence of information patterns that increase with time, i.e. filtrations. Martingale theory is therefore the framework in which the notion of stochastic intensity will be formalized in this book.

Using the martingale point of view, the notion of stochastic intensity receives a rigorous mathematical definition, and, more importantly, the martingale definition of intensity is the basis on which is constructed a martingale calculus which has the same power as Ito calculus for diffusions: it allows a unified treatment of *dynamical point process systems* along the same lines as the theories previously developed for Wiener-driven stochastic systems.

Perhaps one of the most important achievements of the martingale approach is to be found in queueing theory. The traditional tools, such as Markov chains and renewal theory, are useful in design: one first computes the operational characteristics of different policies (priority, routing) and then chooses among the different strategies. This of course is possible if the analytical problem is tractable. The martingale point of view is adapted to problems of a dynamical nature (dynamical assignment of priorities, for instance) where the goal is not to assess different strategies, but to find the best one. Also some traditional topics, such as the output theory of markovian queuing systems, find a new and complete solution using the innovations theory of filtering.

The martingale approach is bound to benefit to the theory of point processes by connecting it to a theory (martingale theory) which is already at an advanced stage of development. Of special interest is the analogy with the martingale methods of the theory of Wiener-driven stochastic systems. Indeed, once the martingale modelling and its corollary the martingale calculus are established, the theories of *filtering* via the innovations method or via the probability of reference, the theory of *likelihood ratios*, and the theory of *dynamic programming* can be developed exactly in parallel with the martingale theory of white noise, which grew out of the works of Girsanov {64}, Duncan {53}, Kailath {81}, Beneš {3}, Fujisaki, Kallianpur, and Kunita {62}, and Davis and Varaiya {45}.

The emphasis has been placed on topics of interest in systems science at large, and the exposition has been set at a level of sophistication which seems to be the relevant one for applications. For instance, we have avoided the more theoretical presentation in terms of the modern theory of stochastic processes. We have deliberately restricted attention to point processes with an intensity, since they can then be introduced in convincing terms as a natural generalization of conditional Poisson processes; also the formulas look much more appealing in this case. However, in order to avoid frustrating the earnest reader, the general case is briefly considered in Chapter VIII, devoted to marked point processes and jump processes.

The material of this book has been taught by the author at the University of California at Berkeley, at the University of Paris IX, and at the Summer School organized yearly by Elja Arjas in Finland. The level of exposition and the inclusion of a large number of exercises with complete detailed solutions make this book usable as a text for graduate students in applied probability, electrical engineering, computer science, and operations research. The prerequisites in probability and random processes are recalled in the Appendices.

Dynamical Point-Process Systems: Historical Sketch

The historical account to follow is only concerned with the martingale approach to dynamical point-process systems; it does not include, for instance, the history of martingale theory. It is not intended to replace the General Bibliography, and therefore the only works cited here are those where a concept, a point of view, or a result of direct interest in this book has been introduced for the first time.

That point processes have something to do with martingales is not new: it was observed long ago that the compensated process $N_t - t$ associated to a standard Poisson process N_t is a martingale, and a square-integrable one. Such an illustration of the definition of a martingale is usually given in textbooks, if given at all, after the Wiener process. The interest of the mathematical community in martingales was centered on continuous martingales which possess exciting path behavior, and point processes only made brief and unnoticed appearances in the theoretical literature devoted to martingales. They were taken care of in the completely different settings of renewal theory or of random measure theory. However, the latter approaches are not dynamical by nature and did not help the applied mathematicians struggling with the notion of stochastic intensity, with models of the kind

$$P[dN_t = 1 | \mathscr{F}_t] = \lambda_t \, dt + o(dt). \qquad (*)$$

A definition such as (*) for the $\mathscr{F}_t$-intensity λ_t of a point process N_t (where $\mathscr{F}_t$ is a history of N_t, i.e., a record of the data available at time t, including the trajectory N_s, $s \in [o, t]$, and increasing with t) not only lacks mathematical precision, but it has no real operational value. The applied literature dealt mostly with doubly stochastic or conditional Poisson processes, a more tractable model from the computational point of view. Using such a model, Snyder derived in {137} the first genuine martingale formula[1] in dynamical point-process systems theory, namely the equivalent of Kushner's filtering formula {96}, with point-process observations replacing white-noise observations. The formal resemblance between Kushner's formula and Snyder's formula was notable, and was found surprising even though Snyder's method was inspired by Kushner's derivation of the white-noise nonlinear filtering equation.

However, it was then clear that martingales should enter the scene of point-process systems, because their role in the nonlinear filtering theory for white-noise observations had already become apparent after the work of Duncan {53}, Zakaï {159}, and Kailath {81}. This role was stressed by Wong, who coined the phrase "martingale calculus," demonstrating its power in systems theory in his textbook {156}.

The fundamental article of Kunita and Watanabe {95} already contained many of the ideas behind the martingale approach to point-process systems, but the first nontrivial statement concerning point process and martingales was given in 1964 by Watanabe {152}: the martingale property of the process

$$N_t - \int_0^t \lambda(s)\,ds,$$

where $\lambda(t)$ is some locally integrable deterministic function, characterizes N_t as a Poisson process with intensity $\lambda(t)$.

The fact that Wiener processes also can be characterized in terms of martingales (Doob's characterization) could only reinforce the feeling awakened by the formal similarity between Kushner's and Snyder's formulas.

But other similarities between Wiener process and Poisson process exist: for instance, Doob has shown in {52} that Ito's stochastic integration theory relative to Wiener-process integrands {72} could also be developed with respect to any integrand which is a square-integrable process with independent increments, in particular a compensated Poisson process. The representatation of martingale Wiener functionals as stochastic Ito integrals also found a counterpart for Poissonian functionals: this result was buried in the statement of a more general theorem in Kunita and Watanabe {95} and revealed in its poissonian avatar by Meyer {107}.

Not all of the aforementioned similarities however are of the same kind. Most of them are purely formal. Only Snyder's formula is at the surface of a deeper concept, but this could have been known only to applied probabilists

---

[1] By "genuine" martingale formula, we mean that the filtering equation involves trajectories, and not merely averages.

familiar with recursive-estimation theory and dynamic-programming theory. The analogy between Wiener and Poisson processes is a fruitful one when it is carried over into an *analogy between Wiener-driven stochastic systems and point-process stochastic systems*. Indeed, stochastic-systems theory deals with information patterns which increase with time (at least when there is no memory limitation), i.e. increasing σ-fields, and this is the reason why martingale concepts are basic to the theory.

The definition which opened the way to the martingale approach was given in {12}: $\mathscr{F}_t$ being a history of a point processes N_t, the locally integrable $\mathscr{F}_t$-progressive process λ_t is called the $\mathscr{F}_t$-intensity of N_t if

$$N_t - \int_0^t \lambda_s \, ds \text{ is an } \mathscr{F}_t\text{-martingale.} \qquad (**)$$

This is at first sight merely a sophisticated way of writing (*), since it implies, for all $0 \leq s \leq t$,

$$E[N_t - N_s | \mathscr{F}_s] = E\left[\int_s^t \lambda_u \, du \,\middle|\, \mathscr{F}_s\right],$$

an equality where the exotic word "martingale" has disappeared.

Watanabe's theorem showed that the definition (**) was not void of content. But it was also shown in {12} that the class of point processes satisfying (**) contains many point processes other than Poisson processes. Indeed, a construction by means of absolutely continuous changes of probability measures inspired by Girsanov's construction of Wiener processes with a drift {64} guarantees the existence of all locally integrable point processes with a stochastic intensity in the sense of (**). Therefore, at the same time as the definition (*) was refined into the mathematically meaningful definition (**), the existence problem could be solved (Serfozo {132} had independently solved it by a different method).

However, existence is a pure mathematician's preoccupation, and the real interest of definition (**) lies in the martingale calculus which it generates. This calculus is based on the following two ingredients: a result of Doléans-Dade and Meyer {50} stating that integration of a predictable process with respect to a martingale of bounded variation yields a martingale (this is the counterpart of Ito's integration theorem), and the ordinary Stieltjes–Lebesgue calculus (the counterpart of Ito's differentiation rule).

The martingale calculus was applied to point-process systems in much the same way as it had already been previously applied to Wiener-driven stochastic systems by Duncan {53, 54}, Zakaï {159}, Kailath {81}, Fujisaki, Kallianpur and Kunita {62}, and Davis and Varaiya {45}. For example, the point-process analogues of Girsanov's formula for absolutely continuous changes of probability measures {64}, the detection formula of Duncan {53, 54}, and the filtering theory of Duncan {53} and Zakaï {159} were given in Brémaud {12}; the innovations theory of Kailath {81} and Fujisaki,

Kallianpur, anb Kunita {62} was extended to point-process systems by Van Schuppen {144}; and the martingale theory of dynamic programming for controlled point processes was developped by Boel and Varaiya {9} along the lines of Davis and Varaiya {45}.

Concurrently the theory of {12} was refined, in particular by Boel, Varaiya, and Wong {10}, who extended Kunita and Watanabe's representation theorem for square-integrable Poisson processes to local point-process martingales. Jacod {76} gave a unified account of the mathematical foundations of the martingale approach to point processes in terms of the dual predictable-projection concept of Dellacherie {47}.

The impact of the martingale approach on queueing theory was at first marginal: filtering formulas were derived by Brémaud {15} for queues, a new proof of a formula of Takacš together with applications of the control theory of Boel and Varaiya {9} to queueing systems were given by Martins-Netto and Wong {102}. The first new result with a typical queueing-theoretical flavor obtained by means of martingale calculus was given in {20} and {22}, where it was proven that some local-balance equations relative to networks of queues in equilibrium were necessary as well as sufficient conditions for the Poisson output property to hold. With respect to queueing theory, an article by Kennedy {89} must be mentioned: it provides queueing formulas (not formulas relative to trajectories) using the observation that certain processes are martingales.

The first significant result concerning information theory and using martingale calculus was contributed by Kabanov {79}, who computed the capacity of a point-process channel, transposing a result Kadota, Zakaï, and Ziv {80} relative to white-noise channels: in particular Kabanov showed that feedback does not increase the capacity of a point-process channel.

One must mention in connection with queueing theory the monograph of Franken, König, Arndt, and Schmidt {60}. Although the basic tool used in {60} is the theory of *stationary* point processes, some interesting connections with the martingale point of view have been established.

Special Notations

$[n]$ denotes the nth reference in the reference list at the end of the current chapter, whereas $\{n\}$ denotes the nth reference of the general bibliography at the end of the book.

For indicator functions we have adopted a non-standard notation. An example will suffice: the notation $1(X \in [0, 1])$ where X is a random variable stands for $1_{\{X \in [0, 1]\}}$; similarly, $1(X(\omega) \in [0, 1])$ stands for $1_{\{X \in [0, 1]\}}(\omega)$.

CHAPTER I

Martingales

1. Histories and Stopping Times
2. Martingales
3. Predictability
4. Square-Integrable Martingales
References
Solutions to Exercises, Chapter I

1. Histories and Stopping Times

We wish to recall a few concepts concerning stochastic processes in a vocabulary that sounds familiar to systems-science theorists who deal with random phenomena in continuous time.[1]

As usual in probability theory, a probability space $(\Omega, \mathscr{F}, P)$ is given, and in most statements we will not mention this omnipresent trinity. Similarly, all equalities between random variables are to be understood P-almost surely, and we will—most of the time—omit the ritual scribble corresponding to this notion (P-a.s.).

The evolution of a random phenomenon continuously observed is represented by a family $(X_t, t \geq 0)$ of random variables, usually taking their values in R^n; such a family is called a *stochastic process*. We will take the liberty of writing X_t instead of $(X_t, t \geq 0)$; of course some confusion must be feared, but it is always avoidable at the price of a negligible amount of mental energy.

---

[1] This section merely introduces the terminology. For a more detailed account of the main results in probability theory and stochastic-processes theory to be used in the sequel, the reader is advised to browse through Appendix A1 and Appendix A2.

In association with such a process X_t, one can define for each $t \geq 0$ a sub-σ-field of $\mathscr{F}$, denoted $\mathscr{F}_t^X$, by

$$\mathscr{F}_t^X = \sigma(X_s, s \in [0, t]). \tag{1.1}$$

In words: $\mathscr{F}_t^X$ is generated by the family of random variables $(X_s, s \in [0, t])$. The family $(\mathscr{F}_t^X, t \geq 0)$ is called the *internal history of the process* X_t. The σ-field $\sigma(X_t, t \geq 0)$ which records all the events linked to process X_t is denoted by $\mathscr{F}_\infty^X$.

Let $t \geq 0$ be fixed, and let Y be a random variable, R^k-valued and $\mathscr{F}_t^X$-measurable. Then, by a well-known technical result on measurability (Appendix A1, T7),

$$Y = \phi(X_s, s \in S), \tag{1.2}$$

where S is a countable subset of $[0, t]$ and ϕ is a borelian function from $R^{n|S|}$ into R^k ($|S|$ = cardinality of S). We will sometimes abbreviate (1.2) to "$Y = \phi(X_0^t)$," a notation which speaks for itself.

Let Y_t be a R^k-valued process such that for each $t \geq 0$, Y_t is $\mathscr{F}_t^X$-measurable. One then says that Y_t is adapted to $\mathscr{F}_t^X$. By the above representation,

$$Y_t = \psi_t(X_0^t), \tag{1.3}$$

for some ψ_t; in other words Y_t depends *causally* on X_t.

Let now T be an $\overline{R}_+$-valued random variable which can be interpreted as the (random) time of occurrence of some phenomenon depending causally upon the process X_t. What we mean by "causally" is the following: at each time $t \geq 0$, the answer to the question: "has the phenomenon already taken place?" depends solely on the observation of the past (at time t) of the process. In other words, the indicator function of the set $\{T \leq t\}$ (this indicator function takes the value 1 if the answer is "yes, phenomenon occurred before t," and the value 0 if the answer is negative) is $\mathscr{F}_t^X$-measurable, or equivalently

$$\{T \leq t\} \in \mathscr{F}_t^X. \tag{1.4}$$

This corresponds to the intuitive notion of causality, since then

$$1(T \leq t) = \phi(X_0^t) \tag{1.5}$$

for some ϕ, where $1(T \leq t)$ is the characteristic function of $\{T \leq t\}$. When T satisfies (1.4) for all $t \geq 0$, it is called an $\mathscr{F}_t^X$-*stopping time* (as we just saw, "X_t-causal time" would be a more vivid appellation).

For future purposes, we need to enlarge the scope of the above notions (internal history, stopping time). For instance, a family $(\mathscr{F}_t, t \geq 0)$ of sub-σ-fields of $\mathscr{F}$ will be called a *history* iff it is increasing (i.e. $\mathscr{F}_t \geq \mathscr{F}_s$ whenever $s \leq t$). Clearly $(\mathscr{F}_t^X, t \geq 0)$ deserves the appellation, since it is increasing. As we did for stochastic processes, we will shorten the notation $(\mathscr{F}_t, t \geq 0)$ to $\mathscr{F}_t$. By definition, $\mathscr{F}_\infty$ also denoted $\bigvee_{t \geq 0} \mathscr{F}_t$ is the smallest σ-field containing all the events of $\mathscr{F}_t$, for all $t \geq 0$.

If the process X_t is such that, for all $t \geq 0$,

$$\mathscr{F}_t \supseteq \mathscr{F}_t^X, \tag{1.6}$$

then $\mathscr{F}_t$ is called *a history of* X_t. Clearly, $\mathscr{F}_t^X$ is the smallest history of X_t (by definition history $\mathscr{G}_t$ is *smaller* than history $\mathscr{F}_t$ if for all $t \geq 0$, $\mathscr{G}_t \subseteq \mathscr{F}_t$). When $\mathscr{G}_t$ is a history of X_t, one also says that X_t is *adapted to* $\mathscr{F}_t$ (for more details, see A1, §7).

In most practical cases the histories $\mathscr{F}_t$ will have the representation $\mathscr{F}_t = \mathscr{F}_t^Z$ for some process Z_t. For instance, let $Z_t = (X_t, x_t)$, where X_t and x_t are R^n-valued and R^k-valued processes respectively. Then $\mathscr{F}_t^Z$ is larger than $\mathscr{F}_t^X$, and (therefore) X_t is adapted to $\mathscr{F}_t^Z$.

A history $\mathscr{F}_t$ is best thought of, from the systems-science point of view, as an *increasing information pattern*; in the above example, the history $\mathscr{F}_t^Z$ summarizes the information associated to the observation of *both* processes X_t and x_t, whereas $\mathscr{F}_t^X$ corresponds to the incomplete observation of Z_t through its components X_t.

For stopping times, the natural generalization of the definition already given for X_t-causal times is: let $\mathscr{F}_t$ be a history; then an $\bar{R}_+$-valued random variable T is called an $\mathscr{F}_t$-*stopping time* iff for all $t \geq 0$

$$\{T \leq t\} \in \mathscr{F}_t. \tag{1.7}$$

The random time T being an $\mathscr{F}_t$-stopping time, the *past* $\mathscr{F}_t$ *at time* T corresponding to the history $\mathscr{F}_t$ is defined as follows (A1, §7):

$$\mathscr{F}_T = \{A \in \mathscr{F}_\infty | A \cap \{T \leq t\} \in \mathscr{F}_t \text{ for all } t \geq 0\}. \tag{1.8}$$

The physical interpretation of $\mathscr{F}_T$ as the past at time T is best understood when $\mathscr{F}_t = \mathscr{F}_t^X$ for some process X_t: indeed, one expects from a sensible definition of $\mathscr{F}_T^X$ as the past at time T that any event A thereof depends only upon $(X_s, s \in [0, t])$ when one knows beforehand that $T \leq t$; and this is exactly what the definition (1.8) tells us.

The *strict past* $\mathscr{F}_{T-}$ is defined in Appendix A2, D3 by

$$\mathscr{F}_{T-} = \sigma(A \cap \{t < T\}, A \in \mathscr{F}_t, t \geq 0). \tag{1.9}$$

It is defined in terms of its generators $A \cap \{t < T\}$, $A \in \mathscr{F}_t$, which, intuitively enough, belong to the strict past at time T, where we take the phrase "strict past" in its ordinary acceptation. It can be shown (see Appendix A2, T4) that if T is an $\mathscr{F}_t$-stopping time, then $\mathscr{F}_{T-} \subset \mathscr{F}_T$ and T is $\mathscr{F}_{T-}$-measurable. In general stopping times and their associated pasts always behave as expected, as is shown in detail in Appendix A2.

2. Martingales

The concept of martingales is linked in a natural way to the concept of increasing information pattern. Indeed, let $\mathscr{F}_t$ be a history—for definiteness $\mathscr{F}_t = \mathscr{F}_t^X$, where X_t is a R^n-valued process. Let now Y be a square-integrable real-valued random variable; suppose one wishes to construct the best quadratic estimate of Y given the information X_0^t. One seeks a square-integrable random variable $\hat{Y}_t$ which is $\mathscr{F}_t^X$-measurable [i.e., $\hat{Y}_t$ is of the form $\psi_t(X_0^t)$ for "some" ψ_t] and such that, for any other square-integrable random variable Z which is $\mathscr{F}_t^X$-measurable [i.e., of the form $\phi(X_0^t)$],

$$E[(Y - \hat{Y}_t)^2] \leq E[(Y - Z)^2].$$

As is well known, the conditional expectation $E[Y|\mathscr{F}_t^X]$ answers the question. Also, if $Y_t = E[Y|\mathscr{F}_t^X]$, then, by the successive conditioning rule [A1, Equation (5.12)], $E[\hat{Y}_t|\mathscr{F}_s^X] = \hat{Y}_s$. In other words, $\hat{Y}_t$ is a $(P, \mathscr{F}_t^X)$-martingale according to Definition D1 to be given in a few lines.

The above elementary fact shows that one does not have to stretch one's mind to encounter martingales in the context of dynamical stochastic systems, especially if one is interested in on-line estimation based on increasing information flow (filtering).

Let us now turn to the basic definitions and some examples.

D1 Definition (Martingale, Submartingale, Supermartingale). A history $\mathscr{F}_t$ is given on the probability space $(\Omega, \mathscr{F}, P)$. A $(P, \mathscr{F}_t)$-*martingale over* $[0, c]$ (where c is some nonnegative real number) is a real-valued stochastic process X_t such that:

(i) X_t is adapted to $\mathscr{F}_t$,
(ii) X_t is P-integrable, i.e. $E[|X_t|] < \infty$, $t \in [0, c]$,
(iii) for all $0 \leq s \leq t \leq c$, $E[X_t|\mathscr{F}_s] = X_s$, P-a.s.

If X_t is a $(P, \mathscr{F}_t)$-martingale over $[0, c]$ for all real $c \geq 0$, then X_t is called a $(P, \mathscr{F}_t)$-martingale.

In the above definition, if the equality sign in (iii) is replaced by $\geq$ [$\leq$], and if the word "martingale" is replaced by "submartingale" ["supermartingale"], then we obtain a definition for $(P, \mathscr{F}_t)$-*submartingales* [$(P, \mathscr{F}_t)$-*supermartingales*].

Processes with Independent Increments

Let X_t be a real-valued process with independent increments; that is to say, for all $0 \leq s \leq t$, $X_t - X_s$ is independent of $\mathscr{F}_s^X$. Suppose moreover that for each $t \geq 0$, X_t is integrable and $E[X_t] = 0$. Then X_t is a $(P, \mathscr{F}_t^X)$-martingale. Indeed, for any $0 \leq s \leq t$,

$$E[X_t|\mathscr{F}_s^X] = E[X_s|\mathscr{F}_s^X] + E[X_t - X_s|\mathscr{F}_s^X] = X_s + 0.$$

E1 Exercise. Suppose in addition that X_t^2 is integrable. Show that $X_t^2 - E[X_t^2]$ is a $(P, \mathscr{F}_t^X)$-martingale and that X_t^2 is a $(P, \mathscr{F}_t^X)$-submartingale.

2. Martingales

If X_t is a Wiener process [i.e. (see A3) a continuous process with independent increments such that for all $0 \le s \le t$, $X_t - X_s$ is a Gaussian random variable with variance $t - s$ and mean 0], then E1 specializes to: X_t as well as $X_t^2 - t$ are $(P, \mathscr{F}_t^X)$-martingales (simply observe that $E[X_t^2] = t$).

From now on we will simplify the writing by omitting P as often as possible in phrases such as "$(P, \mathscr{F}_t)$-martingale." Thus we will be speaking of $\mathscr{F}_t$-martingales, $\mathscr{F}_t$-supermartingales, etc. Of course, whenever the clouds of confusion loom on the horizon, we must swiftly go back to the more complete notation (this will be the case in Chapter VI, when changes of probability measures will be considered). However, one must be warned against the simplification which consists in omitting the mention of the history $\mathscr{F}_t$ (the neophyte who dares indulge in such vice falls into all kinds of traps).

We will now return to examples and see that the Markov property is another source of martingales.

Martingales Associated with Markov Chains

Let X_t be a right-continuous N_+-valued $(P, \mathscr{F}_t)$-Markov chain with left-hand limits (here again, we will simplify this notation into "$\mathscr{F}_t$-Markov chain") which is stable and conservative and admits the Q-matrix $(q_{ij}; i, j \in N_+)$. If f is a nonnegative function from $N_+ \times N_+$ into R_+, then for any $0 \le s \le t$, the Lévy formula holds:

$$E\left[\sum_{s < u \le t} f(X_{u-}, X_u) \mid \sigma(X_s)\right] = E\left[\int_s^t \sum_{j \ne X_u} q_{X_u j} f(X_u, j) \, du \mid \sigma(X_s)\right]. \quad (2.1)$$

By the $\mathscr{F}_t$-Markov property of X_t [conditional independence of $\mathscr{F}_s$ and $(X_u, u \ge s)$ given X_s], conditioning with respect to X_s in (2.1) is equivalent to conditioning with respect to $\mathscr{F}_s$:

$$E\left[\sum_{s < u \le t} f(X_{u-}, X_u) \mid \mathscr{F}_s\right] = E\left[\int_s^t \sum_{j \ne X_u} q_{X_u j} f(X_u, j) \, du \mid \mathscr{F}_s\right]. \quad (2.2)$$

From this, the following exercise is straightforward.

E2 Exercise [5]. Show that if $f: N_+ \times N_+ \to R$ is such that, for all $t \ge 0$,

$$E\left[\int_0^t \sum_{j \ne X_s} q_{X_s j} |f(X_s, j)| \, ds\right] < \infty, \quad (2.3)$$

then

$$\sum_{0 < s \le t} f(X_{s-}, X_s) - \int_0^t \sum_{j \ne X_s} q_{X_s j} f(X_s, j) \, ds \text{ is a } (P, \mathscr{F}_t)\text{-martingale.}$$

In particular, if one selects for f the form $f(i,j) = 1(i = k)1(j = l)$, in which case $\sum_{0 < s \le t} f(X_{s-}, X_s) = N_t(k, l)$ is the number of transitions from k to l in the interval $(0, t])$, then

$$N_t(k, l) - q_{kl} \int_0^t 1(X_s = k)\,ds \text{ is a } (P, \mathscr{F}_t)\text{-martingale.} \qquad (2.4)$$

Here the condition (2.3) is automatically satisfied, since

$$E\left[q_{kl}\int_0^t 1(X_s = k)\,ds\right] = q_{kl}\int_0^t P[X_s = k]\,ds \le q_{kl}t < \infty.$$

Another interesting choice of f is the following: $f(i,j) = 1(j = l)$, in which case $\sum_{0 < s \le t} f(X_{s-}, X_s) = N_t(l)$ counts the number of entrances into state l during the interval $(0, t]$. Then

$$N_t(l) - \int_0^t \sum_{i \ne l} q_{il} 1(X_s = i)\,ds \text{ is a } (P, \mathscr{F}_t)\text{-martingale} \qquad (2.5)$$

provided that

$$\sum_{i \ne l} \int_0^t q_{il} P(X_s = i)\,ds < \infty, \qquad t \ge 0. \qquad (2.6)$$

A special case of interest to queueing theory is the *birth-and-death process* (see A1, §8).

E3 Exercise. Let Q_t be a birth and death process with parameters λ_n and μ_n, and such that $E[Q_0] < \infty$. Show that if $\sum_{n=0}^\infty E[\int_0^t \lambda_n P(Q_s = n)\,ds] < \infty$ for all $t \ge 0$, then

$$A_t - \int_0^t \lambda_{Q_s}\,ds \text{ is a } (P, \mathscr{F}_t^Q)\text{-martingale,} \qquad (2.7)$$

and

$$D_t - \int_s^t \mu_{Q_s} 1(Q_s > 0)\,ds \text{ is a } (P, \mathscr{F}_t^Q)\text{-martingale,} \qquad (2.8)$$

where A_t and D_t count respectively the upward jumps and the downward jumps of Q_t in the interval of time $(0, t]$.

Here is a third source of martingales, to which the whole Chapter VI will be devoted:

Martingales and Radon–Nikodyn Derivatives

Let P and $\tilde{P}$ be two probability measures defined on the same measurable space $(\Omega, \mathscr{F})$, and let $\mathscr{F}_t$ be a history. For each $t \ge 0$, denote by P_t and $\tilde{P}_t$ the restrictions to $\mathscr{F}_t$ of P and $\tilde{P}$ respectively, and suppose that for some

2. Martingales

$c \geq 0$, $\tilde{P}_c$ is absolutely continuous with respect to P_c. Then clearly, for all $t \in [0, c]$, $\tilde{P}_t$ is absolutely continuous with respect to P_t. Define, for each $t \in [0, c]$, the random variable L_t to be the Radon–Nikodym derivative of $\tilde{P}_t$ with respect to P_t:

$$L_t = \frac{d\tilde{P}_t}{dP_t}. \tag{2.9}$$

Then:

E4 Exercise. L_t is a $(P, \mathscr{F}_t)$-martingale over $[0, c]$.

Such Radon–Nikodym-derivative martingales play an important role in communications theory, and we will encounter them in Chapter VI under the name of *likelihood ratios*. In VI, §1 the central role of such likelihood ratios in decision theory is briefly explained.

The following result indicates how to obtain submartingales from a martingale by means of convex transformations:

E5 Exercise. Let M_t be a $(P, \mathscr{F}_t)$-martingale, and let $\phi: R \to R$ be a convex function such that for all $t \geq 0$, $E[|\phi(M_t)|] < \infty$. Then $\phi(M_t)$ is a $(P, \mathscr{F}_t)$-submartingale. Compare with E1.

Here is an almost obvious result that will be needed later in the book.

E6 Exercise. Let X_t be a $(P, \mathscr{F}_t)$-submartingale [or a $(P, \mathscr{F}_t)$-supermartingale] such that $E[X_0] = E[X_c]$ for some $c \geq 0$. Then X_t is a $(P, \mathscr{F}_t)$-martingale over $[0, c]$.

Optional Sampling and Local Martingales

One of the central results of martingale theory, due to Doob, is the *optional-sampling theorem*. It says—roughly—that in the martingale equality $E[M_t|\mathscr{F}_s] = M_s$, the deterministic times s and t can be replaced by $\mathscr{F}_t$-stopping times. More precisely:

T2 Theorem (*Optional Sampling* [3]). *Let M_t be a right-continuous $(P, \mathscr{F}_t)$-submartingale, and let S and T be two finite stopping times such that $S \leq T$. If either one of the two following conditions is satisfied*:

(i) *S and T are bounded by a constant $t_0 < \infty$,*
(ii) *$(M_t, t \geq 0)$ is a uniformly integrable family,[2] then*

$$E[M_T|\mathscr{F}_S] \geq M_S. \tag{2.10}$$

---
[2] See Appendix A1, D23.

This result will not be proven here. It can be found in standard textbooks on stochastic processes such as [6].

Here is a kind of converse to Doob's optional-sampling theorem.

E7 Exercise [4]. Let X_t be a real-valued $\mathscr{F}_t$-progressive process such that for all bounded $\mathscr{F}_t$-stopping times T, X_T is integrable and $E[X_T] = E[X_0]$. Show that X_t is a $(P, \mathscr{F}_t)$-martingale.

For technical reasons, the notion of martingale is not quite sufficient, and one needs a more general concept, that of local martingale.

D3 Definition (Local Martingale [5]). Let X_t be a real-valued stochastic process adapted to a history $\mathscr{F}_t$, and let $(S_n, n \geq 1)$ be an increasing family of $\mathscr{F}_t$-stopping times such that

(i) $\lim_{n \uparrow \infty} S_n = +\infty$,
(ii) for each $n \geq 1$, $X_{t \wedge S_n}$ is a $(P, \mathscr{F}_t)$-martingale.

Then X_t is called a $(P, \mathscr{F}_t)$-*local martingale*. The family $(S_n, n \geq 1)$ is a family of localizing times for X_t.

Of course, we will most of the times use the abbreviation "$\mathscr{F}_t$-local martingale."

Next result shows that a nonnegative local martingale is in fact a supermartingale:

E8 Exercise. Let X_t be a $(P, \mathscr{F}_t)$-local martingale, nonnegative. Then if X_t is integrable, it is a $(P, \mathscr{F}_t)$-supermartingale.

3. Predictability

The notions of measurability and of progressiveness are explained in Appendix A1, §7. For our purpose these notions are not sufficient, and we will need to consider *predictable processes*.

D4 Definition (Predictable σ-Field, Predictable Process [2]). Let $\mathscr{F}_t$ be a history defined on the probability space $(\Omega, \mathscr{F}, P)$, and define $\mathscr{P}(\mathscr{F}_t)$ to be the σ-field over $(0, \infty) \times \Omega$ generated by rectangles of the form

$$(s, t] \times A; \quad 0 \leq s \leq t, \quad A \in \mathscr{F}_s. \tag{3.1}$$

$\mathscr{P}(\mathscr{F}_t)$ is called the $\mathscr{F}_t$-*predictable σ-field* over $(0, \infty) \times \Omega$. A real-valued process X_t such that X_0 is $\mathscr{F}_0$-measurable and the mapping $(t, \omega) \to X_t(\omega)$ defined from $(0, \infty) \times \Omega$ into R is $\mathscr{P}(\mathscr{F}_t)$-measurable is said to be $\mathscr{F}_t$-*predictable*. Any process X_t which is P-indistinguishable from a $\mathscr{F}_t$-predictable

3. Predictability

process is said to be $(P, \mathscr{F}_t)$-predictable, or $\mathscr{F}_t$-predictable when the choice of P is clear.

We recall that X_t and Y_t are said to be *P-indistinguishable* iff $P[X_t = Y_t, t \geq 0] = 1$; see A1, §7.

Comment. The set of generators (3.1) is the set which is traditionally given; however, the set $\{(s, \infty) \times A, A \in \mathscr{F}_s, s \geq 0\}$ is also a set of generators, and a simpler one.

In all practical applications, the $\mathscr{F}_t$-predictable processes to be encountered are processes adapted to $\mathscr{F}_t$ and *left-continuous*.

T5 Theorem *An R^n-valued process X_t adapted to $\mathscr{F}_t$ and left-continuous is $\mathscr{F}$-predictable.*

PROOF. It is enough to do the proof in the case of an R-valued process X_t. Since X_t is left-continuous, for all $(t, \omega) \in (0, \infty) \times \Omega$ we have

$$X_t(\omega) = \lim_{n \uparrow \infty} \left\{ \sum_{q=0}^{n2^n - 1} X_{q/2^n}(\omega) 1(q/2^n < t \leq q + 1/2^n) + X_n(\omega) 1(t > n) \right\}.$$

Now, for any $n \geq 1$, any $0 \leq q \leq n2^n - 1$, the mapping $(t, \omega) \to X_{q/2^n}(\omega) 1(q/2^n < t \leq q + 1/2^n)$ is in $\mathscr{P}(\mathscr{F}_t)$, since $X_{q/2^n}$ is $\mathscr{F}_{q/2^n}$-measurable (and therefore can be approximated by weighted sums of indicators of sets in $\mathscr{F}_{q/2^n}$; see Appendix A1, T3). Similarly, the mapping $(t, \omega) \to X_n(\omega) 1(t > n)$ is in $\mathscr{P}(\mathscr{F}_t)$. Hence $(t, \omega) \to X_t(\omega)$, as a limit of mappings in $\mathscr{P}(\mathscr{F}_t)$ is also in $\mathscr{P}(\mathscr{F}_t)$. □

E9 Exercise. Show that $\mathscr{P}(\mathscr{F}_t)$ is generated by the sets $\{(t, \omega) | t \leq T(\omega)\}$ where T varies over the bounded $\mathscr{F}_t$-stopping times.

The next exercise provides some motivation for the terminology "predictable."

E10 Exercise [1]. Let X_t be an $\mathscr{F}_t$-predictable R-valued process, and let T be an $\mathscr{F}_t$-stopping time. The random variable $X_T 1(T < \infty)$ is $\mathscr{F}_{T-}$-measurable [recall that if X_t is $\mathscr{F}_t$-progressive, the random variable $X_T 1(T < \infty)$ is *a priori* $\mathscr{F}_T$-measurable, but not necessarily $\mathscr{F}_{T-}$-measurable].

In the hierarchy of measurability, predictability stands on top; indeed:

E11 Exercise. Show that if X_t is $\mathscr{F}_t$-predictable, it is $\mathscr{F}_t$-progressive.

For our purpose in this book, all that we need to know about predictability is contained in the above exercises and theorems. We will see how such concept emerges in the theory of stochastic integration in a very natural way.

Integration with Respect to Bounded Variation Martingales

A process X_t is said to be of *bounded variation* if P-almost all its trajectories are of bounded variation over bounded intervals, and of *integrable variation* if moreover $E[\int_0^t |dX_s|] < \infty, t \geq 0$ (the standard usage would be to call such processes "processes of bounded variation over finite intervals" and "processes of *locally* integrable variation").

T6 Theorem (Integration with Respect to Bounded Variation Martingales [2]). *Let M_t be a $(P, \mathscr{F}_t)$-martingale of integrable bounded variation, and let C_t be an $\mathscr{F}_t$-predictable process such that $E[\int_0^1 |C_s| \, |dM_s|] < \infty$. The process $\int_0^t C_s \, dM_s$ is then a $(P, \mathscr{F}_t)$-martingale over $[0, 1]$.*

PROOF. Let $\mathscr{H}$ be the vector space of bounded $\mathscr{F}_t$-progressive processes C_t such that $\int_0^t C_s \, dM_s$ is a $(P, \mathscr{F}_t)$-martingale over $[0, 1]$. Clearly, $\mathscr{H}$ contains the indicators of (t, ω)-sets of the form $(s, t] \times A$, where $0 \leq s \leq t$, $A \in \mathscr{F}_s$, and $\mathscr{H}$ contains the constants. Also it is stable by (t, ω)-pointwise bounded limit. By the monotone class theorem (Appendix A1, T5), $\mathscr{H}$ contains all bounded $\mathscr{F}_t$-predictable processes [recall that the sets $\{(s, t] \times A | 0 \leq s \leq t, A \in \mathscr{F}_s\}$ generate $P(\mathscr{F}_t)$]. □

It remains to extend this to *unbounded* $\mathscr{F}_t$-predictable processes satisfying $E[\int_0^1 |C_s| |dM_s|] < \infty$. This is done by application of the Hahn–Banach extension theorem as follows:

Let L be the Banach space of $\mathscr{F}_t$-predictable processes C_t such that $E[\int_0^1 |C_s| |dM_s|] < \infty$ endowed with the norm $\|C\|_L = E[\int_0^1 |C_s| |dM_s|]$.[3] Denote by L_b the set of bounded processes of L; L_b is dense in L.

Let $\mathscr{M}$ be the Banach space of $(P, \mathscr{F}_t)$-martingales m_t with the norm $\|m\|_\mathscr{M} = E[|m_1|]$.[4]

E12 Exercise. Prove that $\mathscr{M}$ endowed with the norm $\|m\|_\mathscr{M} = E[|m_1|]$ is a Banach space.

The mapping $\phi : L_b \to \mathscr{M}$ defined by $\phi(C_t) = \int_0^t C_s \, dM_s$ is linear and continuous (actually a contraction, since $E[|\int_0^1 C_s \, dM_s|] \leq E[\int_0^1 |C_s| |dM_s|]$). By Banach's extension theorem, it can be extended uniquely into a linear continuous mapping from L into $\mathscr{M}$, still denoted ϕ. It remains to verify that for any $C_t \in L$, $\phi(C)_t = \int_0^t C_s \, dM_s$ for all $t \in [0, 1]$.

---

[3] We identify two processes C_t and C'_t such that $E[\int_0^1 |C_s - C'_s| |dM_s|] = 0$.

[4] We identify two $\mathscr{F}_t$-martingales m_t and m'_t if $m_t = m'_t$, P-a.s., $t \in [0, 1]$.

Let $C_t \in L$. Choose a sequence $(C_t^{(n)}, n \geq 1)$ in L_b such that $\lim_{n \uparrow \infty} \|C - C^{(n)}\|_L = 0$. Since

$$E\left[\left|\int_0^t C_s\, dM_s - \int_0^t C_s^{(n)}\, dM_s\right|\right] \leq E\left[\int_0^1 |C_s - C_s^{(n)}|\,|dM_s|\right], \quad t \in [0, 1],$$

it follows that $\int_0^t C_s^{(n)}\, dM_s \to \int_0^t C_s\, dM_s$ in $L^1(P)$. Now, by definition of ϕ, $\|\phi(C^{(n)}) - \phi(C)\|_\mathcal{M} \to 0$, and therefore by Jensen's inequality $\phi(C^{(n)})_t \to \phi(C)_t$ in $L^1(P)$ for all $t \in [0, 1]$. Since $\phi(C^{(n)})_t = \int_0^t C_s^{(n)}\, dM_s$ by definition, it follows that $\int_0^t C_s^{(n)}\, dM_s \to \phi(C)_t$ in $L^1(P)$. By uniqueness of $L^1(P)$ limits, $\int_0^t C_s\, dM_s = \phi(C)_t$ P-a.s.

The proof of T6 shows how martingales, stochastic integrals, and predictable processes are tightly linked. Indeed, M_t being an arbitrary real-valued process, it is natural to call the process $X_t = 1_A(M_{t \wedge v} - M_{t \vee u})$ the integral of $C_t = 1_A 1(u < t \leq v)$ with respect to M_t, denoted $\int_0^t C_s\, dM_s$; and the martingale property of $\int_0^t C_s\, dM_s$ for such C_t is a direct consequence of the martingale property of M_t.

The role of predictable processes also stems from the fact that for M_t to be a $(P, \mathcal{F}_t)$-martingale, it is necessary and sufficient that

$$E\left[\int_0^\infty C_s\, dM_s\right] = 0$$

for all C_t of the form $C_t = 1_A 1(u < t \leq v)$, where $0 \leq u \leq v$, $A \in \mathcal{F}_u$ (we have just written the definition of a martingale). The elementary $\mathcal{F}_t$-predictable processes $1_A 1(u < t \leq v)$ are therefore part of the definition of martingales, and one should not wonder at their crucial role in the theory of stochastic integration.

4. Square-Integrable Martingales

D7 Definition. A *square-integrable $(P, \mathcal{F}_t)$-martingale over $(0, c]$* is a $(P, \mathcal{F}_t)$-martingale over $[0, c]$ such that

$$E[M_c^2] < \infty. \tag{4.1}$$

A *square-integrable $(P, \mathcal{F}_t)$-martingale* is a $(P, \mathcal{F}_t)$-martingale such that

$$\sup_{t \geq 0} E[M_t^2] < \infty. \tag{4.2}$$

Remark. The condition (4.1) implies that

$$\sup_{t \in [0, c]} E[M_t^2] < \infty.$$

Indeed, by E5, M_t^2 is a $(P, \mathcal{F}_t)$-submartingale. Hence for all $t \in [0, c]$, $E[M_t^2] \leq E[M_c^2]$.

Let $\mathcal{M}^2$ be the set of square-integrable $(P, \mathcal{F}_t)$-martingales over $[0, 1]$. Here we do not distinguish two martingales m_t and m'_t such that

$$m_t = m'_t \quad P\text{-a.s.,} \qquad t \in [0, 1], \tag{4.3}$$

so that $\mathcal{M}^2$ is, strictly speaking, a set of equivalence classes. Then:

E13 Exercise. Show that $\mathcal{M}^2$ is a Hilbert space when endowed with the norm

$$\|m\|_{\mathcal{M}^2} = E[m_1^2]. \tag{4.4}$$

The following theorem tells us that we can replace the deterministic times of (4.2) by stopping times. More presisely:

T8 Theorem. *Let M_t be a right-continuous square-integrable $(P, \mathcal{F}_t)$-martingale. Then*

$$\sup_{T \in \mathcal{T}} E[M_T^2] < \infty, \tag{4.5}$$

where $\mathcal{T}$ is the set of all finite $\mathcal{F}_t$-stopping times.

PROOF. By E5, M_t^2 is a $(P, \mathcal{F}_t)$-submartingale. By Doob's optional-sampling theorem, for all $T \in \mathcal{T}$ and all $t \geq 0$

$$E[M_{T \wedge t}^2] \leq E[M_t^2] \leq \sup_{t \geq 0} E[M_t^2].$$

By Fatou's lemma (A1, §4),

$$E[M_T^2] = E\left[\lim_{t \uparrow \infty} M_{T \wedge t}^2\right] \leq \varliminf_{t \uparrow \infty} E[M_{T \wedge t}^2]$$

Hence

$$E[M_T^2] \leq \sup_{t \geq 0} E[M_t^2]. \qquad \square$$

References

[1] Dellacherie, C. (1972) *Capacités et processus stochastiques*, Springer, Berlin.
[2] Doléans-Dade, C. and Meyer, P. A. (1970) Intégrales stochastiques par rapport aux martingales locales, *Séminaire Proba X*, Lect. Notes in Math. **124**, Springer, Berlin, pp. 77–107.
[3] Doob, J. L. (1953) *Stochastic Processes*, Wiley, New York.
[4] Komatsu, T. (1973) Markov processes associated with certain integrodifferential operators, Osaka J. Math. **10**, pp. 271–303.
[5] Kunita, H. and Watanabe, S. (1967) On square integrable martingales, Nagoya Math. J. **30**, pp. 209–245.
[6] Meyer, P. A. (1966) *Probability and Potential*, Blaisdell, San Francisco.

SOLUTIONS TO EXERCISES, CHAPTER I

E1. First observe that
$$E[X_t^2 - X_s^2 | \mathcal{F}_s^X] = E[(X_t - X_s)^2 | \mathcal{F}_s^X] \tag{1}$$
(indeed, $(X_t - X_s)^2 = X_t^2 + X_s^2 - 2X_t X_s$ and $E[X_t X_s | \mathcal{F}_s^X] = X_s E[X_t | \mathcal{F}_s^X] = X_s^2$), and consequently
$$E[X_t^2 - X_s^2] = E[(X_t - X_s)^2]. \tag{2}$$
Therefore the martingale property of $X_t^2 - E[X_t]$ to be proven,
$$E[X_t^2 - X_s^2 | \mathcal{F}_s^X] = E[X_t^2 - X_s^2], \tag{3}$$
is equivalent to
$$E[(X_t - X_s)^2 | \mathcal{F}_s^X] = E[(X_t - X_s)^2]. \tag{4}$$
The latter equality is just a consequence of the hypothesis of independent increments.

The function $t \to E[X_t^2]$ is increasing. Indeed, by (2), $E[X_t^2] = E[X_s^2] + E[X_t^2 - X_s^2] = E[X_s^2] + E[(X_t - X_s)^2] \geq E[X_s^2]$. Therefore X_t^2 is the sum of the $\mathcal{F}_t^X$-martingale $X_t^2 - E[X_t^2]$ and the increasing function $E[X_t^2]$; hence it is an $\mathcal{F}_t^X$-submartingale.

E2. Define
$$A_t = \sum_{0 < s \leq t} f(X_{s-}, X_s),$$
$$B_t = \int_0^t \sum_{j \neq X_s} q_{X_s j} f(X_s, j) \, ds, \tag{1}$$
$$m_t = A_t - B_t.$$

The condition (2.3) guarantees that m_t is integrable (in fact both A_t and B_t are integrable). Equation (2.2) reads
$$E[A_t - A_s | \mathcal{F}_s] = E[B_t - B_s | \mathcal{F}_s], \tag{2}$$
and what we want to prove is
$$E[A_t - B_t | \mathcal{F}_s] = E[A_s - B_s | \mathcal{F}_s] \quad (= A_s - B_s). \tag{3}$$
The passage from (2) to (3) is allowed because A_t, A_s, B_t, B_s are integrable.

E3. The number of upward jumps in $(0, t]$ is
$$A_t = \sum_{n=0}^{\infty} N_t(n, n+1). \tag{1}$$
From (2.4) and the expression for q_{ij} in A1, Equation (8.20),
$$N_t(n, n+1) - \lambda_n \int_0^t 1(Q_s = n) \, ds \text{ is an } \mathcal{F}_t\text{-martingale}. \tag{2}$$

The summation of (2) for all $n \geq 0$ will yield a martingale if $E[A_t] < \infty$ for all $t \geq 0$, or equivalently

$$\sum_{n \geq 0} \lambda_n \int_0^t P[Q_s = n] \, ds < \infty \tag{3}$$

(indeed, $E[A_t] = \sum_{n \geq 0} E[N_t(n, n+1)]$ and

$$E[N_t(n, n+1)] = E\left[\lambda_n \int_0^t 1(Q_s = n) \, ds\right] = \lambda_n \int_0^t P[Q_s = n] \, ds\right).$$

Hence the result, since $\sum_{n=0}^\infty \lambda_n \int_0^t 1(Q_s = n) \, ds = \int_0^t \lambda_{Q_s} \, ds$.

The number of downward jumps in $(0, t]$ is

$$D_t = \sum_{n=1}^\infty N_t(n, n-1). \tag{4}$$

From (2.4) and A1, Equation (8.20) again,

$$N_t(n, n-1) - \mu_n \int_0^t 1(Q_s = n) \, ds \text{ is an } \mathscr{F}_t\text{-martingale.} \tag{5}$$

The integrability condition $E[D_t] < \infty$ is automatically granted by $E[Q_0] < \infty$ and $E[A_t] < \infty$, since $D_t \leq Q_0 + A_t$. Hence the condition (3) guarantees that summation of (5) over n yields a martingale, namely

$$D_t - \int_0^t \mu_{Q_s} 1(Q_s > 0) \, ds \text{ is an } \mathscr{F}_t\text{-martingale} \tag{6}$$

[observe that $\sum_{n=1}^\infty \mu_n \int_0^t 1(Q_s = n) \, ds = \int_0^t \mu_{Q_s} 1(Q_s > 0) \, ds$].

E4. To be proven: for all $0 \leq s \leq t \leq c$ and all $A \in \mathscr{F}_s$,

$$E[1_A L_t] = E[1_A L_s]. \tag{1}$$

But A is in $\mathscr{F}_s$ and $L_s = d\tilde{P}_s/dP_s$; therefore by Appendix A1, T10,

$$E[1_A L_s] = \tilde{E}[1_A] = \tilde{P}(A). \tag{2}$$

Since $s \leq t$, A is also in $\mathscr{F}_t$; therefore, since $L_t = d\tilde{P}_t/dP_t$,

$$E[1_A L_t] = \tilde{E}[1_A] = \tilde{P}(A), \tag{3}$$

and (1) follows from (2) and (3).

Comment. We have taken for granted that L_t is integrable and adapted to $\mathscr{F}_t$, but conditions (i) and (ii) of D1 should always be carefully checked (at least mentally). In this particular case L_t is nonnegative, and by definition of L_t as a Radon–Nikodym derivative of $\tilde{P}_t$ with respect to P_t,

$$E[L_t] = E[1_\Omega L_t] = \tilde{E}[1_\Omega] = \tilde{P}[\Omega] = 1;$$

hence the integrability. Now for the adaptation of L_t to $\mathscr{F}_t$, we have to look back at Appendix A1, T10 and draw up the following "notational-equivalence table":

Solutions to Exercises, Chapter I

A1, T10	This exercise
μ	P_t
ν	$\tilde{P}_t$
$\mathscr{F}$	$\mathscr{F}_t$
$d\nu/d\mu$	L_t

In A1, T10, one of the conclusion is: $d\nu/d\mu$ is a random variable, i.e, it is $\mathscr{F}$-measurable. Thus, according to the equivalence table above, L_t is $\mathscr{F}_t$-measurable.

E5. Follows directly from Jensen's inequality for conditional expectations [A1, §5, Equation (5.11)]:

$$E[\varphi(X_t)|\mathscr{F}_s] \geq \varphi(E[X_t|\mathscr{F}_s]) = \varphi(X_s).$$

Compare with E1: $x \to x^2$ is convex.

Comment. Here again one should check the integrability and adaptation requirements [(i) and (ii) of D1]. The integrability of $\varphi(X_t)$ is part of the hypothesis. For the measurability: $\varphi(X_t)$ is $\mathscr{F}_t$-measurable because X_t is $\mathscr{F}_t$-measurable (X_t is a $\mathscr{F}_t$-martingale) and φ is borelian [being the sup of borelian functions; see the proof of A1, Equation (5.11)].

E6. If X_t is, say, an $\mathscr{F}_t$-submartingale, $t \to E[X_t]$ is increasing. Therefore if $E[X_0] = E[X_c]$, then $t \to E[X_t]$ is a constant over $(0, c]$. Hence for all $0 \leq s \leq t \leq c$

$$E[X_t] = E[X_s]. \qquad (1)$$

By the definition of submartingale,

$$E[X_t|\mathscr{F}_s] \geq X_s \quad P\text{-a.s.} \qquad (2)$$

The announced result follows from (1) and (2) and the following lemma: if X and Y are two integrable randon variables such that $X \geq Y$, P-a.s., and $E[X] = E[Y]$, then $X = Y$, P-a.s.

E7. Let S and T be two bounded $\mathscr{F}_t$-stopping times; by hypothesis

$$E[X_T] = E[X_S]. \qquad (1)$$

Choose S and T as follows: first fix s and t, $0 \leq s \leq t$, and $A \in \mathscr{F}_s$; then define

$$T(\omega) = t$$

$$S(\omega) = \begin{cases} s & \text{if } \omega \in A, \\ t & \text{if } \omega \in A^c. \end{cases} \qquad (2)$$

We check that S is a stopping time: indeed, for any $v \geq 0$,

$$\{S \leq v\} = \begin{cases} \varnothing & \text{if } 0 \leq v < s, \\ A & \text{if } s \leq v < t, \\ \Omega & \text{if } t \leq v. \end{cases} \qquad (3)$$

In the first and third cases $\{S \leq v\}$ is obviously an event of $\mathscr{F}_v$. In the second case, $\{S \leq v\} = A \in \mathscr{F}_s$, but since $v \geq s$, it is also in $\mathscr{F}_v$.

Let us now write the equality (1) with S and T defined by (2). Since

$$E[X_S] = E[X_s 1_A] + E[X_t 1_{A^c}] \tag{4}$$

and

$$E[X_T] = E[X_t 1_A] + E[X_t 1_{A^c}], \tag{5}$$

it follows that

$$E[X_s 1_A] = E[X_t 1_A],$$

which is the announced result, A being arbitrary in $\mathscr{F}_s$.

E8. Let $(S_n, n \geq 1)$ be a family of localizing times; then for all $n \geq 1$, all $0 \leq s \leq t$, and all $A \in \mathscr{F}_s$,

$$E[X_{t \wedge S_n} | \mathscr{F}_s] = X_{s \wedge S_n}. \tag{1}$$

By Fatou's lemma for conditional expectations, $X_{t \wedge S_n}$ being bounded from below (nonnegative),

$$\lim E[X_{t \wedge S_n} | \mathscr{F}_s] \geq E[\lim X_{t \wedge S_n} | \mathscr{F}_s] = E[X_t | \mathscr{F}_s]. \tag{2}$$

Now by (1)

$$\lim E[X_{t \wedge S_n} | \mathscr{F}_s] = \lim X_{s \wedge S_n} = X_s. \tag{3}$$

Hence, from (2) and (3),

$$E[X_t | \mathscr{F}_s] \leq X_s. \tag{4}$$

E9. $\mathscr{P}(\mathscr{F}_t)$ is generated by the sets of $(0, \infty) \times \Omega$ of the form

$$(s, +\infty) \times A, \quad t \geq s \geq 0, \quad A \in \mathscr{F}_s. \tag{1}$$

Define for each $s \geq 0$, $A \in \mathscr{F}_s$, the $\mathscr{F}_t$-stopping time s_A by

$$s_A(\omega) = \begin{cases} s & \text{if } \omega \in A, \\ +\infty & \text{if } \omega \in A^c \end{cases} \tag{2}$$

(check that s_A is an $\mathscr{F}_t$-stopping time). Then

$$\{(t, \omega) | t \leq s_A(\omega)\}^c = (s, +\infty) \times A, \tag{3}$$

where the complement is taken with respect to $(0, \infty) \times \Omega$.

So far we have shown that the generators of $\mathscr{P}(\mathscr{F}_t)$ of type (1) can be obtained by elementary set operations from sets of the form $\{t \leq T(\omega)\}$ where T is an $\mathscr{F}_t$-stopping time. We conclude by observing that $\{(t, \omega) \in (0, \infty) \times \Omega | t \leq T(\omega)\} = \lim_{\downarrow} \{(t, \omega) \in (0, \infty) \times \ \ )| t \leq T(\omega) \wedge n\}$ for any stopping time T, finite or not.

E10. First verify that the r ılt is true for elementary $\mathscr{F}_t$-predictable X_t of the form

$$X_t(\ ) = 1_A(\omega) 1(u < t), \quad 0 \leq u, \quad A \in \mathscr{F}_u. \tag{1}$$

Indeed, $1_A 1(u < T) 1(T < \infty)$ is in $\mathscr{F}_{T-}$, since $\{u < T\}$ is a generator of $\mathscr{F}_{T-}$ whenever $A \in \mathscr{F}_u$, and $\{T < \infty\} \in \mathscr{F}_{T-}$ (recall that T is $\mathscr{F}_{T-}$-measurable: Appendix A2, T4).

Solutions to Exercises, Chapter I   17

The rest of the proof is a straightforward application of the verification theorem (Appendix A1, T4): let $\mathscr{H}$ be the set of processes X_t such that $X_T 1(T < \infty)$ is $\mathscr{F}_{T-}$-measurable for all $\mathscr{F}_t$-stopping times T, and let $\mathscr{S}$ be the subsets of $(0, \infty) \times \Omega$ of the form $(u, \infty) \times A$, $u \geq 0$, $A \in \mathscr{F}_u$.

E11. It suffices to show that the generators of $\mathscr{P}(\mathscr{F}_t)$,

$$(u, \infty) \times A, \quad u \geq 0, \quad A \in \mathscr{F}_u, \tag{1}$$

are $\mathscr{F}_t$-progressive. This is true because the indicators of such sets are left-continuous and adapted to $\mathscr{F}_t$; see Appendix A1, T32.

E12. One should first prove that $E[|m_1|] = \|m\|_{\mathscr{M}}$ really defines a norm; only the fact that $\|m\|_{\mathscr{M}} = 0$ implies $m_t = 0$, P-a.s., for all $t \in [0, 1]$, is not completely obvious. By Jensen's inequality for conditional expectations [A1, Equation (5.11)], for all $t \in [0, 1]$

$$E[|m_t|] = E[|E[m_1|\mathscr{F}_t]|] \leq E[E[|m_1||\mathscr{F}_t]] = E[|m_1|]. \tag{1}$$

Therefore $E[|m_1|] = 0$ implies $E[|m_t|] = 0$, which in turn implies $m_t = 0$, P-a.s. To show that $\mathscr{M}$ is a Banach space, one must prove the completeness. Let $(m^{(n)}, n \geq 1)$ be a sequence in $\mathscr{M}$ such that

$$\lim_{k, n \uparrow \infty} E[|m_1^{(n)} - m_1^{(k)}|] = 0. \tag{2}$$

By the same Jensen's-inequality argument as above, it follows that for each $t \in [0, 1]$

$$\lim_{k, n \uparrow \infty} E[|m_t^{(n)} - m_t^{(k)}|] = 0. \tag{3}$$

Hence $(m_t^{(n)}, n \geq 1)$ is a Cauchy sequence in $L^1(P, \mathscr{F}_t)$. Therefore there exists a random variable m_t of $L^1(P, \mathscr{F}_t)$ such that

$$\lim m_t^{(n)} = m_t \quad \text{in } L^1(P, \mathscr{F}_t). \tag{4}$$

If we can show that m_t is a $(P, \mathscr{F}_t)$-martingale, then we have completed the proof since $\lim \|m^{(n)} - m\|_{\mathscr{M}} = 0$. For this one must prove that for all $0 \leq s \leq t \leq 1$, $A \in \mathscr{F}_s$,

$$E[1_A m_t] = E[1_A m_s]. \tag{5}$$

This is a consequence of the fact that for each $n \geq 1$, $m_t^{(n)}$ is an $\mathscr{F}_t$-martingale:

$$E[1_A m_t^{(n)}] = E[1_A m_s^{(n)}], \tag{6}$$

and that for each $t \in [0, 1]$, m_t is the $L^1(P, \mathscr{F}_t)$-limit of $m_t^{(n)}$ (hence $\lim E[1_A m_t^{(N)}] = E[1_A m_t]$).

E13. The proof parallels that of E12 [use Jensen's inequality with $\varphi: x \to x^2$ instead of $\varphi: x \to |x|$, and replace $L^1(P, \mathscr{F}_t)$ by $L^2(P, \mathscr{F}_t)$]. In addition, one must observe that $\|m\|_{\mathscr{M}^2}$ is generated by the scalar product $\langle m, M \rangle_{\mathscr{M}^2} = E[m_1 M_1]$.

CHAPTER II

Point Processes, Queues, and Intensities

1. Counting Processes and Queues
2. Watanabe's Characterization
3. Stochastic Intensity, General Case
4. Predictable Intensities
5. Representation of Queues
6. Random Changes of Time
7. Cryptographic Point Processes
References
Solutions to Exercises, Chapter II

1. Counting Processes and Queues

A point processes over the half line $[0, \infty)$ can be viewed in three different ways: as a sequence of nonnegative random variables, as a discrete random measure, or via its associated counting process. We adopt in this book the last point of view. It is quite well adapted to those problems of a dynamical nature where time does not merely play the role of a parameter.

Simple Univariate Point Processes

A realization of a point process over $[0, \infty)$ can be described by a sequence T_n in $[0, \infty]$ such that

$$T_0 = 0,$$
$$T_n < \infty \Rightarrow T_n < T_{n+1}. \tag{1.1}$$

This realization is, by definition, *nonexplosive* iff

$$T_\infty = \lim \uparrow T_n = +\infty. \tag{1.2}$$

1. Counting Processes and Queues

To each realization T_n corresponds a counting function N_t defined by

$$N_t = \begin{cases} n & \text{if } t \in [T_n, T_{n+1}), n \geq 0, \\ +\infty & \text{if } t \geq T_\infty. \end{cases} \quad (1.3)$$

N_t is therefore a right-continuous step function such that $N_0 = 0$, and its jumps are upward jumps of magnitude 1. (See Figure 1.)

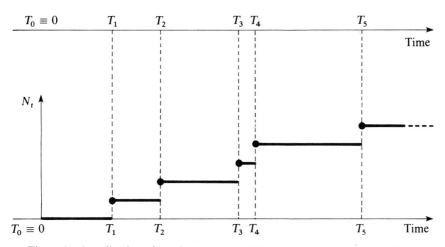

Figure 1 A realization of a point process and the associated counting function.

If the above T_n's are random variables, defined on some probability space $(\Omega, \mathscr{F}, P)$, one then calls the sequence T_n *a point process*. The associated *counting process* N_t is also called a point process, by abuse of notation (an innocuous one, since N_t and T_n obviously carry the same information). *Henceforward, unless explicitly mentioned, attention will be restricted to P-nonexplosive point processes*, that is to say point processes such that, *P*-a.s.,

$$N_t < \infty, \quad t \geq 0 \quad \text{(or equivalently } T_\infty \equiv \infty\text{)}.$$

Moreover, if the condition

$$E[N_t] < \infty, \quad t \geq 0$$

holds, the point process N_t is said to be *integrable*.

Multivariate Point Processes

Let T_n be a point process defined on $(\Omega, \mathscr{F}, P)$, and let $(Z_n, n \geq 1)$ be a sequence of $\{1, 2, \ldots, k\}$-valued random variables, also defined on $(\Omega, \mathscr{F}, P)$. Define for all i, $1 \leq i \leq k$, and all $t \geq 0$:

$$N_t(i) = \sum_{n \geq 1} 1(T_n \leq t) 1(Z_n = i). \quad (1.4)$$

(See Figure 2.)

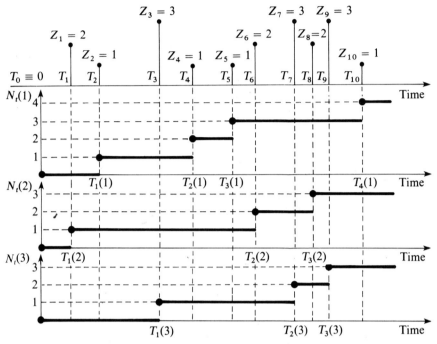

Figure 2 A realization of a 3-variate point process.

Both the k-vector process $N_t = (N_t(1), \ldots, N_t(k))$ and the double sequence $(T_n, Z_n, n \geq 1)$ are called *k-variate point processes*.

The limit $T_\infty \equiv \lim_{n \uparrow} T_n$ is the explosion point of N_t. If for some i ($1 \leq i \leq k$) $N_t(i)$ has an explosion point $T_\infty(i)$, it is necessarily "contained in T_∞" in the sense that on the set $\{T_\infty(i) < \infty\}$ we have $T_\infty(i) = T_\infty$. Note also that the $N_t(i)$'s have *no common jumps*; in general we say that two point processes $N_t(1)$ and $N_t(2)$ defined on $(\Omega, \mathscr{F}, P)$ have no common jumps if $\Delta N_t(1) \Delta N_t(2) = 0$, $t \geq 0$, P-a.s.

Queueing Processes

A (simple) queueing process Q_t is a N_+-valued process, defined on some $(\Omega, \mathscr{F}, P)$, and of the form

$$Q_t = Q_0 + A_t - D_t, \tag{1.5}$$

where A_t and D_t are P-nonexplosive point processes *without common jumps*. The above definition implies that P-a.s., $D_t \leq Q_0 + A_t$, $t \geq 0$, since Q_t is nonnegative. Q_t is called the *state process*; Q_0 is the *initial state*. For each $t \geq 0$, the random variable Q_t is to be interpreted as the number of customers waiting in line or being attended by a server, A_t is the number of arrivals in $(0, t]$, and D_t is the number of departures in $(0, t]$. The processes A_t and D_t

1. Counting Processes and Queues

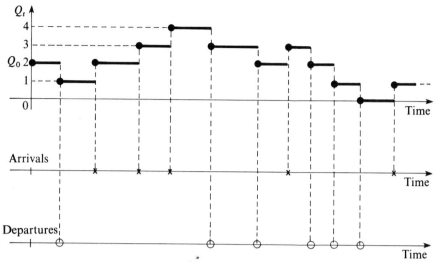

Figure 3 A realization of a simple queueing process.

are called the *arrival* (*or input*) *process* and the *departure* (*or output*) *process* respectively. (See Figure 3.)

E1 Exercise. Let $N_t(1)$ and $N_t(2)$ be two nonexplosive point processes without common jumps and let Q_0 be a nonnegative integer-valued random variable. Define $X_t = Q_0 + N_t(1) - N_t(2)$ and $m_t = \min\{X_s \wedge 0, s \in [0, t]\}$. Show that $Q_t = X_t - m_t$ is a simple queueing process with arrival process $A_t = N_t(1)$ and departure process $D_t = \int_0^t 1(Q_{s-} > 0) \, dN_s(2)$; also $m_t = -\int_0^t 1(Q_{s-} = 0) \, dN_s(2)$.

Doubly Stochastic or Conditional Poisson Processes

A nonmathematical definition of doubly stochastic Poisson processes consists in describing a two-step randomization procedure: for instance, in the first step one draws at random the trajectory of a "driving" process, say Y_t, and once the whole trajectory is selected, one generates a Poisson process of intensity $f(t, Y_t)$. Of course such an intensity is random, since Y_t is a random process, and therefore the point process is Poisson only conditionally with respect to Y_t.

The domain of application of such conditional Poisson models is considerable, and the interested reader will find a host of examples in the book of Snyder {140}, Srinivasan {141} and the monograph of Grandell {65}. For the time being, the above heuristic definition of a doubly stochastic point process will be formalized and extended, namely:

D1 Definition (Doubly Stochastic or Conditional Poisson Processes). Let N_t be a point process adapted to a history $\mathscr{F}_t$, and let λ_t be a nonnegative measurable process [all given on the same probability space $(\Omega, \mathscr{F}, P)$]. Suppose that

$$\lambda_t \text{ is } \mathscr{F}_0\text{-measurable}, \qquad t \geq 0, \tag{1.6}$$

and that

$$\int_0^t \lambda_s \, ds < \infty \quad P\text{-a.s.}, \qquad t \geq 0. \tag{1.7}$$

If for all $0 \leq s \leq t$ and all $u \in R$

$$E[e^{iu(N_t - N_s)} | \mathscr{F}_s] = \exp\left\{(e^{iu} - 1) \int_s^t \lambda_v \, dv\right\}, \tag{1.8}$$

then N_t is called a $(P, \mathscr{F}_t)$-*doubly stochastic Poisson process* or a $(P, \mathscr{F}_t)$-*conditional Poisson process* with the (stochastic) intensity λ_t.

If λ_t is deterministic—in which case the notation "$\lambda(t)$" is used—then N_t is called a $(P, \mathscr{F}_t)$-*Poisson process*.[1] If moreover $\mathscr{F}_t \equiv \mathscr{F}_t^N$, one simply says: N_t is a Poisson process with the intensity $\lambda(t)$. If $\mathscr{F}_t \equiv \mathscr{F}_t^N$, $\lambda(t) \equiv 1$, then N_t is the *standard Poisson process*.

Comments. The conditions (1.6) and (1.8) imply that for all $0 \leq s \leq t$, $N_t - N_s$ is P-independent of $\mathscr{F}_s$ given $\mathscr{F}_0$. Indeed, since λ_t is $\mathscr{F}_0$-measurable, conditioning by $\mathscr{F}_0$ does not change the right-hand side of (1.8), and therefore

$$E[e^{iu(N_t - N_s)} | \mathscr{F}_0] = E[e^{iu(N_t - N_s)} | \mathscr{F}_0 \vee \mathscr{F}_s], \tag{1.8'}$$

which implies the announced conditional independence.

Also (1.8) yields, for all $0 \leq s \leq t$ and all $k \geq 0$,

$$P[N_t - N_s = k | \mathscr{F}_s] = e^{-\int_s^t \lambda_u \, du} \frac{(\int_s^t \lambda_u \, du)^k}{k!} \tag{1.9}$$

(with the usual convention $0! = 1$).

It is clear that, given (1.6), the condition (1.8) is equivalent to both conditions (1.8') and (1.9).

Special Cases

(α) if, in D1, $\lambda_t = \Lambda$ where Λ is some nonnegative $\mathscr{F}_0$-measurable random variable, N_t is called a *homogeneous* doubly stochastic Poisson process.

---

[1] For the sake of typographical economy, we will use the shorter appelations "$\mathscr{F}_t$-doubly stochastic Poisson process," "$\mathscr{F}_t$-Poisson process," "$\mathscr{F}_t$-conditional Poisson process," omitting P in the notation. Also we will call λ_t the "$\mathscr{F}_t$-intensity" instead of the "$(P, \mathscr{F}_t)$-intensity." This is in accord with our previous notational conventions of Chapter I.

(β) if, in D1, $\lambda_t = f(t, Y_t)$ for some appropriately measurable nonnegative function f and for some measurable process Y_t, and if $\mathscr{F}_0$ contains $\mathscr{F}_\infty^Y$, then N_t is called a doubly stochastic Poisson process *driven by* Y_t. This case is the most common in applications.

The terminology "doubly stochastic" was introduced by Cox {37}. Some authors prefer "conditional" (an $\mathscr{F}_t$-conditional Poisson process, a Poisson process "conditioned by Y_t," etc.). We will adopt either one of these terminologies.

In the next section an alternative (and equivalent) definition of doubly stochastic Poisson processes will be given, which has the advantage of a greater generality, since it will be extended to include point processes with a stochastic intensity which are not Poisson or doubly stochastic Poisson processes.

2. Watanabe's Characterization

Let N_t be a doubly stochastic Poisson process with the $\mathscr{F}_t$-intensity λ_t. Multiplying both sides of (1.9) by k and summing over all $k \geq 0$, one obtains, in view of the conditional independence of $N_t - N_s$ and $\mathscr{F}_s$ given $\mathscr{F}_0$ and of the $\mathscr{F}_0$-measurability of λ_t,

$$E[N_t - N_s | \mathscr{F}_s] = E\left[\int_s^t \lambda_u \, du \, \Big| \, \mathscr{F}_s\right]. \tag{2.1}$$

Suppose now that for all $t \geq 0$ $E[\int_0^t \lambda_u \, du] < \infty$ and hence (from (2.1)) $E[N_t] < \infty$. In other words, N_t is integrable. It then follows that the process M_t defined by

$$M_t = N_t - \int_0^t \lambda_s \, ds \tag{2.2}$$

is integrable, and in view of (2.1), a $\mathscr{F}_t$-martingale.

Let $\mathscr{H}$ be the vector space of measurable bounded processes C_t, adapted to $\mathscr{F}_t$, and such that

$$E\left[\int_0^t C_s \, dN_s\right] = E\left[\int_0^t C_s \lambda_s \, ds\right], \qquad t \geq 0.^2$$

By (2.1), $\mathscr{H}$ contains the indicators of sets in the π-system[3]

$$\mathscr{S} = \{(s, t] \times A, 0 \leq s \leq t, A \in \mathscr{F}_s\}$$

---
[2] $\int_0^t C_s \, dN_s = \sum_{n \geq 1} C_{T_n} 1(T_n \leq t)$ and $\int_0^\infty C_s \, dN_s = \sum_{n \geq 1} C_{T_n} 1(T_n < \infty)$.
[3] See Appendix A1, §2 for definitions (§2).

that generates $\mathcal{P}(\mathcal{F}_t)$. It also contains the constants and is stable by (t, ω)-pointwise bounded limits. Therefore by the monotone class theorem (Appendix A1, T4), $\mathcal{H}$ contains all bounded $\mathcal{F}_t$-predictable processes C_t. In summary:

Partial Result. *If N_t is an integrable doubly stochastic Poisson process with the $\mathcal{F}_t$-intensity λ_t, then for all nonnegative $\mathcal{F}_t$-predictable processes C_t*

$$E\left[\int_0^\infty C_s \, dN_s\right] = E\left[\int_0^\infty C_s \lambda_s \, ds\right]. \tag{2.3}$$

It turns out that the equality (2.3) can be used as a definition for the intensity of a doubly stochastic Poisson process. To see this a few preliminary lemmas of a technical nature are needed:

L2 Lemma. *Let N_t be a point process adapted to the history $\mathcal{F}_t$, and let λ_t be a nonnegative process, $\mathcal{F}_t$-progressive and satisfying (1.7). Suppose moreover that (2.3) holds for all nonnegative $\mathcal{F}_t$-predictable processes C_t. Then N_t is P-nonexplosive and M_t defined by (2.2) is an $\mathcal{F}_t$-local martingale.*

PROOF. Let $S_n = \inf\{t \mid \int_0^t \lambda_s \, ds \geq n\}$ if $\{\cdots\} \neq \emptyset$, and $+\infty$ otherwise. For each $n \geq 0$, S_n is an $\mathcal{F}_t$-stopping time, and by (1.7), $S_n \uparrow \infty$, P-a.s. Writing (2.3) with $C_t = 1(t \leq S_n)$, obtain $E[N_{S_n}] = E[\int_0^{S_n} \lambda_s \, ds] \leq n < \infty$ for all $n \geq 1$, and therefore $N_{S_n} < \infty, n \geq 1$, P-a.s., which implies the nonexplosiveness since $S_n \uparrow \infty$, P-a.s.

The rest of the proof will be carried out by the reader in the following exercise:

E2 Exercise. Complete the proof of the local martingale property. Also prove the following:

L3 Lemma. *Under the conditions of Lemma L2, let X_t be an $\mathcal{F}_t$-predictable process such that for all $t \geq 0$*

$$E\left[\int_0^t |X_s|\lambda_s \, ds\right] < \infty. \tag{2.4}$$

Then $Y_t = \int_0^t X_s \, dM_s$ is an $\mathcal{F}_t$-martingale, where $M_t = N_t - \int_0^t \lambda_s \, ds$. If (2.4) is replaced by the weaker condition

$$\int_0^t |X_s|\lambda_s \, ds < \infty \quad \text{P-a.s.,} \quad t \geq 0 \tag{2.5}$$

then $Y_t = \int_0^t X_s \, dM_s$ is an $\mathcal{F}_t$-local martingale.

We now turn to the fundamental characterization theorem:

2. Watanabe's Characterization

T4 Theorem (Characterization of Doubly Stochastic Poisson Processes). *Let N_t be a point process adapted to some history $\mathscr{F}_t$, and let λ_t be a nonnegative measurable process such that for all $t \geq 0$*

(α) λ_t is $\mathscr{F}_0$-measurable,
(β) $\int_0^t \lambda_s \, ds < \infty$, P-a.s.

Then, if the equality

$$E\left[\int_0^\infty C_s \, dN_s\right] = E\left[\int_0^\infty C_s \lambda_s \, ds\right] \tag{2.6}$$

is verified for all nonnegative $\mathscr{F}_t$-predictable processes C_t, N_t is a doubly stochastic Poisson process with the $\mathscr{F}_t$-intensity λ_t.[4]

PROOF. For each $u \in R$, define the process M_t^u by

$$M_t^u = \frac{e^{iuN_t}}{\exp\{(e^{iu}-1)\int_0^t \lambda_s \, ds\}}. \tag{2.7}$$

From the exponential formula (Appendix A4, T4),

$$M_t^u = 1 + \int_0^t (e^{iu}-1)M_{s-}^u (dN_s - \lambda_s \, ds). \tag{2.8}$$

Let $S_n = \inf\{t \mid \int_0^t \lambda_s \, ds \geq n\}$ if $\{\cdots\} \neq \emptyset$, and $+\infty$ otherwise. By L3 and (2.8), $M_{t \wedge S_n}^u$ is an $\mathscr{F}_t$-martingale. Hence

$$E\left[\frac{\exp\{iu(N_{t \wedge S_n} - N_{s \wedge S_n})\}}{\exp\{(e^{iu}-1)\int_{s \wedge S_n}^{t \wedge S_n} \lambda_v \, dv\}} \bigg| \mathscr{F}_s\right] = 1.$$

Now λ_t and S_n are $\mathscr{F}_0$-measurable for all $t \geq 0$, $n \geq 0$. Hence

$$E[\exp\{iu(N_{t \wedge S_n} - N_{s \wedge S_n})\} \mid \mathscr{F}_s] = \exp\left\{(e^{iu}-1)\int_{s \wedge S_n}^{t \wedge S_n} \lambda_v \, dv\right\}.$$

Letting $n \uparrow \infty$ [in which case $S_n \uparrow \infty$, P-a.s., by condition (β)], we obtain (1.8), which is the desired result. □

The version presented by Watanabe in 1964, with a quite different proof, concerns Poisson processes. It is the first important result linking point processes and martingales:

T5 Theorem (Watanabe [9]). *Let N_t be a point process adapted to the history $\mathscr{F}_t$, and let $\lambda(t)$ be a locally integrable nonnegative measurable function. Suppose that $N_t - \int_0^t \lambda(s) \, ds$ is an $\mathscr{F}_t$-martingale. Then N_t is an $\mathscr{F}_t$-Poisson*

---

[4] This result is somewhat puzzling at first sight: indeed, one does not believe *a priori* that the Poisson law, which is a very picturesque law, can be extracted from an equation as flat and dull as (2.6).

process with the intensity $\lambda(t)$, i.e., for all $0 \leq s \leq t$, $N_t - N_s$ is a Poisson random variable with parameter $\int_0^t \lambda(u)\, du$, independent of $\mathscr{F}_s$.

PROOF. As in the beginning of the present paragraph, we can show that (2.3) holds with $\lambda_t = \lambda(t)$. The rest follows from T5. □

The following extension of Watanabe's result will be useful in the analysis of multiple output streams of queueing systems (Chapter V).

T6 Theorem (Multichannel Watanabe's Theorem). *Let $(N_t(1), \ldots, N_t(k))$ be a k-variate point process adapted to some history $\mathscr{F}_t$, and suppose that for all $1 \leq i \leq k$*

$$N_t(i) - \int_0^t \lambda_i(s)\, ds \text{ is an } \mathscr{F}_t\text{-martingale}, \tag{2.9}$$

where $\lambda_i(t)$ is a nonnegative locally integrable measurable function for all $1 \leq i \leq k$. Then the $N_t(i)$, $1 \leq i \leq k$, are independent $\mathscr{F}_t$-Poisson processes with the intensities $\lambda_i(t)$, $1 \leq i \leq k$, respectively.

(One must insist on the assumption that the $\lambda_i(t)$'s are deterministic.)

PROOF. By T5, for all $1 \leq i \leq k$, $N_t(i)$ is an $\mathscr{F}_t$-Poisson process with intensity $\lambda_i(t)$. In particular, for all $0 \leq s \leq t$ and all $1 \leq i \leq k$, $N_t(i) - N_s(i)$ is independent of $\mathscr{F}_s$. Since $\mathscr{F}_t$ contains $\mathscr{F}_t^N(i)$ for all $1 \leq i \leq k$, it follows that for any fixed i and for all $0 \leq s \leq t$, $N_t(i) - N_s(i)$ is independent of all the increments $\{N_v(j) - N_u(j), t \leq u \leq v \text{ or } u \leq v \leq s; j \neq i\}$.

It remains to show that for all $1 \leq i \neq j \leq k$ and for all $0 \leq s \leq t$, $N_t(i) - N_s(i)$ is P-independent of $N_t(j) - N_s(j)$. This can be done as follows, in the case $i = 1$, $j = 2$ (for notational convenience): define, for fixed $u, v \in R$, $X_t = e^{iuN_t(1) + ivN_t(2)}$, and observe that

$$X_t = 1 + \int_0^t (e^{iu} - 1)X_{s-}\, dN_s(1) + \int_0^t (e^{iv} - 1)X_{s-}\, dN_s(2).$$

Subtracting $\int_0^t (e^{iu} - 1)X_{s-}\lambda_1(s)\, ds + \int_0^t (e^{iv} - 1)X_{s-}\lambda_2(s)\, ds$ from both sides of the above equality and using L3, we see that

$$X_t - \int_0^t [(e^{iu} - 1)\lambda_1(s) + (e^{iv} - 1)\lambda_2(s)]X_s\, ds$$

is an $\mathscr{F}_t$-martingale. Hence, for all $0 \leq s \leq t$,

$$E\left[\frac{X_t}{X_s}\right] = 1 + \int_s^t E\left[\frac{X_y}{X_s}\right]((e^{iu} - 1)\lambda_1(y) + (e^{iv} - 1)\lambda_2(y))\, dy.$$

Therefore, recalling the definition of X_t:

$$E[e^{iu(N_t(1)-N_s(1))+iv(N_t(2)-N_s(2))}]$$

$$= \exp\left\{(e^{iu}-1)\int_s^t \lambda_1(y)\,dy + (e^{iv}-1)\int_s^t \lambda_2(y)\,dy\right\}$$

$$= E[e^{iu(N_t(1)-N_s(1))}]E[e^{iv(N_t(2)-N_s(2))}],$$

and the announced result then follows from (Appendix A1, C21). □

E3 Exercise [4]. Devise a direct proof of Watanabe's result T5 inspired by the above proof for T6.

3. Stochastic Intensity, General Case

If we want to define the $\mathscr{F}_t$-intensity of a point process N_t in the general case, a definition such as D1 is not available. However, Theorem T4 suggests that the following definition is appropriate:

D7 Definition (Stochastic Intensity [1]). Let N_t be a point process adapted to some history $\mathscr{F}_t$, and let λ_t be a nonnegative $\mathscr{F}_t$-progressive process such that for all $t \geq 0$

$$\int_0^t \lambda_s\,ds < \infty \quad \text{P-a.s.} \tag{3.1}$$

If for all nonnegative $\mathscr{F}_t$-predictable processes C_t, the equality

$$E\left[\int_0^\infty C_s\,dN_s\right] = E\left[\int_0^\infty C_s\lambda_s\,ds\right] \tag{3.2}$$

is verified, then we say: N_t admits the $(P,\mathscr{F}_t)$-intensity (or $\mathscr{F}_t$-intensity) λ_t.

E4 Exercise. Let N_t be a point process with the $\mathscr{F}_t$-intensity λ_t. Show that if λ_t is $\mathscr{G}_t$-progressive for some history $\mathscr{G}_t$ such that $\mathscr{F}_t^N \subseteq \mathscr{G}_t \subseteq \mathscr{F}_t$, $t \geq 0$, then λ_t is also the $\mathscr{G}_t$-intensity of N_t.

T8 Theorem (Integration Theorem). If N_t admits the $\mathscr{F}_t$-intensity λ_t (where $\int_0^t \lambda_s\,ds < \infty$. P-a.s., $t \geq 0$), then N_t is P-nonexplosive and

(α) $M_t = N_t - \int_0^t \lambda_s\,ds$ is an $\mathscr{F}_t$-local martingale;
(β) if X_t is an $\mathscr{F}_t$-predictable process such that $E[\int_0^t |X_s|\lambda_s\,ds] < \infty$, $t \geq 0$, then $\int_0^t X_s\,dM_s$ is an $\mathscr{F}_t$-martingale;
(γ) if X_t is an $\mathscr{F}_t$-predictable process such that $\int_0^t |X_s|\lambda_s\,ds < \infty$, P-a.s., $t \geq 0$, then $\int_0^t X_s\,dM_s$ is an $\mathscr{F}_t$-local martingale.

PROOF. T8 is just a concatenation of L2 and L3. □

Some examples were given after Definition D1 in Chapter 1 [the *transition counting process* $N_t(i, j)$ in a Markov chain, the *passage* (through state i) *counting process* $N_t(i)$] that yielded a martingale relation of type (α) in T8. This martingale relation in fact characterizes $\lambda_t[q_{ij}1(X_t = i)$ for $N_t(i, j)$, $q_{X_t,j}$ for $N_t(i)]$ as the intensity of the corresponding processes $[N_t(i, j), N_t(i)]$. Indeed:

T9 Theorem (Martingale Characterization of Intensity). *Let N_t be a non-explosive point process adapted to $\mathscr{F}_t$, and suppose that for some nonnegative $\mathscr{F}_t$-progressive process λ_t and for all $n \geq 1$,*

$$N_{t \wedge T_n} - \int_0^{t \wedge T_n} \lambda_s \, ds \text{ is a } (P, \mathscr{F}_t)\text{-martingale}. \qquad (3.3)$$

Then N_t is the $\mathscr{F}_t$-intensity of N_t.

PROOF. Repeat the arguments at the beginning of §2 to obtain

$$E\left[\int_0^{t \wedge T_n} C_s \, dN_s\right] = E\left[\int_0^{t \wedge T_n} C_s \lambda_s \, ds\right]$$

for all bounded $\mathscr{F}_t$-predictable processes C_t, and let T_n go to ∞. □

Remark. For all $n \geq 1$ and all $0 \leq s \leq t$, we have $E[N_{t \wedge T_n} - N_{s \wedge T_n} | \mathscr{F}_s] = E[\int_{s \wedge T_n}^{t \wedge T_n} \lambda_u \, du | \mathscr{F}_s]$, and therefore, letting $n \uparrow \infty$,

$$E[N_t - N_s | \mathscr{F}_s] = E\left[\int_s^t \lambda_u \, du \Big| \mathscr{F}_s\right], \qquad (3.4)$$

which reminds one of a more classical definition of the intensity. In particular, if λ_t is right-continuous and bounded, it follows from (3.4), by application of the Lebesgue averaging theorem and the Lebesgue dominated-convergence theorem successively, that

$$\lim_{t \downarrow s} \frac{1}{t - s} E[N_t - N_s | \mathscr{F}_s] = \lambda_s \quad P\text{-a.s.} \qquad (3.5)$$

E5 Exercise. Suppose N_t has the $\mathscr{F}_t$-intensity λ_t, and let $\mathscr{G}_t$ be a history such that $\mathscr{G}_\infty$ is independent of $\mathscr{F}_t$ for all $t \geq 0$. Show that λ_t is also the $\mathscr{F}_t \vee \mathscr{G}_t$-intensity of N_t.

Parameters of a Queueing Process

Let Q_t be a simple queueing process adapted to $\mathscr{F}_t$ and with input process A_t and output process D_t. Whenever $Q_{t-} = 0$, there is no departure at time t; therefore for any non-negative process C_t

$$\int_0^\infty C_s \, dD_s = \int_0^\infty C_s 1(Q_{s-} > 0) \, dD_s. \qquad (*)$$

3. Stochastic Intensity, General Case

Suppose now that D_t has the $\mathscr{F}_t$-intensity μ_t. Then for any nonnegative $\mathscr{F}_t$-predictable process C_t

$$E\left[\int_0^\infty C_s\, dD_s\right] = E\left[\int_0^\infty C_s \mu_s\, ds\right].$$

In particular,

$$E\left[\int_0^\infty C_s 1(Q_{s-} > 0)\, dD_s\right] = E\left[\int_0^\infty C_s 1(Q_{s-} > 0)\mu_s\, ds\right],$$

and therefore by (*)

$$E\left[\int_0^\infty C_s\, dD_s\right] = E\left[\int_0^\infty C_s 1(Q_s > 0)\mu_s\, ds\right],$$

which reads: $\mu_t 1(Q_t > 0)$ is the $\mathscr{F}_t$-intensity of D_t. Thus

T10 Theorem [2]. *Let D_t be the departure process of a simple queueing process adapted to $\mathscr{F}_t$; then if D_t admits the $\mathscr{F}_t$-intensity μ_t, it also admits the $\mathscr{F}_t$ intensity $\mu_t 1(Q_t > 0)$.*

D11 Definition (Parameters of a Queueing Process). Let Q_t be a simple queueing process adapted to $\mathscr{F}_t$ and with input and output processes A_t and D_t respectively. If A_t and D_t have the $\mathscr{F}_t$-intensities λ_t and $\mu_t 1(Q_t > 0)$ respectively, then one says Q_t has the $\mathscr{F}_t$-*parameters* λ_t and μ_t, where λ_t is the arrival intensity, and μ_t is the service potential.

E6 Exercise. The queue Q_t corresponding to the birth-and-death process of (Appendix A1, §8) has the $\mathscr{F}_t$-parameters λ_{Q_t} and μ_{Q_t}.

E7 Exercise. Let Q_t be a $M/M/1$ queue with $\mathscr{F}_t^Q$-parameters λ and μ (i.e., take $\lambda_n \equiv \lambda$ and $\mu_n \equiv \mu$ in E6). Define θ_t to be the *last regeneration of service before time t*, i.e., if $(T_n, n \geq 1)$ is the sequence of jump times of Q_t,

$$\theta_t = \begin{cases} 0 & \text{if } t < T_1 \text{ and } Q_0 > 0 \\ \sup\{T_n, n \geq 1 | T_n \leq t \text{ and either} \\ \qquad \Delta Q_{T_n} = -1, Q_{T_n} > 0 \text{ or } \Delta Q_{T_n} = 1, Q_{T_n} = 0\} & \text{if } Q_t > 0 \\ t & \text{if } Q_t = 0. \end{cases}$$

Define ξ_t the *elapsed service at time t* by

$$\xi_t = t - \theta_t. \tag{3.7}$$

Find for all real numbers u an evolution equation for $\exp(iu\xi_t)$ in terms of the processes Q_t and D_t. Find an evolution equation for $\phi_t(u) = E[\exp(iu\xi_t)]$ in terms of $p_0(t) = P[Q_t = 0]$. Solve for $\phi_t(u)$. Find the limiting distribution for ξ_t [i.e. $\lim_{t \uparrow \infty} \phi_t(u)$] when $P[Q_t = n] = \rho^n(1 - \rho)$, where $\rho = \lambda/\mu$ is assumed to be strictly less than 1 [in which case $\rho_n = \rho^n(1 - \rho)$ is the equilibrium distribution of Q_t].

First-Order Equivalent Queues

E8 Exercise. Let Q_t be a simple queue with $\mathscr{F}_t$-parameters λ_t and μ_t where λ_t and μ_t are bounded. Define $p_n(t)$, $\lambda_n(t)$, and $\mu_n(t)$ by

$$p_n(t) = P[Q_t = n],$$
$$\lambda_n(t) = E[\lambda_t | Q_t = n], \qquad (3.8)$$
$$\mu_n(t) = E[\mu_t | Q_t = n].$$

Show that for all $n \geq 0$

$$p_n(t) = p_n(0) + \int_0^t (p_{n-1}(s)\lambda_{n-1}(s)\mathbf{1}(n > 0)$$
$$- p_n(s)(\lambda_n(s) + \mu_n(s)\mathbf{1}(n > 0))$$
$$+ p_{n+1}(s)\mu_{n+1}(s))\,ds. \qquad (3.9)$$

Comment. The queue Q_t is said to be *first-order equivalent* to a (nonhomogeneous) markovian queue $\tilde{Q}_t$ with parameters $\lambda_{\tilde{Q}_t}(t)$ and $\mu_{\tilde{Q}_t}(t)$: indeed, if Q_t and $\tilde{Q}_t$ have the same initial distribution, then they have the same distribution at any time $t \geq 0$. In particular, if Q_t is in equilibrium, i.e. if $p_n(t) \equiv p_n$, $t \geq 0$, $n \geq 0$, then it is equivalent to a birth-and-death process with parameters λ_n and μ_n determined by

$$\lambda_n p_n = \mu_{n+1} p_{n+1}, \qquad (3.10)$$

and moreover, if $p_n > 0$ for all $n \geq 0$, then λ_n can be chosen to be independent of n (Poissonian input).

E9 Exercise {120}. Let N_t be a Poisson process with intensity λ and let T be a fixed time (the terminal time). Let $\mathscr{C}$ be the class of all $\mathscr{F}_t^N$-stopping times τ bounded by T. Find $\tau^* \in \mathscr{C}$ such that $E[N_{\tau^*}(T - \tau^*)] \geq E[N_\tau(T - \tau)]$ for all $\tau \in \mathscr{C}$. [Comment: N_t can be interpreted as follows: it is the counting process of a flow of goods, or items, entering a warehouse. For any given item sojourning a time x in the warehouse, a fee proportional to x must be paid. At time T, the owner of the goods will take them back to the factory or dump them, so that there are no storage expenses after T. At an intermediary time τ, the owner of the goods has the option of removing the N_τ items present in the warehouse, thus saving an amount of money proportional to $N_\tau(T - \tau)$. The time τ is chosen according to the observation of N_t and cannot anticipate on the future observations, therefore it has to be a $\mathscr{F}_t^N$-stopping time.]

4. Predictable Intensities

At this point the reader might be puzzled by the fact that in the proof of T10, two $\mathscr{F}_t$-intensities for D_t were exhibited, namely μ_t and $\mu_t \mathbf{1}(Q_t > 0)$. This habit of speaking of *the* $(P, \mathscr{F}_t)$-intensity is subject to criticism. However, it is harmless, and we will stick to it. On the other hand, we want to show that if "the" intensity is constrained to be predictable, it is essentially unique. Also one can always find such a predictable version of the intensity.

4. Predictable Intensities

T12 Theorem (Uniqueness of Predictable Intensities). *Let N_t be a point process adapted to the history $\mathscr{F}_t$, and let λ_t and $\tilde{\lambda}_t$ be two $\mathscr{F}_t$-intensities of N_t which are $\mathscr{F}_t$-predictable. Then*

$$\lambda_t(\omega) = \tilde{\lambda}_t(\omega) \quad P(d\omega)\, dN_t(\omega)\text{-a.e.} \tag{4.1}$$

In particular, P-a.s.,

$$\lambda_{T_n} = \tilde{\lambda}_{T_n} \quad \text{on } \{T_n < \infty\}, \quad n \geq 1, \tag{4.2}$$

and

$$\lambda_t(\omega) = \tilde{\lambda}_t(\omega) \quad \lambda_t(\omega)\, dt \text{ and } \tilde{\lambda}_t(\omega)\, dt\text{-a.e.} \tag{4.3}$$

Also, P-a.s.,

$$\lambda_{T_n} > 0 \quad \text{on } \{T_n < \infty\}, \quad n \geq 1. \tag{4.4}$$

PROOF. Apply (3.2) to $C_s = 1(\lambda_s > \tilde{\lambda}_s)1(s \leq a)$ to obtain

$$E\left[\int_0^a 1(\lambda_s > \tilde{\lambda}_s)\lambda_s\, ds\right] = E\left[\int_0^a 1(\lambda_s > \tilde{\lambda}_s)\tilde{\lambda}_s\, ds\right],$$

which implies, since a is arbitrary,

$$1(\lambda_t(\omega) > \tilde{\lambda}_t(\omega)) = 0 \quad P(d\omega)\lambda_t(\omega)\, dt\text{- or } P(d\omega)\tilde{\lambda}_t(\omega)\, dt\text{-a.e.}$$

Similarly,

$$1(\lambda_t(\omega) < \tilde{\lambda}_t(\omega)) = 0 \quad P(d\omega)\lambda_t(\omega)\, dt\text{- or } P(d\omega)\tilde{\lambda}_t(\omega)\, dt\text{-a.e.}$$

(4.1) then follows, since *by definition* of the intensities λ_t and $\tilde{\lambda}_t$,

$$P(d\omega)\, dN_t(\omega) = P(d\omega)\lambda_t(\omega)\, dt = P(d\omega)\tilde{\lambda}_t(\omega)\, dt \quad \text{on } \mathscr{P}(\mathscr{F}_t). \tag{4.5}$$

The equalities (4.2) and (4.3) follow by Fubini's theorem (Appendix A4, T13).

To prove (4.4), apply (3.2) to

$$C_t = 1(\lambda_t = 0)1(T_{n-1} < t \leq T_n)$$

to obtain

$$E[1(\lambda_{T_n} = 0)1(T_n < \infty)] = E\left[\int_{T_{n-1}}^{T_n} 1(\lambda_t = 0)\lambda_t\, dt\right] = 0,$$

which implies the announced result. □

T13 Theorem (Existence of Predictable Versions of the Intensity). *Let N_t be a point process with an $\mathscr{F}_t$-intensity λ_t. Then one can find an $\mathscr{F}_t$-intensity $\tilde{\lambda}_t$ that is $\mathscr{F}_t$-predictable.*

PROOF. It suffices to define $\tilde{\lambda}_t(\omega)$ as the Radon–Nikodym derivative (Appendix A1, T10) of the restriction to $\mathscr{P}(\mathscr{F}_t)$ of $P(d\omega)\lambda_t(\omega)\, dt$ with respect to the restriction to $\mathscr{P}(\mathscr{F}_t)$ of $P(d\omega)\, dt$. Indeed, by that definition,

$$E\left[\int_0^\infty C_s\, dN_s\right] = E\left[\int_0^\infty C_s \lambda_s\, ds\right] = E\left[\int_0^\infty C_s \tilde{\lambda}_s\, ds\right]$$

for all nonnegative $\mathscr{F}_t$-predictable processes C_t, and $\tilde{\lambda}_t$ is $\mathscr{F}_t$-predictable, the mapping $(t, \omega) \to \tilde{\lambda}_t(\omega)$ being $\mathscr{P}(\mathscr{F}_t)$-measurable. □

Remark. Since $P(d\omega)\lambda_t(\omega)\, dt = P(d\omega)\, dN_t(\omega)$ on $\mathscr{P}(\mathscr{F}_t)$, one observes that, by definition of $\tilde{\lambda}_t$, $\tilde{\lambda}_t$ is just the Radon–Nikodym derivative

$$\tilde{\lambda}_t(\omega) = \left(\frac{dP\, dN_u}{dP\, du}\right)(t, \omega) \quad \text{on } \mathscr{P}(\mathscr{F}_t).$$

From now on, *the* $(P, \mathscr{F}_t)$-*intensity is the predictable version.*

T14 Theorem (Change of History for Intensities). *Let N_t be a point process with the $\mathscr{F}_t$-intensity λ_t. Let $\mathscr{G}_t$ be a history of N_t smaller than $\mathscr{F}_t$, i.e.,*

$$\mathscr{F}_t^N \subseteq \mathscr{G}_t \subseteq \mathscr{F}_t, \quad t \geq 0. \tag{4.6}$$

Then N_t admits a $\mathscr{G}_t$-intensity μ_t defined by

$$\mu_t(\omega) = \left(\frac{\lambda_u\, dP\, du}{dP\, du}\right)(t, \omega) \quad \text{on } \mathscr{P}(\mathscr{G}_t). \tag{4.7}$$

PROOF. By the definition of Radon–Nikodym derivatives, (4.7) is equivalent to

$$E\left[\int_0^\infty C_s \lambda_s\, ds\right] = E\left[\int_0^\infty C_s \mu_s\, ds\right]$$

for all nonnegative $\mathscr{G}_t$-predictable processes C_t. Such processes C_t are *a fortiori* $\mathscr{F}_t$-predictable, and therefore, by definition of the $\mathscr{F}_t$ intensity λ_t,

$$E\left[\int_0^\infty C_s\, dN_s\right] = E\left[\int_0^\infty C_s \lambda_s\, ds\right].$$

Hence for all nonnegative $\mathscr{G}_t$-predictable processes C_t,

$$E\left[\int_0^\infty C_s\, dN_s\right] = E\left[\int_0^\infty C_s \mu_s\, ds\right],$$

and this defines μ_t as the $\mathscr{G}_t$-intensity of N_t. □

Comment. Theorem T14 can be quoted in loose terms as follows: *If N_t admits the $\mathscr{F}_t$-intensity λ_t and if $\mathscr{G}_t$ is a history of N_t which is smaller than $\mathscr{F}_t$, then $E[\lambda_t | \mathscr{G}_t]$ is the $\mathscr{G}_t$-intensity of N_t.*

We will explain in a few lines what is wrong with such a statement from a pure mathematician's point of view. First we wish to give a pseudo-proof of the above avatar of T14.

PSEUDO-PROOF OF T14. One must show that:

$$E\left[\int_0^\infty C_s\, dN_s\right] = E\left[\int_0^\infty C_s E[\lambda_s | \mathscr{G}_s]\, ds\right]$$

4. Predictable Intensities

for all nonnegative $\mathscr{G}_t$-predictable processes C_t. But since $\mathscr{G}_t$-predictable $\Rightarrow$ $\mathscr{F}_t$-predictable when $\mathscr{G}_t$ is smaller than $\mathscr{F}_t$, $E[\int_0^\infty C_s dN_s] = E[\int_0^\infty C_s \lambda_s ds]$ by definition of λ_t as the $\mathscr{F}_t$-intensity of N_t. It therefore suffices to show that

$$E\left[\int_0^\infty C_s \lambda_s \, ds\right] = E\left[\int_0^\infty C_s E[\lambda_s | \mathscr{G}_s] \, ds\right].$$

This follows by application of Fubini's theorem, since

$$E\left[\int_0^\infty C_s \lambda_s \, ds\right] = \int_0^\infty E[C_s \lambda_s] \, ds = \int_0^\infty E[C_s E[\lambda_s | \mathscr{G}_s]] \, ds$$

(for the last equality, use the fact that C_s is $\mathscr{G}_s$-measurable), and

$$E\left[\int_0^\infty C_s E[\lambda_s | \mathscr{G}_s] \, ds\right] = \int_0^\infty E[C_s E[\lambda_s | \mathscr{G}_s]] \, ds. \qquad \square$$

Let us now appraise the validity of the above simplified proof. In fact, the proof would be completely valid if one were granted that process $E[\lambda_t | \mathscr{G}_t]$ is measurable (Fubini then applies, just as was done). More precisely, if for each $t \geq 0$, one could find a version μ_t of the conditional expectation $E[\lambda_t | \mathscr{G}_t]$ such that the process μ_t were $\mathscr{G}_t$-progressive, then the pseudo-proof would acquire the status of proof, and the simplified statement of T14 would become a well-formulated statement. [Recall that it was imposed in the definition of intensity that the intensity must be progressive. The $\mathscr{F}_t$-progressiveness requirement for the $\mathscr{F}_t$-intensity λ_t of a point process N_t serves two technical purposes: first it allows one to speak about integrals such as $\int_0^t \lambda_s ds$, since progressiveness implies measurability, and secondly $\int_0^t \lambda_s ds$ is then $\mathscr{F}_t$-measurable (see E10); the latter is needed in statements such as "$N_t - \int_0^t \lambda_s ds$ is an $\mathscr{F}_t$-martingale" because an $\mathscr{F}_t$-martingale must be, by definition, adapted to $\mathscr{F}_t$.]

The correct proof of T14 circumvents all these measurability problems, but it does so in a rather abstract way. Fortunately, in practical situations, those situations where actual computations are performed, one will always find versions μ_t of $E[\lambda_t | \mathscr{G}_t]$ such that the process μ_t is adapted to $\mathscr{G}_t$ and right- (left-) continuous, and therefore $\mathscr{G}_t$-progressive ($\mathscr{G}_t$-predictable).

E10 Exercise. Show that if a nonnegative process X_t is $\mathscr{F}_t$-progressive, then $\int_0^t X_s ds$ is adapted to $\mathscr{F}_t$.

The next theorem describes the statistics of the imbedded "mark process" $(Z_n, n \geq 1)$ in terms of the history $(\mathscr{F}_n, n \geq 1)$, where $\mathscr{F}_n \equiv \mathscr{F}_{T_n-}$:

T15 Theorem. *Let $(T_n, Z_n, n \geq 0)$ be an m-variate point processes, and let $N_t(i)$, $1 \leq i \leq m$, be its associated counting processes (see §1). Let $\mathscr{F}_t$ be a history of the form*

$$\mathscr{F}_t = \mathscr{F}_0 \vee \left(\bigvee_{i=1}^m \mathscr{F}_t^N(i)\right), \qquad (4.8)$$

where $\mathscr{F}_t^N(i)$ is the internal history of $N_t(i)$, and suppose that for each $1 \leq i \leq m$, $N_t(i)$ admits the $\mathscr{F}_t$-intensity $\lambda_t(i)$. Then for all $n \geq 1$

$$\frac{\lambda_{T_n}(i)}{\lambda_{T_n}} = P[Z_n = i | \mathscr{F}_{T_n-}] \quad on \; \{T_n < \infty\}, \tag{4.9}$$

where λ_t is the $\mathscr{F}_t$-intensity of $N_t = \sum_{i=1}^m N_t(i)$, i.e. $\lambda_t = \sum_{i=1}^m \lambda_t(i)$.

PROOF. Let A be any set in $\mathscr{F}_{T_n-}$. By (Appendix A2, T31) there exists an $\mathscr{F}_t$-predictable bounded process X_t such that $X_{T_n} = 1_A$ on $\{T_n < \infty\}$ and X_t is null outside $(T_{n-1}, T_n]$. If we apply the integration theorem T8 to $C_t = \lambda_t X_t$, we obtain

$$E\left[\int_0^\infty \lambda_s X_s \, dN_s(i)\right] = E\left[\int_0^\infty \lambda_s X_s \lambda_s(i) \, ds\right] = E\left[\int_0^\infty \lambda_s(i) X_s \, dN_s\right]. \tag{4.10}$$

But, by definition of the symbols involved,

$$E\left[\int_0^\infty \lambda_s X_s \, dN_s(i)\right] = E[\lambda_{T_n} 1_A 1(T_n < \infty) 1(Z_n = i)] \tag{4.11}$$

and

$$E\left[\int_0^\infty \lambda_s(i) X_s \, dN_s\right] = E[\lambda_{T_n}(i) 1_A 1(T_n < \infty)]. \tag{4.12}$$

Thus, (4.10) can be read [taking into account the $\mathscr{F}_{T_n-}$-measurability of $\lambda_{T_n} 1(T_n < \infty)$ and $\lambda_{T_n}(i) 1(T_n < \infty)$; see (I, E10)]

$$\lambda_{T_n} P[Z_n = i | \mathscr{F}_{T_n-}] 1(T_n < \infty) = \lambda_{T_n}(i) 1(T_n < \infty).$$

Then (4.9) follows by division by λ_{T_n} (which is allowed in view of T12). □

Comments. Roughly speaking, $\lambda_t(i)/\sum_{j=1}^m \lambda_t(j)$ is the probability of having a jump of $N_t(i)$ at time t, given $\mathscr{F}_{t-}$ and given that there is a jump of one of the $N_t(j)$'s at time t. A heuristic proof of T15 goes as follows:

$$P[dN_t(i) = 1 | \mathscr{F}_{t-}, dN_t = 1] = \frac{P[dN_t(i) = 1, dN_t = 1 | \mathscr{F}_{t-}]}{P[dN_t = 1 | \mathscr{F}_{t-}]}$$

$$= \frac{P[dN_t(i) = 1 | \mathscr{F}_{t-}]}{P[dN_t = 1 | \mathscr{F}_{t-}]}$$

$$= \frac{\lambda_t(i) \, dt}{\lambda_t \, dt} = \frac{\lambda_t(i)}{\lambda_t}$$

(where it has been observed that $dN_t(i) = 1$ implies $dN_t = 1$).

The result of T15 should be compared with the interpretation of q_{ij}/q_i in a Markov chain with Q-matrix $\{q_{ij}; i,j \in E\}$ as the probability of jumping to state j at time t, knowing that the chain indeed jumps at time t, and was in state i at the time "just before" t.

5. Representation of Queues

Standard and Feedback Representations of a Queue

For students of queueing systems, a feedback queue is a queue in which part or totality of the flow of customers leaving the service facility is recycled into the system. This corresponds to the schematic picture in Figure 4.

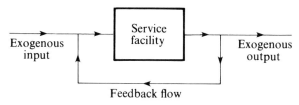

Figure 4   A feedback queue.

From a mathematical point of view, one should prefer to speak of a *feedback representation* of a queue, or of a queue *in feedback form*, rather than of a feedback queue. However, we will not maintain this subtle distinction all the way through, and we will very often adopt the practitioner's terminology, less precise but more physically intuitive. But let us see first what is meant by a feedback representation.

First, recall that in the definition of a queue,

$$Q_t = Q_0 + A_t - D_t, \tag{5.1}$$

the point processes A_t and D_t were supposed without common jumps, a very natural restriction if one interprets A_t and D_t as arrivals and departures from a service station in which a service facility is located (the distinction between service station and service facility being precisely that once a customer leaves the station, he does not immediately come back, whereas in a service facility the service of a customer may be interrupted according to some rule, say a priority rule, and this customer is recycled into the waiting line of the service facility without leaving the station; see Figure 5).

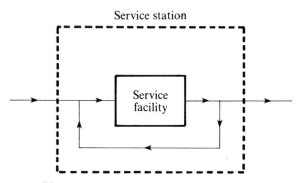

Figure 5   Service station and service facility.

A representation (A_t, D_t) where A_t and D_t have no common jumps is called a *representation in standard form* or a *standard representation* of the queue Q_t. Suppose now that Q_t is represented as

$$Q_t = Q_0 + E_t - S_t, \qquad (5.2)$$

where E_t and D_t are nonexplosive point processes that *do possess common jumps*. Define F_t by

$$F_t = \sum_{0 < s \leq t} \Delta E_s \, \Delta S_s. \qquad (5.3)$$

This process counts the number of jumps common to E_t and S_t in the interval $(0, t]$. Also define

$$\begin{aligned} A_t &= E_t - F_t, \\ D_t &= S_t - F_t. \end{aligned} \qquad (5.4)$$

Then (A_t, D_t) a standard representation of Q_t. The representation (E_t, S_t, F_t) is called a *feedback representation* of Q_t with *feedback process* F_t. Generally the modeler will have a physical reason for calling F_t the feedback process. However, one should note that starting from a standard representation of a queue Q_t one can obtain arbitrarily many feedback representations of the type (5.2), (5.3), (5.4): it suffices to choose an arbitrary nonexplosive point process F_t, and then define E_t and S_t by (5.4).

We now proceed to the construction of a feedback queue of the Jackson type. In such a queue, a customer leaving the service facility is recycled with a probability p into the waiting line independently of the previous history (in the worldly acceptation of the word history) of the queue. This recycling procedure is sometimes called a *Bernoulli switch*. Such queues were studied by Jackson {74} in the more general context of markovian networks where a customer leaving the service station i with probability $1 - p$ is routed towards other stations of the network with probabilities depending upon the label of the station and upon the label of the target station (see Figure 6). The probabilities $p_i, p_{ij}, p_{ik}, \ldots$ constitute a *routing procedure*.

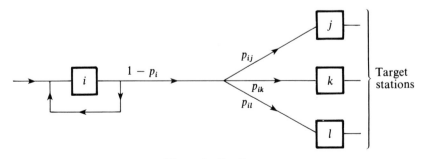

Figure 6 Routings.

5. Representation of Queues

Such networks are very idealized models of what actually happens in computer networks; still, they are useful. (See Kleinrock {91} for a discussion of computer network modeling.)

In this section, we will be concerned with just one station, our general aim being to provide tools for the analysis of queueing systems, rather than to make an inventory of results. We will go back to networks in Chapter V.

Construction of a $G/M/1$ Queue

We now present a method for constructing a feedback queue and calculating its arrival intensity and service potential. The following paragraphs of this section are just meant as illustrations of the martingale view of intensity. More general considerations of queueing systems will be given in Chapter V.

Consider two point processes $N_t(1)$ and $N_t(2)$ and a nonnegative random variable Q_0, mutually independent. Suppose that $N_t(1)$ admits the $\mathscr{F}_t^N(1)$-intensity λ_t and that $N_t(2)$ is $\mathscr{F}_t^N(2)$-Poisson with the intensity μ. By the independence assumption, $N_t(1)$ and $N_t(2)$ also have the $\mathscr{F}_t$-intensities λ_t and μ, where

$$\mathscr{F}_t = \mathscr{F}_t^N(1) \vee \mathscr{F}_t^N(2) \vee \sigma(Q_0)$$

(see E5). A queueing process Q_t is constructed from $N_t(1)$ and $N_t(2)$ as in E1:

$$A_t = N_t(1),$$

$$D_t = \int_0^t 1(Q_{s-} > 0) \, dN_s(2), \tag{5.5}$$

$$Q_t = Q_0 + A_t - D_t.$$

The $\mathscr{F}_t^Q$-parameters of this queue can be calculated as follows: First, since

$$\mathscr{F}_t^Q \subset \mathscr{F}_t$$

and since $N_t(1) = A_t$ has the $\mathscr{F}_t$-intensity λ_t, which is $\mathscr{F}_t^Q$-measurable (for it is $\mathscr{F}_t^N(1)$-measurable, i.e., $\mathscr{F}_t^A$-measurable), the $\mathscr{F}_t^Q$-rate of arrival of A_t is λ_t (see E4). Now let C_t be a $\mathscr{F}_t^Q$-predictable (hence $\mathscr{F}_t$-predictable) nonnegative process. The process $C_t 1(Q_{t-} > 0)$ is also such a process. Therefore

$$E\left[\int_0^\infty C_s 1(Q_{s-} > 0) \, dN_s(2)\right] = E\left[\int_0^\infty C_s 1(Q_{s-} > 0) \mu \, ds\right].$$

Since

$$E\left[\int_0^\infty C_s 1(Q_{s-} > 0) \, dN_s(2)\right] = E\left[\int_0^\infty C_s \, dD_s\right],$$

it follows, by Definition D1, that $\mu 1(Q_{t-} > 0)$ is the $\mathscr{F}_t$-intensity of D_t. We have therefore constructed a $G/M/1$ queue (here we use an extension of Kendall's notational system).

We now turn to the slightly more complex construction of a jacksonian feedback queue. The basic steps of the construction are the same as above.

Construction of a Feedback Queue

One wishes to model a waiting line with exponential service (parameter μ) and "external" Poisson arrivals (rate λ) in which a customer who has just completed service either leaves the system or is recycled. The decision on recycling is taken independently of the past, with *a priori* probability p. We are going to give a precise mathematical construction of such a feedback queue.

One starts by defining formally the arrival and departure processes A_t and D_t of a queueing process Q_t. As in the previous construction, let $N_t(1)$ and $N_t(2)$ be two Poisson processes with intensities λ and μ respectively; in addition an i.i.d. sequence X_n is given on the same probability space $(\Omega, \mathscr{F}, P)$ with the following statistics:

$$P[X_n = 1] = 1 - P[X_n = 0] = p, \qquad p \in (0, 1). \tag{5.6}$$

Assume that the processes $N_t(1)$, $N_t(2)$, the sequence X_n, and Q_0 are mutually independent. Define

$$\mathscr{F}_n^X = \sigma(X_1, \ldots, X_n), \qquad \mathscr{F}_\infty^X = \sigma(X_1, X_2, \ldots) \tag{5.7}$$

and

$$\mathscr{G}_t = \mathscr{F}_t^N(1) \vee \mathscr{F}_t^N(2) \vee \mathscr{F}_\infty^X \vee \sigma(Q_0). \tag{5.8}$$

Then $N_t(1)$ and $N_t(2)$ have $\mathscr{G}_t$-intensity λ and μ respectively. Now, define

$$\begin{aligned} A_t &= N_t(1), \\ D_t &= \sum_{0 < T_n(2) \le t} (1 - X_n) 1(Q_{T_n(2)-} > 0), \\ Q_t &= Q_0 + N_t(1) - D_t. \end{aligned} \tag{5.9}$$

In particular $Q_t = Q_0 + A_t - D_t$. The process $A_t = N_t(1)$ is interpreted as the *exogenous stream of arrivals*: by (5.9), the nth jump $T_n(2)$ of $N_t(2)$ is counted as a departure if and only if the queue is not empty at $T_n(2)-$ and $X_n = 0$; the *feedback stream* is represented by

$$F_t = \sum_{0 < T_n(2) \le t} X_n 1(Q_{T_n(2)-} > 0). \tag{5.10}$$

D_t is the actual output or departure process of the whole service system. The *service facility output stream* (see Figure 7) is defined by

$$S_t = \sum_{0 < T_n(2) \le t} 1(Q_{T_n(2)-} > 0). \tag{5.11}$$

5. Representation of Queues

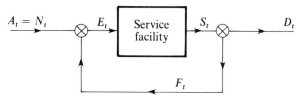

Figure 7 The flows in the feedback queue.

Similarly the service facility input stream is defined by

$$E_t = F_t + A_t. \tag{5.12}$$

Clearly, (5.1), (5.2), (5.3), and (5.4) are verified: Q_t admits the standard representation (A_t, D_t) and the feedback representation (E_t, S_t, F_t).

It should now be checked that the queue so constructed corresponds to the intuitive description of a feedback queue. First, if we denote by τ_n the sequence of jumps corresponding to the intermediary output $N_t(2)$, we must verify that the sequence $\tau_{n+1} - \tau_n$ is i.i.d., exponentially distributed (parameter μ), and independent of the input A_t. This verification, which is left to the reader, should be intuitively obvious in view of the independence assumptions and the "absence of memory" of exponential random variables. Next, D_t can be written as

$$D_t = \sum_{0 < \tau_n \le t} (1 - X_n), \tag{5.13}$$

and therefore D_t is obtained from the intermediary output stream as follows: a coin of bias p is tossed at time τ_n, and if the result X_n is 0 (say "heads"), the corresponding customer leaves the system (i.e. is counted in the output stream of the total system); otherwise he is recycled. By assumption, X_n is independent of the input $A_t = N_t(1)$; it is also independent of Q_{τ_n-}, since Q_{τ_n-} depends only upon the processes $N_t(1)$ and $N_t(2)$ and upon $X_n, \ldots, X_{n-1}$. Therefore D_t is obtained by random cancellation of points on the intermediary output, and each cancellation is independent of the strict past of the queue and of past cancellations.

E11 Exercise. Show that the $\mathscr{F}_t^Q \vee \mathscr{F}_t^F$-parameters of Q_t are λ and $\mu(1-p)1(Q_t > 0)$. Show that the $\mathscr{F}_t^Q \vee \mathscr{F}_t^F$-intensities of S_t, F_t, and E_t are respectively $\mu 1(Q_t > 0)$, $\mu p 1(Q_t > 0)$, and $\lambda + \mu p 1(Q_t > 0)$. Can you say that E_t has the $\mathscr{F}_t^Q$-intensity $\lambda + \mu 1(Q_t > 0)$? Can you say that F_t has the $\mathscr{F}_t^Q$-intensity $\mu p 1(Q_t > 0)$?

We will now give a different construction of the $M/M/1$ feedback queue which will be extended to Markovian queueing networks in Chapter VI.

The basic ingredients are three mutually independent Poisson processes $\bar{A}_t$, $\bar{D}_t$, and $\bar{F}_t$ of respective intensities λ, $\mu(1-p)$, and μp, and a nonnegative integer-valued random variable Q_0 independent of these three processes. We then define Q_t, F_t, S_t, E_t, A_t, and D_t by

$$Q_t = Q_0 + \bar{A}_t - \int_0^t 1(Q_{s-} > 0) \, d\bar{D}_s,$$

$$F_t = \int_0^t 1(Q_{s-} > 0) \, d\bar{F}_s,$$

$$S_t = \int_0^t 1(Q_{s-} > 0) \, d\bar{D}_s + \int_0^t 1(Q_{s-} > 0) \, d\bar{F}_s, \qquad (5.14)$$

$$E_t = \bar{A}_t + \int_0^t 1(Q_{s-} > 0) \, d\bar{F}_s,$$

$$A_t = \bar{A}_t,$$

$$D_t = \int_0^t 1(Q_{s-} > 0), \, d\bar{D}_s.$$

If we let $\mathscr{G}_t$ be the history

$$\mathscr{G}_t = \sigma(Q_0) \vee \mathscr{F}_t^{\bar{A}} \vee \mathscr{F}_t^{\bar{D}} \vee \mathscr{F}_t^{\bar{F}}, \qquad (5.15)$$

it can be shown as in E11 that the $\mathscr{G}_t$-intensities of A_t, D_t, E_t, S_t, and F_t are respectively λ, $\mu(1-p)1(Q_t > 0)$, $\lambda + \mu p 1(Q_t > 0)$, $\mu 1(Q_t > 0)$, $\mu p 1(Q_t > 0)$.

It is clear that we have obtained a queueing process Q_t in feedback form with feedback flow F_t. The number p admits the same interpretation as above, by virtue of T15. Indeed, let us consider the bivariate point process (D_t, F_t) with basic point process $S_t = D_t + F_t$. With respect to the history $\mathscr{F}_t = \mathscr{F}_t^D \vee \mathscr{F}_t^F$, D_t, F_t, and S_t admit the predictable intensities $\mu(1-p)h_t$, $\mu p h_t$, and μh_t, where h_t is a predictable version of $P(Q_t > 0 | \mathscr{F}_t^D \vee \mathscr{F}_t^F)$. Denoting by T_n the basic points of S_t, and associating to T_n the random variable X_n which has two values: $X_n = 0$ if T_n is a jump time of D_t; $X_n = 1$ if T_n is a jump time of F_t, we have by T15

$$P(X_n = 0 | \mathscr{F}_{T_n-}) = 1 - p, \qquad P(X_n = 1 | \mathscr{F}_{T_n-}) = p \quad \text{on } \{T_n < \infty\}.$$

6. Random Changes of Time

In the present section, the effect of a change of clock on the intensity of a point process will be examined. In particular it will be shown how to transform a given non-Poisson process into a standard Poisson process. It turns out that there are many changes of clock which achieve such a transformation, and that not all preserve the information content of the point process, in a sense that will be made precise at the end of the section and in the next section.

6. Random Changes of Time

The basic result is:

T16 Theorem [7, 8]. *Let N_t be a point process with the $\mathscr{F}_t$-intensity λ_t and the $\mathscr{G}_t$-intensity $\tilde{\lambda}_t$, where $\mathscr{F}_t$ and $\mathscr{G}_t$ are histories of N_t such that*

$$\mathscr{F}_t^N \subset \mathscr{G}_t \subset \mathscr{F}_t. \tag{6.1}$$

Suppose that

$$N_\infty = \infty \quad P\text{-a.s.} \tag{6.2}$$

Define for each t the $\mathscr{G}_t$-stopping time $\tau(t)$ by

$$\int_0^{\tau(t)} \tilde{\lambda}_s \, ds = t. \tag{6.3}$$

Then the point process $\tilde{N}_t$ defined by

$$\tilde{N}_t = N_{\tau(t)} \tag{6.4}$$

is a standard Poisson process (intensity 1).

PROOF. We must first show that (6.3) indeed defines $\tau(t)$ for all t, that is to say, we must show that

$$\int_0^\infty \tilde{\lambda}_s \, ds = \infty \quad P\text{-a.s.} \tag{6.5}$$

$\square$

E12 Exercise. Prove the following:

L17 Lemma [6]. *Let N_t be a point process with the $\mathscr{F}_t$-intensity λ_t. The two conditions*

(α) $N_\infty = \infty$, P-a.s.,
(β) $\int_0^\infty \lambda_s \, ds = \infty$, P-a.s.

are equivalent.

E13 Exercise. Show that for fixed $t \geq 0$, $\tau(t)$ defined by (6.3) is a $\mathscr{G}_t$-stopping time, and that $\mathscr{F}_t^{\tilde{N}} \subset \mathscr{F}_{\tau(t)}^N \subset \mathscr{G}_{\tau(t)}$.

In order to prove T16, it suffices show that for all nonnegative $\mathscr{G}_t$-predictable processes $\tilde{C}_t$ of the form

$$\tilde{C}_t(\omega) = 1_A(\omega) 1_{(a,b]}(t), \quad 0 \leq a \leq b, \quad A \in \mathscr{G}_{\tau(a)}, \tag{6.6}$$

the equality

$$E\left[\int_0^\infty \tilde{C}_s \, d\tilde{N}_s\right] = E\left[\int_0^\infty \tilde{C}_s \, ds\right] \tag{6.7}$$

holds. One observes that

$$E\left[\int_0^\infty \tilde{C}_s\, d\tilde{N}_s\right] = E\left[1_A \int_a^b d\tilde{N}_s\right] = E[1_A(\tilde{N}_b - \tilde{N}_a)] = E[1_A(N_{\tau(b)} - N_{\tau(a)})], \tag{6.8}$$

and therefore

$$E\left[\int_0^\infty \tilde{C}_s\, d\tilde{N}_s\right] = E\left[\int_0^\infty C_s\, dN_s\right], \tag{6.9}$$

where

$$C_t = 1_A 1(\tau(a) < t \leq \tau(b)). \tag{6.10}$$

If one agrees that C_t is a $\mathscr{G}_t$-predictable process, then by the definition of intensity,

$$E\left[\int_0^\infty C_s\, dN_s\right] = E\left[\int_0^\infty C_s \tilde{\lambda}_s\, ds\right]. \tag{6.11}$$

The right-hand side of (6.11) is nothing but

$$E\left[1_A \int_{\tau(a)}^{\tau(b)} \tilde{\lambda}_s\, ds\right] = E[1_A(b - a)],$$

by the definition of $\tau(t)$; now $E[1_A(b - a)] = E[\int_0^\infty \tilde{C}_s\, ds]$. The above chain of equalities yields (6.7).

The last point to be proven is left as an exercise:

E14 Exercise. Show that C_t given by (6.10), where $A \in \mathscr{F}_t^{\tilde{N}}$ and $\tau(t)$ is defined by (6.3), is $\mathscr{F}_t^N$-predictable.

EXAMPLE. Suppose that on $(\Omega, \mathscr{F}, P)$ are defined a point process N_t and a nonnegative random variable Λ with distribution $F(d\lambda)$. Define the history $\mathscr{F}_t$ of N_t by

$$\mathscr{F}_t = \sigma(\Lambda) \vee \mathscr{F}_t^N, \tag{6.12}$$

and suppose that N_t has the $\mathscr{F}_t$-intensity Λ. In other words, N_t is a homogeneous doubly stochastic Poisson process with intensity Λ.

Take, in the statement of T16,

$$\mathscr{G}_t = \mathscr{F}_t^N. \tag{6.13}$$

Anticipating a result of Chapter VI (VI, E9),

$$\hat{\lambda}_t = E[\Lambda | \mathscr{F}_t^N] = \frac{\int_0^\infty \lambda^{N_t+1} e^{-\lambda t} F(d\lambda)}{\int_0^\infty \lambda^{N_t} e^{-\lambda t} F(d\lambda)}. \tag{6.14}$$

In order to obtain more explicit formulas, we specialize to the situation where Λ takes only two equiprobable values a and b:

$$P[\Lambda = a] = P[\Lambda = b] = \tfrac{1}{2}, \tag{6.15}$$

in which case

$$\hat{\lambda}_t = \frac{1 + (b/a)^{N_t+1}e^{(a-b)t}}{1 + (b/a)^{N_t}e^{(a-b)t}}. \tag{6.16}$$

If we let $\tilde{T}_n$ be the sequence associated to $\tilde{N}_t$, we have, since $\tau(\tilde{T}_n) = T_n$,

$$\int_{T_n}^{T_{n+1}} \hat{\lambda}_s \, ds = \tilde{T}_{n+1} - \tilde{T}_n, \tag{6.17}$$

or, more explicitly (after easy calculations),

$$\tilde{T}_{n+1} - \tilde{T}_n = f(n, T_{n+1}) - f(n, T_n), \tag{6.18}$$

where

$$f(n, t) = at - \log(1 + (b/a)^{n+1}e^{(a-b)t}). \tag{6.19}$$

Equation (6.18) gives an explicit algorithm for inverting the change of time $T_n \to \tilde{T}_n$.

Let us now perform the change of time of T16 with

$$\mathcal{G}_t = \mathcal{F}_t = \sigma(\Lambda) \vee \mathcal{F}_t^N. \tag{6.20}$$

In this case

$$\tau(t) = \frac{t}{\Lambda} \tag{6.21}$$

and the resulting point process $\tilde{N}_t$ is also Poisson, rate 1.

There is however a fundamental difference between the two above changes of time. Indeed, in the first case, where $\mathcal{G}_t = \mathcal{F}_t^N$, the information $\Lambda(\omega)$ can be recovered exactly from the sequence $\tilde{T}_n(\omega)$: in order to do this, it suffices to invert $T_n(\omega) \to \tilde{T}_n(\omega)$ by means of (6.18) and (6.19), and then to extract $\Lambda(\omega)$:

$$\Lambda(\omega) = \lim_{t \uparrow \infty} \frac{N_t(\omega)}{t} \tag{6.22}$$

(this is nothing but the law of large numbers).

In the case where $\mathcal{G}_t = \sigma(\Lambda) \vee \mathcal{F}_t^N$, $\tilde{N}_t(\omega)$ does not contain any information about $\Lambda(\omega)$. Indeed:

E15 Exercise. Show that $\tilde{N}_t = N_{t/\Lambda}$ is independent of Λ.

We will now apply the above remarks to the problem of complete security in cryptographic systems.

7. Cryptographic Point Processes

The problem of cryptography consists of sending messages generated by a *stochastic source* to a *receiver* (friend) in such a way that a potential *interceptor* (foe) cannot retrieve the information content of the messages. Thus the original message has to be coded (*encrypted*) before it is sent through the *wiretaped channel*. The interceptor is granted intelligence and he is supposed to know the general principle of the coding scheme, except for a few parameters which constitute the *key* of the coding procedure. The key is transmitted to the authorized receiver by means of a secure courier.

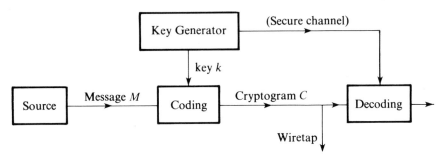

Figure 8   Classical coding-decoding system.

Usually the coding procedures consist of complex algebraic transformations of large stretches of the original message. The output then appears to be random. It is, in fact, not random since the transformations have to be invertible. Because they are invertible the receiver who has the key can reconstruct the original sequence. The interceptor cannot do so because the key is not accessible to him, he has the choice between a very large number of possible transformations $(U_k, k \in \mathcal{K})$ where $\mathcal{K}$ is the set of keys among which the key k_0 in operation is chosen.

Denoting the message by M, the encoded message or *cryptogram*, is $C = U_{k_0}(M)$. The receiver then just decodes the cryptogram: $M = U_{k_0}^{-1}(C)$.

The general problem of cryptoanalysis is to recover the message M from the cryptogram C. In fact the interceptor is supposed to continuously wiretap the line in order to have access to *all* messages M which go through. He will therefore aim at obtaining the key k_0 from the message M. He then makes the assumption that a key k_0 is used several times before it becomes obsolete. Such assumption is realistic because the procedures of key transmission and of changes of keys at both the encoder and decoder locations are very costly.

In summary, the designer of a cryptographic system must provide *key security*. We will not give details about classical cryptography, referring the reader to the fundamental article of Shannon {135} where it is explained that the algebraic procedures do not in general provide the users with complete key security, unless prohibitive dimensions of the key space are technologically acceptable.

In this subsection we will give an alternative method of coding, which is not algebraic but stochastic in the sense that mapping from M to C is not one-to-one. Such a method is completely key secure in the sense of classical cryptoanalysis, as will be now explained by means of an example.

7. Cryptographic Point Processes

A Specific Procedure for Coding and Decoding a Digital Message

Let a_n be a sequence of iid random variables such that:

$$P(a_n = 0) = P(a_n = 1) = \tfrac{1}{2}, \qquad n \in Z_+. \tag{7.1}$$

We are going to transform a_n into a sequence of iid random numbers S'_n where:

$$P(S'_n \leq t) = 1 - \exp(-t), \qquad n \in Z_+, \; t \in R_+ \tag{7.2}$$

in such a way as to preserve information about the original sequence a_n (if this requirement were not forced upon us we would just draw independent samples from an exponential law of fixed parameter).

Thus doing, we will have given a way of encoding a random sequence a_k, each a_k being coded into a block of random numbers, of random length:

$$a_{k+1} \to (S'_{n_k+1}, \ldots, S'_{n_{k+1}}).$$

The decoding is then made with the help of a deterministic transformation:

$$(S'_n) \to ((S_n), (n_k))$$

and of a bayesian test on the observations:

$$S'_{n_k+1}, \ldots, S'_{n_{k+1}}.$$

The following algorithms are proposed. We first describe the coding algorithm into which the message a_n is fed.

Coding Algorithm

Step 1: Choose a triple (a, b, T) of strictly positive real numbers. This triple is called the key. Set $K = 0$, $c = 0$, $N_0 = 0$, $n_0 = 0$.
Step 2: $K = K + 1$, $N_K = N_{K-1} + n_{K-1}$.
Step 3: If $a_K = 0$ set $c = a$, if $a_K = 1$ set $c = b$.
Step 4: Generate a sequence of iid r.v. (U_n) drawn from the exponential law of parameter c.
Step 5: Calculate $n_K = \inf(n/U_1 + \cdots + U_n \geq T)$.
Step 6: Construct $(U'_n, 1 \leq n \leq n_K)$ from $(U_n, 1 \leq n \leq n_K)$ with the help of the following transformation:

$$U'_n = f(n, U_0 + U_1 + \cdots + U_n) - f(n, U_0 + \cdots + U_{n-1})$$

where: $f(n, t) = at - \log(1 + (b/a)^n \exp(-(b-a)t))$.
Step 7: Set: $S'_{N_{K-1}+1} = U'_1, \ldots, S'_{N_{K-1}+n_K} = U'_{N_K}$.
Step 8: Go to Step 2.

Comments. This program generates $S'_1, S'_2, \ldots$ from $a_1, a_2, \ldots$. It first considers a_1, then generates iid exponential r.v.s of parameter a if $a_1 = 0$, of parameter b if $a_1 = 1$. It selects a random number of these r.v.s, this number being determined by step 5. Then it performs on this block of r.v.s the change of time described in §1. These new r.v.s are transmitted. Then a_2 is examined, and so on.

Each block $(S'_{N_K-1+1}, \ldots, S'_{N_K-1+n_K})$ consists of iid exponential r.v.s with parameter 1. All the blocks being built independently if a_n is iid, the resulting sequence S'_n is a sequence of iid exponential r.v.s with parameter 1.

We now proceed to the decoding algorithm into which are fed the cryptogram S'_n and the key (a, b, T).

Decoding Algorithm

Step 1: Set $K = 0$, $n_K = 0$, $N_K = 0$.
Step 2: $K = K + 1$, $N_K = N_K + n_K$.
Step 3: Define $U'_1, U'_2, \ldots$ by $U'_j = S'_{N_K+j}$.
Step 4: Transform sequence U'_n into sequence U_n by inverting the transformation of step 6 of the encoding algorithm.
Step 5: Set $n_K = \inf(n/U_1 + \cdots + U_n \geq T)$.
Step 6: Perform on $(U_1, \ldots, U_{n_K})$ a bayesian test to discriminate between the two hypotheses: $H_0: (U'_1, \ldots, U'_n)$ is iid exponential, parameter a; and $H_1: (U'_1, \ldots, U'_{n_K})$ is iid, exponential, parameter b, each hypothesis having a priori probability $\frac{1}{2}$. If the test decides in favor of H_0, print $a_K = 0$, and if in favor of H_1, print $a_K = 1$.
Step 7: Go to step 2.

Comments. Steps 3, 4, 5 separate the blocks of S'_ns corresponding to different a'_ks. Step 4 undoes the change of time of the coding algorithm (note that a, b has to be fed into the algorithm to perform the change of time; note also, that T has to be fed into it so as to recover the block dimensions).

Remark. Step 6 is a bayesian test, and therefore the decoding is not error-free. However the probability of error P_E can be analytically computed, and it can be checked that P_E depends on $a' = a/T$ and $b' = b/T$, and that, as intuition tells, $P_E(a', b')$ goes to zero as $|a' - b'|$ goes to infinity (we leave this verification to the reader).

Another Coding Scheme, Using Wiener Processes

It so happens that the above situation is not unique; indeed, we have seen that an "anonymous" process such as a standard Poisson process can actually transport a nonzero quantity of information. But the other "anonymous" process, the noise process "par excellence," the Wiener process, can perform the same trick; it is then known as the "innovations process." Let us recall some facts and definitions (see also Appendix A3).

Let W_t be a continuous process and $\mathscr{F}_t$ a history of W_t. If, for all $0 < s < t$:

(a) $W_t - W_s$ is P-independent of $\mathscr{F}_s$,
(b) $W_t - W_s$ is a centered gaussian r.v. with variance $t - s$,

then W_t is called $a(P, \mathscr{F}_t)$ Wiener process. If the process Y_t has the form:

$$Y_t = \int_0^t S_u \, du + W_t, \qquad (7.3)$$

where W_t is a $(P, \mathscr{F}_t)$ Wiener process and S_t is a signal process, measurable adapted to $\mathscr{F}_t$ and such that: $E[\int_0^t S_u \, du] < \infty$, then Y_t is called a $(P, \mathscr{F}_t)$ Wiener process with a drift.

In communication theory, S_t is a signal and W_t a channel noise which is not accessible, neither to the transmitter nor to the receiver. The problem of the receiver is to calculate $\hat{S}_t(\omega) = E[S_t | \mathscr{F}_t^Y](\omega)$ from the observation $Y_t(\omega)$ the joint statistics of W_t and S_t being supposed known. We note that the transmitter is not able to calculate $S_t(\omega)$ since $Y_t(\omega)$ is not accessible to him.

Now if we change our point of view, and suppose that W_t is not a channel noise, but instead purposely generated and added to S_t by the transmitter T itself, then T is able to generate $\hat{S}_t(\omega)$ when knowing the joint statistics of W_t and S_t, since T knows $Y_t(\omega)$ in this new situation. Let then T construct the innovations $\hat{W}_t$ of Y_t:

$$\hat{W}_t = Y_t - \int_0^t (S_u - \hat{S}_u) \, du \qquad (7.4)$$

and transmit $\hat{W}_t$ instead of Y_t. It is known {81} that $\hat{W}_t$ is a $(P, \mathscr{F}_t^Y)$ brownian motion, and it is conjectured that in nonpathological situations

$$\mathscr{F}_t^W = \mathscr{F}_t^Y \quad P\text{-a.s.} \qquad (7.5)$$

In fact, this conjecture has been proven in the special case where S_t is P-independent of W_t and bounded (Clark {34}). We will suppose that this situation prevails. What the verification of the conjecture (7.5) means is that the receptor does not loose information about S_t when sent the innovation $\hat{W}_t$ instead of the original observation Y_t as long as he knows the joint statistics of W_t and S_t and that what he is receiving is the innovations process of Y_t. On the other hand, an interceptor who has the same knowledge as the receiver except for the joint statistics of W_t and S_t—say he knows that W_t is Wiener, but knows nothing about S_t—, is not able to recover S_t. And if such an interceptor tries to establish statistics about S_t, he will not be able to do so, because the statistics of his observation $\hat{W}_t$ does not depend upon the statistics of S_t.

References

[1] Brémaud, P. (1972) A martingale approach to point processes, Ph.D. Thesis, Memo ERL-M-345, Dept. of EECS, Univ. of Calif., Berkeley.
[2] Brémaud, P. (1974) The martingale theory of point processes over the real line admitting an intensity, *Proc. Coll. on Control Theory at IRIA*, Lect. Notes in Op. Res. and Math. Syst. **107**, Springer, Berlin, pp. 519–542.

[3] Brémaud, P. (1975) On the information carried by a point process, Cahiers du CETHEDEC **45**, pp. 43–70.
[4] Brémaud, P. (1975) An extension of Watanabe's characterization theorem, *J. Appl. Probab.* **8**, 2, pp. 396–399.
[5] Brémaud, P. (1978) Streams in a $M/M/1$ feedback queue, *Z. für Wahrscheinlichkeitstheorie* **45**, pp. 21–33.
[6] Liptser, R. and Shiryayev, A. (1978) *Statistics of Random Processes, II: Applications*, Springer, New York.
[7] Meyer, P. A. (1969) Démonstration simplifiée d'un théorème de Knight, in *Séminaire Proba V*, Lect. Notes in Math. **191**, Springer, Heidelberg.
[8] Papangelou, F. (1972) Integrability of expected increments of point processes and a related change of scale, *Trans. Amer. Math. Soc.* **165**, pp. 483–506.
[9] Watanabe, S. (1964) Additive functionals of Markov processes and Lévy Systems, *Jap. J. Math.*, **34**, pp. 53–79.

SOLUTIONS TO EXERCISES, CHAPTER II

E1. For t to be a point of decrease of m_t, it is necessary and sufficient that (i) $X_{t-} = m_{t-}$ and (ii) $\Delta N_t(2) = 1$; moreover the decrease is then of one unit. Hence

$$\Delta m_t = -1(X_{t-} = m_{t-}) \Delta N_t(2), \tag{1}$$

i.e, since by definition $Q_t = X_t - m_t$,

$$m_t = -\int_0^t 1(Q_{s-} = 0) \, dN_s(2). \tag{2}$$

Note that by the definition of m_t, $X_t \geq m_t$, and hence $Q_t \geq 0$. Also

$$Q_t = X_t - m_t = Q_0 + N_t(1) - N_t(2) + \int_0^t 1(Q_{s-} = 0) \, dN_s(2), \tag{3}$$

or

$$Q_t = Q_0 + N_t(1) - \int_0^t 1(Q_{s-} > 0) \, dN_s(2) \tag{4}$$

(since either $Q_{s-} = 0$ or $Q_{s-} > 0$). Defining $D_t = \int_0^t 1(Q_{s-} > 0) \, dN_s(2)$ and $A_t = N_t(1)$, one observes that D_t and A_t have no common jumps, since $N_t(1)$ and $N_t(2)$ have no common jumps.

E2. Write (2.3) with $C_t = 1_A 1(t \leq T_n) 1(a < t \leq b)$, $0 \leq a \leq b$, $A \in \mathscr{F}_a$, to obtain

$$E[1_A(N_{b \wedge T_n} - N_{a \wedge T_n})] = E\left[1_A \int_{a \wedge T_n}^{b \wedge T_n} \lambda_s \, ds\right] \qquad A \in \mathscr{F}_a. \tag{1}$$

In particular for $A = \Omega$, $a = 0$, $b = t$,

$$E[N_{t \wedge T_n}] = E\left[\int_0^{t \wedge T_n} \lambda_s \, ds\right]. \tag{2}$$

Hence since $N_{t \wedge T_n} \leq n$, $E[\int_0^{t \wedge T_n} \lambda_s \, ds] < \infty$. Moving terms from one side to the other in (1) is therefore licit (no $\infty - \infty$ form); for instance

$$E\left[1_A\left(N_{b \wedge T_n} - \int_0^{b \wedge T_n} \lambda_s \, ds\right)\right] = E\left[1_A\left(N_{a \wedge T_n} - \int_0^{a \wedge T_n} \lambda_s \, ds\right)\right]. \tag{3}$$

Solutions to Exercises, Chapter II   49

Because a, b, A are arbitrary modulo the requirements $0 \le a \le b$, $A \in \mathcal{F}_a$, (3) implies that $N_{t \wedge T_n} - \int_0^{t \wedge T_n} \lambda_s \, ds$ is an $\mathcal{F}_t$-martingale (one need only check the integrability, which was done, and the adaptation to $\mathcal{F}_t$, which is clear). Now, since $T_n \uparrow \infty$, P-a.s. (as seen in the text), $N_t - \int_0^t \lambda_s \, ds$ is an $\mathcal{F}_t$-local martingale (by the definition of local martingales).

Proof of Lemma L3. First part: Apply (2.3) with

$$C_t = X_t 1(a < t \le b) 1_A, \qquad (4)$$

where $0 \le a \le b$ and $A \in \mathcal{F}_a$, to obtain

$$E\left[1_A \int_a^b X_s \, dN_s\right] = E\left[1_A \int_a^b X_s \lambda_s \, ds\right]. \qquad (5)$$

By (2.4) and (5) applied with $|X_t|$ replacing X_t, for all $t \ge 0$

$$E\left[\int_0^t |X_s| \, dN_s\right] = E\left[\int_0^t |X_s| \lambda_s \, ds\right] < \infty. \qquad (6)$$

Hence, the jumbling of (5) is permitted, and yields

$$E\left[1_A \int_0^b X_s(dN_s - \lambda_s \, ds)\right] = E\left[1_A \int_0^a X_s(dN_s - \lambda_s \, ds)\right], \qquad (7)$$

which says that $\int_0^t X_s(dN_s - \lambda_s \, ds)$ is an $\mathcal{F}_t$-martingale, A being arbitrary in $\mathcal{F}_a$.

Second part: Apply E2 with X_t replaced by $X_t 1(t \le S_n)$, where

$$S_n = \begin{cases} \inf\left\{t \,\Big|\, \int_0^t |X_s| \lambda_s \, ds = n\right\} & \text{if } \{\cdots\} \ne \emptyset, \\ +\infty & \text{otherwise.} \end{cases} \qquad (8)$$

This yields

$$\int_0^{t \wedge S_n} X_s(dN_s - \lambda_s \, ds) \text{ is an } \mathcal{F}_t\text{-martingale.} \qquad (9)$$

Now, in view of (2.5), S_n defined by (8) goes to infinity P-a.s. as n goes to infinity, hence $\int_0^t X_s(dN_s - \lambda_s \, ds)$ is an $\mathcal{F}_t$-local martingale.

E3. Write

$$e^{iuN_t} = 1 + \int_0^t (e^{iu} - 1) e^{iuN_{s^-}} \, dN_s. \qquad (1)$$

Add $\int_0^t (e^{iu} - 1) e^{iuN_s} \lambda(s) \, ds$ to both sides of (1), and obtain

$$m_t = e^{iuN_t} - (e^{iu} - 1) \int_0^t e^{iuN_s} \lambda(s) \, ds \text{ is an } \mathcal{F}_t\text{-martingale} \qquad (2)$$

[since $(e^{iu} - 1) \int_0^t e^{iuN_{s^-}} (dN_s - \lambda(s) \, ds)$ is an $\mathcal{F}_t$-martingale]. Write the martingale property, i.e, for all $0 \le s \le t$, $A \in \mathcal{F}_s$,

$$E[1_A(m_t - m_s)] = 0, \qquad (3)$$

that is to say,

$$E[1_A(e^{iuN_t} - e^{iuN_s})] = E\left[1_A(e^{iu} - 1) \int_s^t e^{iuN_v} \lambda(v) \, dv\right]. \qquad (4)$$

Divide both sides of (4) by e^{iuN_s} to obtain, after rearrangement and interchange of integral signs,

$$E[1_A e^{iu(N_t - N_s)}] = P(A) + (e^{iu} - 1) \int_s^t E[1_A e^{iu(N_v - N_s)}] \lambda(v)\, dv. \tag{5}$$

Define

$$H(t, s, A) = E[1_A e^{iu(N_t - N_s)}]. \tag{6}$$

Then (5) reads

$$H(t, s, A) = P(A) + (e^{iu} - 1) \int_s^t H(v, s, A) \lambda(v)\, dv. \tag{7}$$

Hence, observing that $H(s, s, A) = P(A)$,

$$H(t, s, A) = E[1_A e^{iu(N_t - N_s)}] = P(A) \exp\left\{(e^{iu} - 1) \int_s^t \lambda(v)\, dv\right\}. \tag{8}$$

In particular, taking $A = \Omega$,

$$E[e^{iu(N_t - N_s)}] = \exp\left\{(e^{iu} - 1) \int_s^t \lambda(v)\, dv\right\}, \tag{9}$$

which expresses that $N_t - N_s$ is Poisson, parameter $\int_s^t \lambda(v)\, dv$. Now expand (8) in powers of e^{iun} and identify the factors:

$$E[1_A 1(N_t - N_s = n)] = P(A) \exp\left\{-\int_s^t \lambda(v)\, dv\right\} \frac{(\int_s^t \lambda(v)\, dv)^n}{n!} \tag{10}$$

$$= P(A) P(N_t - N_s = n).$$

Since A is arbitrary in $\mathscr{F}_s$, this implies that $N_t - N_s$ is independent of $\mathscr{F}_s$.

E4. Clearly a $\mathscr{G}_t$-predictable process is an $\mathscr{F}_t$-predictable process whenever the history $\mathscr{G}_t$ is smaller than the history $\mathscr{F}_t$. Hence, if (3.2) holds for all nonnegative $\mathscr{F}_t$-predictable processes C_t, it is also verified, *a fortiori*, for all nonnegative $\mathscr{G}_t$-predictable processes C_t.

E5. By T9 and Remark 1 thereafter, it suffices to show that for all $n \geq 0$

$$N_{t \wedge T_n} - \int_0^{t \wedge T_n} \lambda_s\, ds \text{ is an } \mathscr{F}_t \vee \mathscr{G}_\infty\text{-martingale} \tag{1}$$

and to apply E4 (indeed, $\mathscr{F}_t \vee \mathscr{G}_t \subseteq \mathscr{F}_t \vee \mathscr{G}_\infty$). Now, by hypothesis,

$$N_{t \wedge T_n} - \int_0^{t \wedge T_n} \lambda_s\, ds \text{ is an } \mathscr{F}_t\text{-martingale}, \tag{2}$$

(1) therefore follows from (2) and from the general result: if m_t is a $\mathscr{F}_t$-martingale and if $\mathscr{A}$ is a σ-field which is independent of $\mathscr{F}_t$ for all $t > 0$, then m is an $\mathscr{F}_t \vee \mathscr{A}$-martingale (indeed, $E[m_t | \mathscr{F}_s \vee \mathscr{A}] = E[m_t | \mathscr{F}_s]$, since $\mathscr{A}$ is independent of $\mathscr{F}_s$ and m_t).

E6. This is just Definition D11 combined with I, E3.

E7. When $Q_t = 0$, then $\zeta_t = 0$. Also, when $Q_t > 0$, ζ_t increases smoothly (at the same rate as t), except at a departure, where it falls to 0. Hence

$$d\exp(iu\zeta_t) = +iu\exp(iu\zeta_t)1(Q_t > 0)\,dt + (1 - \exp(iu\zeta_{t-}))\,dD_t, \quad (1)$$

or

$$\exp(iu\zeta_t) = \exp(iu\zeta_0) + iu\int_0^t \exp(iu\zeta_s)1(Q_s > 0)\,ds - \int_0^t (1 - \exp(iu\zeta_{s-}))\,dD_s. \quad (2)$$

Now $\mu 1(Q_t > 0)$ is the $\mathscr{F}_t^Q$-intensity of D_t, and $1 - \exp(iu\zeta_{t-})$ is $\mathscr{F}_t^Q$-predictable (it is left-continuous and adapted to $\mathscr{F}_t^Q$), and therefore by the definition of intensity

$$E\left[\int_0^t (1 - \exp(iu\zeta_{s-}))\,dD_s\right] = E\left[\int_0^t (1 - \exp(iu\zeta_s))\mu 1(Q_s > 0)\,ds\right]. \quad (3)$$

Also $1_t(Q_t > 0) = 1 - 1(Q_t = 0)$, and $Q_t = 0$ implies $\zeta_t = 0$; therefore

$$E\left[\int_0^t (1 - \exp(iu\zeta_{s-}))\,dD_s\right] = E\left[\int_0^t (1 - \exp(iu\zeta_s))\mu\,ds\right] \quad (4)$$

and

$$E\left[\int_0^t \exp(iu\zeta_s)1(Q_s > 0)\,ds\right] = E\left[\int_0^t (\exp(iu\zeta_s) - 1(Q_s = 0))\,ds\right]. \quad (5)$$

Therefore, taking expectations of both sides of (2) and using (4) and (5),

$$\Phi_t(u) = \Phi_0(u) + (iu - \mu)\int_0^t \Phi_s(u)\,ds + \int_0^t (\mu - iup_0(s))\,ds. \quad (6)$$

The solution of (6) is easily computed:

$$\Phi_t(u) = \left(\Phi_0(u) + \int_0^t (\mu - iup_0(s))\exp\{(\mu - iu)s\}\,ds\right)\exp\{(iu - \mu)t\}. \quad (7)$$

If the queue is in equilibrium, $p_0(t) = p_0 = 1 - \rho$ and therefore

$$\Phi_t(u) = \Phi_0(u)\exp\{(iu - \mu)t\} + \frac{\mu - iup_0}{\mu - iu}. \quad (8)$$

Hence

$$\lim_{t\uparrow\infty}\Phi_t(u) = \frac{\mu - iup_0}{\mu - iu} = p_0 + (1 - p_0)\frac{\mu}{\mu - iu}. \quad (9)$$

Note that the distribution corresponding to the characteristic function $p_0 + (1 - p_0)\mu/(\mu - iu)$ is a mixture of the distribution centered at 0 (weight p_0) and the exponential distribution with mean $1/\mu$ (weight $1 - p_0$); this is quite expected.

E8. Start from the identity

$$1(Q_t = n) = 1(Q_0 = n) + \sum_{0 < s \le t} 1(Q_s = n) - 1(Q_{s-} = n), \quad (1)$$

[valid because $1(Q_t = n)$ is a step process]. Thus, separating downward and upward jumps,

$$1(Q_t = n) = 1(Q_0 = n) + \int_0^t [1(Q_s = n) - 1(Q_{s-} = n)] \, dD_s$$

$$+ \int_0^t [1(Q_s = n) - 1(Q_{s-} = n)] \, dD_s. \tag{2}$$

Now:

$$1(Q_s = n) \, dA_s = 1(Q_{s-} = n-1) 1(n > 0) \, dA_s, \tag{3}$$

since if there is an arrival at time s, $Q_s = Q_{s-} + 1$, and also Q_s is then strictly positive. Also, for similar reasons,

$$1(Q_s = n) \, dD_s = 1(Q_{s-} = n+1) \, dD_s. \tag{4}$$

Now $1(Q_{t-} = n-1) 1(n > 0) - 1(Q_{t-} = n)$ and $1(Q_{t-} = n+1) - 1(Q_{t-} = n)$ are $\mathscr{F}_t^Q$-predictable; therefore, by the integration theorem T8 and the definition of $\mathscr{F}_t^Q$-parameters,

$$E[1(Q_t = n)] = E[1(Q_0 = n)]$$

$$+ E\left[\int_0^t [1(Q_s = n-1) 1(n > 0) - 1(Q_s = n)] \lambda_s \, ds\right]$$

$$+ E\left[\int_0^t [1(Q_s = n+1) - 1(Q_s = n)] \mu_s 1(Q_s > 0) \, ds\right]. \tag{5}$$

By Fubini's theorem and the definition of $\lambda_n(t)$ and $\mu_n(t)$,

$$E\left[\int_0^t 1(Q_s = k) \lambda_s \, ds\right] = \int_0^t E[1(Q_s = k) \lambda_s] \, ds = \int_0^t p_k(s) \lambda_k(s) \, ds. \tag{6}$$

Similarly

$$E\left[\int_0^t 1(Q_s = k) \mu_s 1(Q_s > 0)\right] = \int_0^t p_k(s) \mu_k(s) 1(k > 0) \, ds. \tag{7}$$

Hence (5) can be rewritten as

$$p_n(t) = p_n(0) + \int_0^t \{p_{n-1}(s) \lambda_{n-1}(s) 1(n > 0) - p_n(s)[\lambda_n(s) + \mu_n(s) 1(n > 0)]$$

$$+ p_{n+1}(s) \mu_{n+1}(s)\} \, ds. \tag{8}$$

E9. Use Stieltjes's rule of integration by parts

$$N_\tau(T - \tau) = \int_0^\tau (T - s) \, dN_s - \int_0^\tau N_s \, ds. \tag{1}$$

The intensity of N_t is λ, thus:

$$E\left[\int_0^\tau (T - s) \, dN_s\right] = E\left[\int_0^\tau (T - s) \lambda \, ds\right] \tag{2}$$

(apply II, T8) with $C_t = 1(t \le \tau)(T-t)$). Hence:

$$E[N_\tau(T-\tau)] = E\left[\int_0^\tau (\lambda(T-s) - N_s)\,ds\right]. \qquad (3)$$

The integrand $\lambda(T-t) - N_t$ is decreasing and takes the value $\lambda T > 0$ at time $t = 0$. It is therefore clear that the optimal τ^* is the first time at which this integrand becomes negative. Figure 9 describes τ^*.

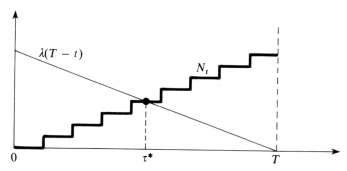

Figure 9   The optimal stopping time.

E10. It suffices to show, for any $t \ge 0$, that if $(s, \omega) \to Y_s(\omega)$ is a mapping defined on $[0, t] \times \Omega$ which is bounded and $\mathscr{B}([0, t]) \otimes \mathscr{F}_t$-measurable, then $\int_0^t Y_s\,ds$ is $\mathscr{F}_t$-measurable.

Let $\mathscr{H}$ be vector space of such mappings, and let $\mathscr{S}$ be the π-system (see A1, §2) for a definition) consisting of the sets in $[0, t] \times \Omega$ of the form

$$C \equiv [u, v] \times A, \qquad 0 \le u \le v \le t, \quad A \in \mathscr{F}_t. \qquad (1)$$

Clearly if $Y_t(\omega) = 1_C(t, \omega)$, then Y_t belongs to $\mathscr{H}$, since

$$\int_0^t Y_s\,ds = (v-u)1_A. \qquad (2)$$

The rest follows from Appendix A1, T4.

E11. By E6, $N_t(1)$ and $N_t(2)$ have the $(P, \mathscr{G}_t)$-intensities λ and μ. Now $A_t = N_t(1)$, $\mathscr{G}_t \supset \mathscr{F}_t^Q \vee \mathscr{F}_t^F$, and λ is $\mathscr{F}_t^Q \vee \mathscr{F}_t^F$-measurable (being a constant); therefore A_t has the $\mathscr{F}_t^Q \vee \mathscr{F}_t^F$-intensity λ, by E4.

In order to compute the $\mathscr{F}_t^Q \vee \mathscr{F}_t^F$-intensity of D_t we must compute, for all nonnegative $\mathscr{F}_t^F \vee \mathscr{F}_t^Q$-predictable processes C_t,

$$E\left[\int_0^\infty C_s\,dD_s\right]. \qquad (1)$$

But this quantity is also

$$E\left[\sum_{n>0} C_{T_n(2)}1(Q_{T_n(2)-} > 0)(1 - X_n)1(T_n(2) < \infty)\right]. \qquad (2)$$

By (I, E10), $C_{T_n(2)}1(T_n(2) < \infty)$ is $(\mathscr{F}^Q \vee \mathscr{F}^F)_{T_n(2)-}$-measurable, and therefore $C_{T_n(2)}1(T_n(2) < \infty)1(Q_{T_n(2)-} > 0)$ is independent of X_n. Thus the quantity in (2) is just

$$\sum_{n>0} E[(1 - X_n)]E\left[\sum_{n>0} C_{T_n(2)}1(Q_{T_n(2)-} > 0)1(T_n(2) < \infty)\right]$$

$$= (1 - p)E\left[\int_0^\infty C_s 1(Q_{s-} > 0)\, dN_s(2)\right]. \quad (3)$$

Now, since μ is the $\mathscr{F}_t^Q \vee \mathscr{F}_t^F$-intensity of $N_t(2)$,

$$E\left[\int_0^\infty C_s 1(Q_{s-} > 0)\, dN_s(2)\right] = E\left[\int_0^\infty C_s \mu 1(Q_s > 0)\, ds\right]. \quad (4)$$

Hence, combining (1), (2), (3), and (4),

$$E\left[\int_0^\infty C_s\, dD_s\right] = E\left[\int_0^\infty C_s \mu(1 - p)1(Q_s > 0)\, ds\right],$$

and therefore, by definition of intensity, $\mu(1 - p)1(Q_t > 0)$ is the $\mathscr{F}_t^Q \vee \mathscr{F}_t^F$-intensity of D_t.

Similar manipulations yield the $\mathscr{F}_t^Q \vee \mathscr{F}_t^F$-intensities of S_t and F_t: $\mu 1(Q_t > 0)$ and $\mu p 1(Q_t > 0)$ respectively. As for the $\mathscr{F}_t^Q \vee \mathscr{F}_t^F$-intensity of E_t, since $E_t = A_t + F_t$, it is the sum of the $\mathscr{F}_t^Q \vee \mathscr{F}_t^F$-intensities of A_t and F_t: $\lambda + \mu p 1(Q_t > 0)$.

For E_t, the answer is yes, by E4. For F_t it is no, because $\mathscr{F}_t^Q$ is *not* a history of F_t, and therefore one cannot speak of the $\mathscr{F}_t^Q$-intensity of F_t.

E12. First we show that $N_\infty = \infty$, P-a.s., implies $\int_0^\infty \lambda_s\, ds = \infty$, P-a.s. Suppose that it were not true; then there would exist a finite number c such that

$$P\left[\int_0^\infty \lambda_s\, ds \leq c\right] > 0. \quad (1)$$

By definition of the intensity,

$$E\left[\int_0^\infty 1\left(\int_0^s \lambda_u\, du < c\right) dN_s\right] = E\left[\int_0^\infty 1\left(\int_0^s \lambda_u\, du < c\right) \lambda_s\, ds\right]. \quad (2)$$

The right-hand side of (2) is bounded above by c, whereas the left-hand side is bounded below by

$$E\left[1\left(\int_0^\infty \lambda_s\, ds \leq c\right) N_\infty\right], \quad (3)$$

a quantity which is infinite in view of (1) and of the hypothesis $N_\infty = \infty$; hence a contradiction.

The converse part follows the same lines; we assume that $\int_0^\infty \lambda_s\, ds = \infty$, P-a.s., and we suppose with a view to contradiction that there exists a finite number c such that

$$P[N_\infty \leq c] > 0. \quad (4)$$

By definition of the intensity,

$$E\left[\int_0^\infty 1(N_s \leq c)\, dN_s\right] = E\left[\int_0^\infty 1(N_s \leq c)\lambda_s\, ds\right]. \tag{5}$$

The left-hand side of (5) is finite (bounded above by c), and the right hand-side is infinite, being bounded below by

$$E\left[1(N_\infty \leq c)\int_0^\infty \lambda_s\, ds\right]. \tag{6}$$

E13. Fix t and u. Since

$$\{\tau(t) \leq u\} = \left\{\int_0^u \tilde{\lambda}_s\, ds \leq t\right\} \tag{1}$$

and since $\int_0^u \tilde{\lambda}_s\, ds$ is $\mathcal{G}_u$-measurable, it follows that $\{\tau(t) \leq u\} \in \mathcal{G}_u$. This being true for all $u \geq 0$, $\tau(t)$ is indeed a $\mathcal{G}_t$-stopping time.

In order to show that $\mathcal{F}^N_{\tau(t)} \subset \mathcal{G}_{\tau(t)}$, we will prove a more general result: if $\mathcal{F}_t$ and $\mathcal{G}_t$ are two histories such that $\mathcal{F}_t \subset \mathcal{G}_t$, and if τ is an $\mathcal{F}_t$-stopping time (hence a $\mathcal{G}_t$-stopping time), then

$$\mathcal{F}_\tau \subset \mathcal{G}_\tau. \tag{2}$$

This is true because if $A \in \mathcal{F}_\tau$, then by definition of $\mathcal{F}_\tau$,

$$A \cap \{\tau \leq t\} \in \mathcal{F}_t, \qquad t \geq 0. \tag{3}$$

But $\mathcal{F}_t \subset \mathcal{G}_t$; therefore

$$A \cap \{\tau \leq t\} \in \mathcal{G}_t, \qquad t \geq 0. \tag{4}$$

That is to say, $A \in \mathcal{G}_\tau$, by definition of $\mathcal{G}_\tau$. Since N_t is $\mathcal{F}^N_t$-progressive, $N_{\tau(t)}$ is adapted to $\mathcal{F}^N_{\tau(t)}$ by Appendix A1, T1. Hence $\mathcal{F}^{\tilde{N}}_t \subset \mathcal{F}^N_{\tau(t)}$. In fact it is shown in Appendix A2 that $\mathcal{F}^N_{\tau(t)} = \mathcal{F}^{\tilde{N}}_t$, but we do not need to show this for the proof of T16.

E14. Since C_t is left-continuous, it suffices to show that it is adapted to $\mathcal{G}_t$. For this it suffices to show that for any $c \geq 0$, $A \in \mathcal{G}_{\tau(c)}$, the set

$$A \cap \{t \leq \tau(c)\} \tag{1}$$

is in $\mathcal{G}_t$. But this is just the definition of $\mathcal{G}_{\tau(t)}$.

E15. First show that $\tilde{N}_t = N_{t/\Lambda}$ has the $\sigma(\Lambda) \vee \mathcal{F}^{\tilde{N}}_t$-intensity 1. Therefore by Watanabe's theorem $\tilde{N}_t - \tilde{N}_s$ is independent of $\sigma(\Lambda) \vee \mathcal{F}^{\tilde{N}}_s$ for all $0 \leq s \leq t$. In particular $\tilde{N}_t$ is independent of $\sigma(\Lambda)$ for all $t \geq 0$.

CHAPTER III

Integral Representation of Point-Process Martingales

1. The Structure of Internal Histories
2. Regenerative Form of the Intensity
3. The Representation Theorem
4. Hilbert-Space Theory of Poissonian Martingales
5. Useful Extensions
References
Solutions to Exercises, Chapter III

1. The Structure of Internal Histories

This chapter is devoted to the proof of a structural result concerning the representation of martingales relative to the internal history of point processes. It turns out that such martingales have the form of a Stieltjes integral with respect to the fundamental martingale. More precisely, if m_t is a $(P, \mathscr{F}_t^N)$-martingale, where $\mathscr{F}_t^N$ is the internal history of a point process N_t admitting the $(P, \mathscr{F}_t^N)$-intensity λ_t, then

$$m_t = m_0 + \int_0^t K_s(dN_s - \lambda_s \, ds), \tag{A}$$

where K_t is a $\mathscr{F}_t^N$-predictable process such that for all $t \geq 0$, $\int_0^t |K_s|\lambda_s \, ds < \infty$, P-a.s. A similar representation theorem for martingales is also available in a quite different context: let m_t be a square-integrable $(P, \mathscr{F}_t^W)$-martingale, where $\mathscr{F}_t^W$ is the internal history of a $(P, \mathscr{F}_t^W)$-Wiener process[1] W_t; then

$$m_t = m_0 + \int_0^t \phi_s \, dW_s, \tag{B}$$

---
[1] I.e., W_t is a continuous process such that for all $0 \leq s \leq t$, $W_t - W_s$ is a gaussian random variable with mean 0 and variance $t - s$, P-independent of $\mathscr{F}_s^W$.

where ϕ_t is a $\mathscr{F}_t^W$-progressive process such that for all $t \geq 0$, $E[\int_0^t \phi_s^2\, ds] < \infty$, and the symbol $\oint$ denotes Ito stochastic integration (see A3). The representation (B) for "Wiener martingales" plays a crucial role in the theory of Wiener-driven stochastic systems, a brief summary of which is given in Appendix A3. The representation (A) for point-process martingales is an equally important result, and we will use it in the present text to develop the innovations theory of filtering based on point-process observations (Chapter IV) and the representation of likelihood ratios as martingale exponentials (Chapter VI).

There are various proofs available for the representation (A); we will give here the most elementary one, which unfortunately turns out to be very computational. However, being elementary, it has the advantage of not requiring a background in martingale theory beyond Doob's optional sampling theorem (I T2).

The proof takes three steps: first we elucidate the structure of the internal histories $\mathscr{F}_t^N$, then we describe P in terms of the intensity, and after that we apply Doob's optional-sampling theorem, from which the result falls. As for the structure of internal histories, all the results are intuitively fairly obvious; however, their proofs are not so straightforward, and we have relegated most of them to Appendix A2, §3.

The setting is that of m-variate point processes. We recall the terminology: $(T_n, Z_n, n \geq 1)$ is a sequence of random variables defined on $(\Omega, \mathscr{F})$, $(T_n, n \geq 1)$ being a point process; Z_0 is a constant, say δ, and the sequence $(Z_n, n \geq 1)$ is $\{1, \ldots, m\}$-valued. The explosion point T_∞ is defined by $T_\infty = \lim \uparrow T_n$. For each $1 \leq i \leq m$, the point process $N_t(i)$ is defined by

$$N_t(i) = \sum_{n \geq 1} 1(T_n \leq t) 1(Z_n = i), \tag{1.1}$$

and the internal history of $(N_t(1), \ldots, N_t(m))$ is denoted by $\mathscr{G}_t$:

$$\mathscr{G}_t = \bigvee_{i=1}^m \mathscr{F}_t^N(i). \tag{1.2}$$

For each $n \geq 0$, define

$$S_{n+1} = \begin{cases} T_{n+1} - T_n & \text{if } T_n < \infty, \\ +\infty & \text{if } T_n = \infty. \end{cases} \tag{1.3}$$

The following exercises are easy:[2]

E1 Exercise.

(α) For each $n \geq 0$, T_n is a $\mathscr{G}_t$-stopping time.
(β) For each $n \geq 0$, $\sigma(T_0, Z_0, \ldots, T_n, Z_n) \subset \mathscr{G}_{T_n}$.
(γ) For each $t \geq 0$,

$$\mathscr{G}_t = \sigma(Z_n 1(T_n \leq s); 0 \leq s \leq t, n \geq 0). \tag{1.4}$$

---

[2] Exercises E1, E2, and E4 will not be solved at the end of the chapter, since they are solved by T23, T24, and T34, respectively, in A2. We have set the corresponding results as exercises because they do not require tricks, in contrast with the results stated as theorems in the current section.

E2 Exercise. Show that the (right-continuous) process $1(t \geq T_\infty)$ is $\mathscr{G}_t$-predictable, and so is any process X_t of the form

$$X_t(\omega) = \sum_{n \geq 0} f^{(n)}(t, \omega) 1(T_n(\omega) < t \leq T_{n+1}(\omega)) 1(0 < t < \infty) \tag{1.5}$$

where for each $n \geq 0$, the mapping $(\omega, t) \to f^{(n)}(t, \omega)$ is $\mathscr{G}_{T_n} \otimes \mathscr{B}_+$-measurable.

We now state a few other results which are just as intuitive as the above ones, but a little harder to prove. The proofs are more technical than difficult and are given in Appendix A2; they can be skipped without harm in a first reading.

In all the statements of this chapter, recall that $\mathscr{G}_t = \bigvee_{i=1}^m \mathscr{F}_t^N(i)$.

T1 Theorem [3, 7]. *The history $\mathscr{G}_t$ is right continuous, i.e., for all $t \geq 0$, $\mathscr{G}_t = \bigwedge_{h>0} \mathscr{G}_{t+h}$.*

PROOF. See Appendix A2, T25. □

Comment. The right continuity of $\mathscr{G}_t$ follows from the fact that the R^m-valued process $Y_t = (N_t(1), \ldots, N_t(m))$ possesses the following property: for all $\omega \in \Omega$, all $t \geq 0$, there exists a strictly positive real number $\varepsilon(t, \omega)$ such that $Y_s(\omega) = Y_t(\omega)$ for all $s \in [t, t + \varepsilon(t, \omega))$ (see [7] and Appendix A2, T26).

T2 Theorem. *The following equalities hold for all $n \geq 0$:*

$$\mathscr{G}_{T_n} = \sigma(T_0, Z_0, \ldots, T_n, Z_n), \tag{1.6}$$

$$\mathscr{G}_{T_n-} = \sigma(T_0, Z_0, \ldots, T_{n-1}, Z_{n-1}, T_n). \tag{1.7}$$

Also

$$\mathscr{G}_{T_\infty} = \mathscr{G}_{T_\infty-} = \sigma(T_n, Z_n, n \geq 0) = \mathscr{G}_\infty. \tag{1.8}$$

PROOF. See Appendix A2, T30. □

Remark. If we apply T2 to the case of a *univariate* point process N_t with internal history $\mathscr{G}_t = \mathscr{F}_t^N$, we find that

$$\mathscr{F}_{T_n}^N = \mathscr{F}_{T_n-}^N. \tag{1.9}$$

However, this is not true in the general case where $\mathscr{G}_{T_n} = \mathscr{G}_{T_n-} \vee \sigma(Z_n)$.

T3 Theorem [3, 7]. *For any $\mathscr{G}_t$-stopping time S,*

$$\mathscr{G}_S = \sigma(N_{t \wedge S}(i), 1 \leq i \leq m, t \geq 0). \tag{1.10}$$

PROOF. See Appendix A2, T28. □

Comment. Theorem T3 says that "the stopped history is the history of the stopped process."

The statement of the next result requires the definition of the *trace* of a σ-field.

D4 Definition. Let $\mathscr{G}$ be a σ-field on Ω, and let A be a subset of Ω. The trace of $\mathscr{G}$ on A, denoted $\mathscr{G} \cap A$, is, by definition, the σ-field on A formed of the sets $B \cap A$ where $B \in \mathscr{G}$.

In the statement of T5, the convention $T_{\infty+1} = \infty$ is adopted.

T5 Theorem [5]. *Let S be a finite $\mathscr{G}_t$-stopping time. Then for all $n \in \overline{N}_+$,*

$$\mathscr{G}_S \cap \{T_n \leq S < T_{n+1}\} = \mathscr{G}_{T_n} \cap \{T_n \leq S < T_{n+1}\}. \quad (1.11)$$

PROOF. See Appendix A2, T32. □

Comment. Theorem T5 can be read as follows: "If you already know that $T_n \leq S < T_{n+1}$, then $\mathscr{G}_{T_n}$ contains the same information as $\mathscr{G}_S$."

T6 Theorem [5]. *Let S be a $\mathscr{G}_t$-stopping time. Then there exists a sequence $(R_n, n \geq 0)$ of non-negative $\mathscr{G}_{T_n}$-measurable random variables such that, for all $n \in \overline{N}_+$:*

$$S \wedge T_{n+1} = (T_n + R_n) \wedge T_{n+1} \quad \text{on } \{S \geq T_n\}. \quad (1.12)$$

PROOF. See Appendix A2, T33. □

E3 Exercise. What is the representation of $S = T_k$?

E4 Exercise [5]. Show that in order for the process X_t to be $\mathscr{G}_t$-predictable, it is *necessary* (and sufficient as was seen in Exercise E2) that it admit the representation

$$X_t(\omega) = \sum_{n \geq 1} f^{(n)}(t, \omega) 1(T_n(\omega) < t \leq T_{n+1}(\omega))$$
$$+ f^{(\infty)}(t, \omega) 1(T_\infty(\omega) < t) 1(t < \infty) \quad \text{on } \{t > 0\}, \quad (1.13)$$

where for each $n \in \overline{N}_+$, the mapping $(\omega, t) \to f^{(n)}(t, \omega)$ is $\mathscr{G}_{T_n} \otimes \mathscr{B}_+$-measurable.

2. Regenerative Form of the Intensity

From the Probability to the Intensity

The notations are those of §1; that is to say, an m-variate point process $(T_n, Z_n, n \geq 1)$ or $(N_t(1), \ldots, N_t(m))$ is given, and its internal history $\bigvee_{i=1}^{m} \mathscr{F}_t^N(i)$ is denoted by $\mathscr{G}_t$. Also the interoccurrence times $T_{n+1} - T_n$

are denoted by S_{n+1}. It is supposed moreover that the conditional distributions of S_{n+1} given the past $\mathscr{G}_{T_n} = \sigma(T_0, Z_0, \ldots, T_n, Z_n)$ admit densities. More precisely, for all $n \geq 0$, all i, $1 \leq i \leq m$, and all $A \in \mathscr{B}_+$,

$$P[S_{n+1} \in A, Z_{n+1} = i | \mathscr{G}_{T_n}](\omega) = \int_A g^{(n+1)}(\omega, x, i) \, dx$$
$$= G^{(n+1)}(\omega, A, i), \tag{2.1}$$

where the mapping $(\omega, x) \to g^{(n+1)}(\omega, x, i)$ from $\Omega \times R_+$ into R_+ is $\mathscr{G}_{T_n} \otimes \mathscr{B}_+$-measurable. Also, for all $n \geq 0$ and all $A \in \mathscr{B}_+$,

$$P[S_{n+1} \in A | \mathscr{G}_{T_n}](\omega) = \int_A g^{(n+1)}(\omega, x) \, dx = G^{(n+1)}(\omega, A), \tag{2.2}$$

where $g^{(n+1)}$ is defined by

$$g^{(n+1)}(\omega, x) = \sum_{i=1}^{m} g^{(n+1)}(\omega, x, i). \tag{2.3}$$

Note that:

$$P[S_{n+1} = \infty | \mathscr{G}_{T_n}] = 1 - \int_{R_+} g^{(n+1)}(x) \, dx. \tag{2.4}$$

Our goal is to give an explicit form for the $(P, \mathscr{G}_t)$-intensity $\lambda_t(i)$ of $N_t(i)$ in terms of the conditional densities $g^{(n+1)}(x, i)$, $1 \leq i \leq m$.

If we agree not to bother about mathematical rigor, this is an easy task: indeed, for each $1 \leq i \leq m$, $\lambda_t(i)$ satisfies (*begin heuristics*)

$$\lambda_t(i) \, dt = P[dN_t(i) = 1 | \mathscr{G}_t].$$

Suppose that t lies between T_n and T_{n+1}; then the past history $\mathscr{G}_t$ consists of $\mathscr{G}_{T_n}$ and of the information $S_{n+1} \geq t$. Also the event $\{dN_t(i) = 1\}$ is identical, when t is between T_n and T_{n+1}, to the event $\{S_{n+1} \in (t, t+dt], Z_{n+1} = i\}$. Hence, on $\{T_n \leq t < T_{n+1}\}$,

$$P[dN_t(i) = 1 | \mathscr{G}_t] = P[S_{n+1} \in (t, t+dt] - T_n,$$
$$Z_{n+1} = i | S_{n+1} \geq t - T_n, \mathscr{G}_{T_n}],$$

where $(t, t+dt] - T_n$ represents the interval $(t - T_n, t + dt - T_n]$. Hence by Bayes' rule,

$$P[dN_t(i) = 1 | \mathscr{G}_t]$$
$$= \frac{P[S_{n+1} \in (t, t+dt] - T_n, Z_{n+1} = i, S_{n+1} \geq t - T_n | \mathscr{G}_{T_n}]}{P[S_{n+1} \geq t - T_n | \mathscr{G}_{T_n}]}.$$

Now $\{S_{n+1} \in (t, t+dt] - T_n, S_{n+1} \geq t - T_n\} = \{S_{n+1} \in (t, t+dt] - T_n\}$. Therefore

$$P[dN_t(i) = 1 | \mathscr{G}_t] = \frac{P[S_{n+1} \in (t, t+dt] - T_n, Z_{n+1} = i | \mathscr{G}_{T_n}]}{P[S_{n+1} \geq t - T_n | \mathscr{G}_{T_n}]},$$

2. Regenerative Form of the Intensity

and therefore, on $\{T_n \leq t < T_{n+1}\}$,

$$\lambda_t(i) \, dt = \frac{g^{(n+1)}(t - T_n, i) \, dt}{1 - \int_0^{t - T_n} g^{(n+1)}(x) \, dx}$$

(end heuristics).
The above calculations will now be formalized.

T7 Theorem [8, 5]. *Under the assumptions of the present section, if for each $1 \leq i \leq m$, we define the processes $\lambda_t(i)$ and $A_t(i)$ by*

$$\lambda_t(i) = \sum_{n \geq 0} \frac{g^{(n+1)}(t - T_n, i)}{1 - \int_0^{t - T_n} g^{(n+1)}(x) \, dx} \mathbf{1}(T_n \leq t < T_{n+1}) \tag{2.5}$$

and

$$A_t(i) = \int_0^t \lambda_s(i) \, ds, \tag{2.6}$$

then $N_{t \wedge T_n}(i) - A_{t \wedge T_n}(i)$ is a $(P, \mathscr{G}_t)$-martingale for each $n \geq 0$.

Comments

(α) In T7, one does not make the assumption that T_n is nonexplosive. If T_n were nonexplosive, then the conclusion of T7 would read: $N_t(i) - A_t(i)$ is a $(P, \mathscr{G}_t)$-local martingale.

(β) One can show, as we have seen already in II, §2, that if $N_{t \wedge T_n}(i) - A_{t \wedge T_n}(i)$ is a $(P, \mathscr{G}_t)$-martingale, then

$$E\left[\int_0^{T_n} C_s \, dN_s(i)\right] = E\left[\int_0^{T_n} C_s \lambda_s(i) \, ds\right] \tag{2.7}$$

for all nonnegative $\mathscr{G}_t$-predictable process C_t. Now if we assume that T_n is nonexplosive, the equality (2.7) extends to

$$E\left[\int_0^\infty C_s \, dN_s(i)\right] = E\left[\int_0^\infty C_s \lambda_s(i) \, ds\right], \tag{2.8}$$

and therefore $\lambda_t(i)$ is the $(P, \mathscr{G}_t)$-intensity of $N_t(i)$, in the sense of II, D7.

PROOF OF T7. By Komatsu's lemma (I, E7), it suffices to show that

$$E[N_{S \wedge T_n}(i)] = E[A_{S \wedge T_n}(i)] \tag{2.9}$$

for all $n \geq 0$ and all finite $\mathscr{G}_t$-stopping times S. Such a stopping time S admits, by T6, the representation

$$S \wedge T_{n+1} = (T_n + R_n) \wedge T_{n+1} \quad \text{on } S \geq T_n$$

(with the convention $T_{\infty+1} = \infty$), where R_n is a $\mathscr{G}_{T_n}$-measurable nonnegative random variable. Therefore

$$E[A_{S \wedge T_n}(i)] = E\left[\sum_{j=0}^{n-1} \int_0^{R_j \wedge S_{j+1}} \frac{g^{(j+1)}(s, i)\, ds}{G^{(j+1)}([s, \infty])} 1(S \geq T_j)\right]$$

$$= E\left[\sum_{j=0}^{n-1} E\left[\int_0^{R_j \wedge S_{j+1}} \frac{g^{(j+1)}(s, i)\, ds}{G^{(j+1)}([s, \infty])} 1(S \geq T_j) \Big| \mathscr{G}_{T_j}\right]\right]$$

$$= E\left[\sum_{j=0}^{n-1} E\left[\int_0^{R_j \wedge S_{j+1}} \frac{g^{(j+1)}(s, i)\, ds}{G^{(j+1)}([s, \infty])} \Big| \mathscr{G}_{T_j}\right] 1(S \geq T_j)\right],$$

where we have used the fact that $\{S \geq T_j\}$ belongs to $\mathscr{G}_{T_j}$ (Appendix A2, T7). But

$$E\left[\int_0^{R_j \wedge S_{j+1}} \frac{g^{(j+1)}(s, i)\, ds}{G^{(j+1)}([s, \infty])} \Big| \mathscr{G}_{T_j}\right]$$

$$= \int_0^\infty g^{(j+1)}(u)\, du \int_0^{R_j \wedge u} \frac{g^{(j+1)}(s, i)}{G^{(j+1)}([s, \infty])}\, ds.$$

By Fubini's theorem, this last quantity equals

$$\int_0^{R_j} \frac{g^{(j+1)}(s, i)}{G^{(j+1)}([s, \infty])}\, ds \int_s^\infty g^{(j+1)}(u)\, du$$

or $\int_0^{R_j} g^{(j+1)}(s, i)\, ds$ since $\int_s^\infty g^{(j+1)}(u)\, du = G^{(j+1)}([s, \infty])$. The conclusion of the above chain of inequalities is:

$$E[A_{S \wedge T_n}(i)] = E\left[\sum_{j=0}^{n-1} \int_0^{R_j} g^{(j+1)}(s, i)\, ds\, 1(S \geq T_j)\right]. \tag{$*$}$$

Now

$$E[N_{S \wedge T_n}(i)] = E\left[\sum_{j=0}^{n-1} (N_{S \wedge T_{j+1}}(i) - N_{S \wedge T_j}(i)) 1(S \geq T_j)\right]$$

$$= E\left[\sum_{j=0}^{n-1} 1(R_j \geq S_{j+1}) 1(S \geq T_j) 1(Z_{j+1} = i)\right]$$

$$= E\left[\sum_{j=0}^{n-1} P[S_{j+1} \leq R_j, Z_{j+1} = i | \mathscr{G}_{T_j}] 1(S \geq T_j)\right],$$

where we have again used the fact that $\{S \geq T_j\} \in \mathscr{G}_{T_j}$. Therefore

$$E[N_{S \wedge T_n}(i)] = E\left[\sum_{j=0}^{n-1} \int_0^{R_j} g^{(j+1)}(s, i)\, ds\, 1(S \geq T_j)\right], \tag{$**$}$$

and (2.9) follows from $(*)$ and $(**)$. $\square$

2. Regenerative Form of the Intensity

Remark. Theorem T7 remains valid if we take for the definition of $\lambda_t(i)$

$$\lambda_t(i) = \sum_{n \geq 0} \frac{g^{(n+1)}(t - T_n, i)}{1 - \int_0^{t-T_n} g^{(n+1)}(x)\, dx} 1(T_n < t \leq T_{n+1}) 1(T_n < \infty). \quad (2.10)$$

The difference between (2.5) and (2.10) occurs at the T_n's, and therefore both expressions lead to the same $A_t(i)$. The version (2.10) of $\lambda_t(i)$ is $\mathcal{G}_t$-predictable by E2.

Given the "shape" of the intensity, it is possible to recover the conditional densities. More precisely:

E5 Exercise. Let N_t be a point process with the $(P, \mathcal{F}_t^N)$-intensity λ_t of the form

$$\lambda_t(\omega) = \sum_{n \geq 0} \lambda^{(n)}(t, \omega) 1(T_n(\omega) < t \leq T_{n+1}(\omega)) \quad \text{on } t \in (0, \infty), \quad (2.11)$$

where for each $n \geq 0$ the mapping $(\omega, t) \to \lambda^{(n)}(t, \omega)$ is $\mathcal{F}_{T_n}^N \otimes \mathcal{B}_+$ measurable. Let $G^{(n+1)}(\cdot, \omega)$ be a regular version of the conditional probability distribution of S_{n+1} given $\mathcal{F}_{T_n}^N$. Show that for each $n \geq 0$, $G^{(n+1)}(\cdot, \omega)$ admits a density, that is to say,

$$G^{(n+1)}(A, \omega) = \int_A g^{(n+1)}(x, \omega)\, dx, \quad A \in \mathcal{B}_+$$

for some mapping $(\omega, x) \to g^{(n+1)}(x, \omega)$ which is $\mathcal{F}_{T_n}^N \otimes \mathcal{B}_+$-measurable, and show that

$$\lambda^{(n)}(t, \omega) = \frac{g^{(n+1)}(t - T_n(\omega), \omega)}{G^{(n+1)}([t - T_n(\omega), \infty], \omega)}. \quad (2.12)$$

E6 Exercise. Let Q_t be a simple queue with a service-time sequence which is i.i.d. and independent of the arrival process. Let $F(x) = \int_0^x f(u)\, du$ be the common distribution function of the service times. Show that the $(P, \mathcal{F}_t^Q)$-service potential of Q_t is μ_t given by

$$\mu_t = \frac{f(t - \theta_t)}{1 - \int_0^{t-\theta_t} f(u)\, du}, \quad (2.13)$$

where θ_t is the last regeneration of service before t (see II, E7 for a definition).

The Uniqueness Theorem

E7 Exercise. Let N_t be a point process, and let λ_t be a process of the form (2.11). Suppose that the two probabilities P_1 and P_2 [defined on the same $(\Omega, \mathcal{F})$ as N_t and λ_t] are such that N_t admits the $(P_i, \mathcal{F}_t^N)$-intensity λ_t, $i = 1, 2$. Show that $P_1 \equiv P_2$ on $\mathcal{F}_\infty^N$.

The last exercise states a uniqueness theorem which says very roughly: to a given intensity corresponds one probability measure at most. Here is the precise statement in the general case.

T8 Theorem (Uniqueness [5]). *Let P_1 and P_2 be two probability measures on $(\Omega, \mathscr{F})$, and let $(N_t(1), \ldots, N_t(m))$ be an m-variate point process such that for some $\mathscr{G}_t$-predictable processes $\lambda_t(1), \ldots, \lambda_t(m)$, $N_t(i)$ admits the $(P_j, \mathscr{G}_t)$-intensity $\lambda_t(i)$, for all i, $1 \leq i \leq m$, and for $j = 1, 2$. Then P_1 and P_2 coincide on the events of $\mathscr{G}_\infty$.*

PROOF. We do the case $m = 1$ [notation $N_t(1) = N_t$, $\lambda_t(1) = \lambda_t$, $\mathscr{G}_t = \mathscr{F}_t^N$]. One needs only recall that λ_t has the form (2.11) (by E4) and the fact that λ_t can always be taken to be 0 after T_∞. The rest of the proof then follows from E7. □

3. The Representation Theorem

Existence

T9 Theorem (Integral Representation of Point Process Martingales [1, 2]). *Let $(N_t(1), \ldots, N_t(m))$ be an m-variate point process on $(\Omega, \mathscr{F}, P)$, P-non-explosive, and let $\mathscr{G}_t$ be its internal history. Suppose that for each i ($1 \leq i \leq m$), $N_t(i)$ admits the $(P, \mathscr{G}_t)$-predictable intensity $\lambda_t(i)$. Let now M_t be a right-continuous $(P, \mathscr{G}_t)$-martingale of the form $M_t = E[M_\infty | \mathscr{G}_t]$, where M_∞ is some P-integrable random variable. Then for each $t \geq 0$*

$$M_t = M_0 + \sum_{i=1}^{m} \int_0^t H_s(i)(dN_s(i) - \lambda_s(i)\, ds) \quad \text{P-a.s.}, \tag{3.1}$$

where for each $1 \leq i \leq m$, $H_t(i)$ is a $\mathscr{G}_t$-predictable process satisfying

$$\int_0^t |H_s(i)| \lambda_s(i)\, ds < \infty \quad \text{P-a.s.}, \quad t \geq 0. \tag{3.2}$$

PROOF. The proof will be carried out in three steps after some notational preliminaries:

Preliminaries. We will use the form (2.10) of $\lambda_t(i)$:

$$\lambda_t(i) = \sum_{n \geq 0} \frac{g^{(n+1)}(t - T_n, i)}{G^{(n+1)}([t - T_n, \infty])} 1(T_n < t \leq T_{n+1}) 1(T_n < \infty). \tag{3.3}$$

The $H_t(i)$'s will be found to be given by

$$H_t(i) = \sum_{n \geq 0} \left(f^{(n)}(t - T_n, i) - \frac{\sum_{i=1}^{m} \int_{t - T_n}^{\infty} f^{(n)}(s, i) g^{(n+1)}(s, i)\, ds}{G^{(n+1)}([t - T_n, \infty])} \right)$$
$$\times 1(T_n < t \leq T_{n+1}) 1(T_n < \infty), \tag{3.4}$$

where for each $n \geq 0$ and each $1 \leq i \leq m$, $(t, \omega) \to f^{(n)}(t, i, \omega)$ is a mapping which is $\mathscr{G}_{T_n} \otimes \mathscr{B}_+$-measurable and such that

$$M_{T_{n+1}}(\omega) = f^{(n)}(S_{n+1}(\omega), Z_{n+1}(\omega), \omega) \quad \text{on } \{T_{n+1} < \infty\}. \tag{3.5}$$

3. The Representation Theorem

E8 Exercise. Prove the existence of $f^{(n)}$ such that (3.5) holds.

Also we will use the following representation for M_t, between the jumps:

E9 Exercise. Prove that for each $t \geq 0$ and each $n \geq 0$, there exists a mapping $\omega \to h_t^{(n)}(\omega)$ which is $\mathcal{G}_{T_n}$-measurable and such that

$$M_t(\omega)1(T_n(\omega) \leq t < T_{n+1}(\omega))$$
$$= h_t^{(n)}(\omega)1(T_n(\omega) \leq t < T_{n+1}(\omega)) \quad \text{on } \{T_n < \infty\}. \quad (3.6)$$

Step 1. Apply Doob's optional sampling theorem. The martingale M_t is uniformly integrable (see Appendix A1, T27); we can therefore apply Doob's optional sampling theorem (I, T2) to $S = T_{n+1} \wedge t$ and $T = T_{n+1}$: for any set A in $\mathcal{G}_{t \wedge T_{n+1}}$, we have $E[1_A M_{t \wedge T_{n+1}}] = E[1_A M_{T_{n+1}}]$, and therefore $E[1_A M_t 1(t < T_{n+1})] = E[1_A M_{T_{n+1}} 1(t < T_{n+1})]$.

Now, for any $B \in \mathcal{G}_{t \wedge T_{n+1}}$, $A = B \cap \{T_n \leq t\}$ is also in $\mathcal{G}_{t \wedge T_{n+1}}$, since $\{T_n \leq t\} = \{T_n \leq t \wedge T_{n+1}\}$ is in $\mathcal{G}_{t \wedge T_{n+1}}$ (see remark after T7 of Appendix A2). Therefore, for all $B \in \mathcal{G}_{t \wedge T_{n+1}}$, we have $E[1_B M_t 1(T_n \leq t < T_{n+1})] = E[1_B M_{T_{n+1}} 1(T_n \leq t < T_{n+1})]$. Now, by T5, for any $C \in \mathcal{G}_{T_n}$, the event $C \cap \{T_n \leq t < T_{n+1}\}$ has the form $B \cap \{T_n \leq t < T_{n+1}\}$ for some $B \in \mathcal{G}_{t \wedge T_{n+1}}$ (observe that $\{T_n \leq t < T_{n+1}\} = \{T_n \leq t \wedge T_{n+1} < T_{n+1}\}$). Therefore, for all $C \in \mathcal{G}_{T_n}$ such that $C \subset \{T_n < \infty\}$

$$E[1_C M_t 1(T_n \leq t < T_{n+1})] = E[1_C M_{T_{n+1}} 1(T_n \leq t < T_{n+1})]. \quad (3.7)$$

Step 2. Between the jumps. By the representation (3.6) for M_t,

$$E[1_C M_t 1(T_n \leq t < T_{n+1})] = E[1_C h_t^{(n)} 1(T_n \leq t < T_{n+1})]$$
$$= E[1_C h_t^{(n)} 1(T_n \leq t) 1(S_{n+1} > t - T_n)].$$

Hence, since $1_C h_t^{(n)} 1(T_n \leq t)$ is $\mathcal{G}_{T_n}$-measurable,

$$E[1_C M_t 1(T_n \leq t < T_{n+1})]$$
$$= E[1_C h_t^{(n)} G^{(n+1)}([t - T_n, \infty]) 1(T_n \leq t)]. \quad (3.8)$$

Recalling the representation (3.5) for $M_{T_{n+1}}$, we have

$$E[1_C M_{T_{n+1}} 1(T_n \leq t < T_{n+1})]$$
$$= E[1_C f^{(n)}(S_{n+1}, Z_{n+1}) 1(T_n \leq t < T_{n+1})]$$
$$= E[1_C f^{(n)}(S_{n+1}, Z_{n+1}) 1(T_n \leq t) 1(S_{n+1} > t - T_n)].$$

Hence, since $1_C 1(T_n \leq t)$ is $\mathcal{G}_{T_n}$-measurable,

$$E[1_C M_{T_{n+1}} 1(T_n \leq t < T_{n+1})]$$
$$= E\left[1_C 1(T_n \leq t) \left\{\sum_{i=1}^m \int_{t-T_n}^\infty f^{(n)}(s, i) g^{(n+1)}(s, i)\, ds\right\}\right]. \quad (3.9)$$

By combining (3.7), (3.8), and (3.9), we have the equality

$$h_t^{(n)}1(T_n \leq t) = \frac{\sum_{i=1}^m \int_{t-T_n}^\infty f^{(n)}(s, i)g^{(n+1)}(s, i)\, ds}{G^{(n+1)}([t - T_n, \infty])} 1(T_n \leq t).$$

Since $M_{T_n} = h_{T_n}^{(n)} = \sum_{i=1}^m \int_0^\infty f^{(n)}(s, i)g^{(n+1)}(s, i)\, ds$, it follows from the above equality that on $\{T_n \leq t < T_{n+1}\} \cap \{T_n < \infty\}$

$$M_t - M_{T_n} = \frac{\sum_{i=1}^m \int_{t-T_n}^\infty f^{(n)}(s, i)g^{(n+1)}(s, i)\, ds}{G^{(n+1)}([t - T_n, \infty])}$$
$$- \sum_{i=1}^m \int_0^\infty f^{(n)}(s, i)g^{(n+1)}(s, i)\, ds. \tag{3.10}$$

We will now show that on $\{T_n \leq t < T_{n+1}\} \cap \{T_n < \infty\}$

$$M_t - M_{T_n} = \sum_{i=1}^m \int_{T_n}^t H_s(i)\lambda_s(i)\, ds. \tag{3.11}$$

One remarks on $\{T_n \leq t < T_{n+1}\} \cap \{T_n < \infty\}$

$$H_t(i) = h^{(n)}(t - T_n, i),$$

$$\lambda_t(i) = \frac{g^{(n+1)}(t - T_n, i)}{G^{(n+1)}([t - T_n, \infty])},$$

where

$$h^{(n)}(t, i) = f^{(n)}(t, i) - \frac{\sum_{i=1}^m \int_t^\infty f^{(n)}(s, i)g^{(n+1)}(s, i)\, ds}{G^{(n+1)}([t, \infty])}.$$

Letting

$$a(t) = G([s, \infty]),$$
$$b(t) = \sum_{i=1}^m \int_t^\infty f^{(n)}(s, i)g^{(n+1)}(s, i)\, ds,$$

we can write (3.10) as

$$M_t - M_{T_n} = \frac{b(t - T_n)}{a(t - T_n)} - \frac{b(0)}{a(0)}.$$

Integration by parts yields (see Appendix A4, T2)

$$\frac{b(s)}{a(s)} - \frac{b(0)}{a(0)} = \int_0^s \frac{1}{a(u)}\, db(u) - \int_0^s \frac{b(u)}{a^2(u)}\, da(u).$$

3. The Representation Theorem

Taking into account the definition of $a(t)$ and $b(t)$, the right-hand side of the above equality is equal to

$$-\sum_{i=1}^{m} \int_{0}^{s} \frac{g^{(n+1)}(u, i) f^{(n)}(u, i)}{G^{(n+1)}([u, \infty])} du$$
$$+ \int_{0}^{s} \frac{\sum_{i=1}^{m} g^{(n+1)}(u, i) \, du}{G^{(n+1)}([u, \infty])} \int_{u}^{\infty} \sum_{i=1}^{m} f^{(n)}(v, i) g^{(n+1)}(v, i) \, dv,$$

or, according to the definition of $h^{(n)}(t)$,

$$-\sum_{i=1}^{m} \int_{0}^{s} \frac{g^{(n+1)}(u, i) h^{(n)}(u, i)}{G^{(n+1)}([u, \infty])} du.$$

This terminates the proof of (3.11).

Step 3. At the jumps. By (3.11), it follows that there exists a P-modification of M_t (still denoted M_t) which is continuous in the intervals (T_n, T_{n+1}), right-continuous at the T_n's, and with left-hand limits at the T_{n+1}'s; also, by (3.5) and (3.10),

$$M_{T_{n+1}} - M_{T_{n+1}-} = f^{(n)}(S_{n+1}, Z_{n+1}) - \frac{\sum_{i=1}^{m} \int_{S_{n+1}}^{\infty} f^{(n)}(s, i) g^{(n+1)}(s, i) \, ds}{G^{(n+1)}([S_{n+1}, \infty])}.$$

Therefore, by the definition of $H_t(i)$ [see (3.4)],

$$\Delta M_{T_{n+1}} = H(T_{n+1}, Z_{n+1}). \tag{3.12}$$

The proof is now terminated, since the combination of (3.11) and (3.12) yields the representation (3.1). □

Uniqueness

The representation of point-process martingales is unique in the following sense:

T10 Theorem (Uniqueness of the Martingale Representation). *Let $(N_t(1), \ldots, N_t(m))$ be an m-variate point process on $(\Omega, \mathcal{F}, P)$ with the $(P, \mathcal{G}_t)$-intensity $(\lambda_t(1), \ldots, \lambda_t(m))$, where $\mathcal{G}_t$ is the internal history $\bigvee_{i=1}^{m} \mathcal{F}_t^N(i)$. Let M_t be a $(P, \mathcal{G}_t)$- martingale with the representations*

$$M_t = M_0 + \sum_{i=1}^{m} \int_{0}^{t} H_s^{(k)}(i)(dN_s(i) - \lambda_s(i) \, ds), \qquad k = 1, 2,$$

where for $k = 1, 2$ and $1 \leq i \leq m$, $H_t^{(k)}(i)$ is a $\mathcal{G}_t$-predictable process such that for all $t \geq 0$

$$\sum_{i=1}^{m} \int_{0}^{t} H_s^{(k)}(i) \lambda_s(i) \, ds < \infty \quad P\text{-a.s.}, \qquad k = 1, 2.$$

Then for all $1 \leq i \leq m$

$$H_t^{(1)}(i, \omega) = H_t^{(2)}(i, \omega), \quad (3.13)$$

the above equality holding $P(d\omega) \, dN_t(i, \omega)$-*almost everywhere or* (*equivalently*) $P(d\omega)\lambda_t(i, \omega) \, dt$-*almost everywhere.*

PROOF. Clearly (3.13) holds $P(d\omega) \, dN_t(i, \omega)$-a.e., since on $\{T_n < \infty\}$

$$H_{T_n(i)}^{(k)}(i) = \Delta M_{T_n(i)}, \quad k = 1, 2.$$

It also holds $P(d\omega)\lambda_t(i, \omega) \, dt$-a.e., since the $H_t^{(m)}(i)$'s are $\mathcal{G}_t$-predictable and on $\mathcal{P}(\mathcal{G}_t)$, $P(d\omega) \, dN_t(i, \omega) = P(d\omega)\lambda_t(i, \omega) \, dt$ [by the definition of $\lambda_t(i)$ as the $(P, \mathcal{G}_t)$-intensity of $N_t(i)$]. □

The following exercise should be obvious from the proof of T10:

E10 Exercise. The notation is that of T9 and T10. Show that there is no continuous $(P, \mathcal{G}_t)$-martingale other than the constant martingales (i.e. $M_t \equiv M_0$, $t \geq 0$, P-a.s.).

Square-Integrable Point Process Martingales

The structure of square-integrable martingales is obtained as a corollary of T9:

T11 Theorem (Representation for Square-Integrable Martingales). *Let* $(N_t(i), \ldots, N_t(m))$ *be an m-variate point process over* $(\Omega, \mathcal{F}, P)$ *with the* $(P, \mathcal{G}_t)$-*intensity* $(\lambda_t(1), \ldots, \lambda_t(m))$. *If* M_t *is a right-continuous* $(P, \mathcal{F}_t)$-*martingale, square-integrable (i.e.* $\sup_{t \geq 0} [M_t^2] < \infty$), *then it admits the representation* (3.1) *where the* $H_t(i)$'s *are* $\mathcal{G}_t$-*predictable processes satisfying*

$$\sum_{i=1}^m E\left[\int_0^t |H_s(i)|^2 \lambda_s(i) \, ds\right] < \infty, \quad t \geq 0. \quad (3.14)$$

Comment. T11 is the same as T9 except for the integrability condition (3.14).

PROOF. We do the case $m = 1$ with the obvious notational simplifications: For each $n \geq 1$, let U_n, V_n be the $\mathcal{G}_t$-stopping time defined by

$$U_n = \begin{cases} \inf\{t / M_{t-} + \int_0^t \lambda_s \, ds \geq n\} & \text{if } \{\cdots\} \neq \emptyset, \\ +\infty & \text{otherwise,} \end{cases} \quad (3.15)$$

$$V_n = \begin{cases} \inf\{t / \int_0^t H_s^2 \lambda_s \, ds \geq n\} & \text{if } \{\cdots\} \neq \emptyset, \\ +\infty & \text{otherwise.} \end{cases} \quad (3.16)$$

3. The Representation Theorem

By Appendix A4, T2, integration by parts yields, for all t such that $\int_0^t H_s^2 \lambda_s \, ds < \infty$,

$$M_t^2 = M_0^2 + \int_0^t 2M_{s-} \, dM_s + \sum_{0 < s \le t} (\Delta M_s)^2,$$

or

$$M_t^2 = M_0^2 + \int_0^t (2M_{s-} + H_s) H_s (dN_s - \lambda_s \, ds) + \int_0^t H_s^2 \lambda_s \, ds.$$

In particular

$$M_{t \wedge U_n \wedge V_m}^2 = M_0^2 + \int_0^{t \wedge U_n \wedge V_m} (2M_{s-} + H_s) H_s (dN_s - \lambda_s \, ds)$$

$$+ \int_0^{t \wedge U_n \wedge V_m} H_s^2 \lambda_s \, ds. \tag{3.17}$$

$\square$

E11 Exercise. Show that the process $\int_0^{t \wedge U_n \wedge V_m} (2M_{s-} + H_s) H_s (dN_s - \lambda_s \, ds)$ is a $(P, \mathscr{G}_t)$-martingale with mean 0. Therefore by (3.17)

$$E[M_{t \wedge U_n \wedge V_m}^2] = E\left[\int_0^{t \wedge U_n \wedge V_m} H_s^2 \lambda_s \, ds\right] + E[M_0^2]. \tag{*}$$

Since M_{t-} is a finite left-continuous process, $\lim U_n = +\infty$. Let V be the limit of the sequence V_n. The inequality

$$\sup_{T \in \mathscr{T}} E[M_T^2] < \infty, \tag{**}$$

where $\mathscr{T}$ is the set of finite $\mathscr{G}_t$-stopping times, has been proven in I, T7. By (*) and (**),

$$\int_0^{t \wedge V} H_s^2 \lambda_s \, ds < \infty \quad P\text{-a.s.}$$

But, by the definition of the V_n's, this is possible only if $V = \infty$, P-a.s. Letting $S_n = U_n \wedge V_n$, (*) and (**) yield

$$E\left[\int_0^{t \wedge S_n} H_s^2 \lambda_s \, ds\right] = E[M_{t \wedge S_n}^2] \le \sup_{t \in \mathscr{T}} E[M_T^2] < \infty. \tag{3.18}$$

It then suffices to let n go to infinity, and the proof is terminated, since $S_n \uparrow \infty$, P-a.s.

Case of Discrete Time

For pedagogical purposes and in order to demythify the proof of the martingale representation theorem in continuous time, we will now give the analogous result to T9 in discrete time. A *discrete-time point process* is a sequence

$(X_n, n \geq 0)$ of $\{0, 1\}$-valued random variables with $X_0 \equiv 0$. If we define, for all $n \geq 0$, $\mathscr{F}_n^X = \sigma(X_0, X_1, \ldots, X_n)$, then the $(P, \mathscr{F}_n^X)$-intensity of the point process $(X_n, n \geq 0)$ is a discrete time process $(\lambda_n, n \geq 0)$ given by $\lambda_0 \equiv 0$ and $\lambda_n = E[X_n | \mathscr{F}_{n-1}^X]$ for all $n \geq 1$. Note that $0 \leq \lambda_n \leq 1$, P-a.s., $n \geq 0$. Also note that λ_n is $\mathscr{F}_{n-1}^X$-measurable; one says $(\lambda_n, n \geq 0)$ is $\mathscr{F}_n^X$-predictable.

E12 Exercise. Let $(M_n, n \geq 0)$ be a $(P, \mathscr{F}_n^X)$-martingale, that is to say, a real-valued discrete-time process, adapted to $\mathscr{F}_n^X$, integrable, and such that for all $n \geq 1$, $E[M_n | \mathscr{F}_{n-1}^X] = M_{n-1}$. Then for all $n \geq 1$,

$$M_n - M_{n-1} = H_n(X_n - \lambda_n), \tag{3.19}$$

where $(H_n, n \geq 0)$ is an $\mathscr{F}_n^X$-predictable process such that for all $n \geq 0$

$$E[|H_n|\lambda_n(1 - \lambda_n)] < \infty. \tag{3.20}$$

Comment. The equality (3.19) also reads

$$M_n = M_0 + \sum_{j=1}^{n} H_j(X_j - \lambda_j). \tag{3.21}$$

In the above form, the analogy with (3.1) is more apparent.

4. Hilbert-Space Theory of Poissonian Martingales

One can prove the representation result for *poissonian* martingales which are *square-integrable* in a much more elegant way than in the general case. The interest of the proof to follow is its adaptability to the case of brownian (Wiener) martingales; however, this section can be omitted since it has no direct bearing on the rest of the book.

The notation is the same as that of §3. In particular $\mathscr{G}_t$ is the internal history of the m-variate point processes considered.

T12 Theorem [6]. *Suppose that the m-variate point process $(N_t(1), \ldots, N_t(m))$ is Poisson, i.e., for each i, $1 \leq i \leq m$, $N_t(i)$ is a $(P, \mathscr{G}_t)$-Poisson process with the intensity $\lambda(t, i)$, where $\lambda(t, i)$ is a measurable real-valued function, nonnegative and locally integrable $[\int_0^t \lambda(s, i) \, ds < \infty, t \geq 0]$. Let M_t be a square-integrable $(P, \mathscr{G}_t)$-martingale over $[0, 1]$. Then*

$$M_t = M_0 + \sum_{i=1}^{m} \int_0^t H_s(i)(dN_s(i) - \lambda(s, i) \, ds), \quad t \in [0, 1], \tag{4.1}$$

where for each, $1 \leq i \leq m$, $H_t(i)$ is a $\mathscr{G}_t$-predictable process such that

$$E\left[\int_0^1 |H_s(i)|^2 \lambda(s, i) \, ds\right] < \infty. \tag{4.2}$$

4. Hilbert-Space Theory of Poissonian Martingales

This result is just a particular case of T11. However, we will develop a completely different proof based on Hilbert-space methods. We need a sequence of lemmas and preliminaries.

We introduce two Hilbert spaces. First we have the Hilbert space H of R^m-valued processes $C_t = (C_t(1), \ldots, C_t(m))$ such that for each $1 \leq i \leq m$, $C_t(i)$ is $\mathcal{G}_t$-predictable, and such that

$$\|C\|_H^2 = E\left[\sum_{i=1}^m \int_0^1 |C_s(i)|^2 \lambda(s, i)\, ds\right] < \infty.$$

As usual, we consider equivalence classes, not distinguishing two processes C and C' if $\|C - C'\|_H = 0$. The scalar product that makes H a Hilbert space is

$$\langle C, C' \rangle_H = E\left[\sum_{i=1}^m \int_0^1 C_s(i) C_s'(i) \lambda(s, i)\, ds\right]. \tag{4.3}$$

Let H_b be the subset of H consisting of bounded elements of H. Clearly H is dense in H.

E13 Exercise. For any $C \in H$, $\|C\|_H^2 = \frac{1}{2} E[\sum_{i=1}^m \{\int_0^1 |C_s(i)|^2 \lambda(s,i)\, ds + \int_0^1 |C_s(i)|^2\, dN_s(i)\}]$.

The second Hilbert space of interest is $\mathcal{M}^2$, already described in I, E13. It is the set (of equivalence classes) of square-integrable $(P, \mathcal{G}_t)$-martingales over $[0, 1]$ with the scalar product

$$\langle m, m' \rangle_{\mathcal{M}^2} = E[m_1 m_1']. \tag{4.4}$$

We refer to I, E13, where it is shown that $\langle m, m \rangle_{\mathcal{M}^2} = \|m\|_{\mathcal{M}^2}$ defines a norm that makes $\mathcal{M}^2$ a Hilbert space. We call $\mathcal{M}_0^2$ the Hilbert subspace of $\mathcal{M}^2$ consisting of the 0-mean martingales of $\mathcal{M}^2$. A mapping ϕ with domain H is defined as follows: for each $C \in H$, $\phi(C)$ is a process defined over $[0, 1]$ by

$$\phi(C)_t = \sum_{i=1}^m \int_0^t C_s(i)\, dM_s(i), \tag{4.5}$$

where $M_t(i) = N_t(i) - \int_0^t \lambda(s, i)\, ds$.

E14 Exercise. Show that $\phi(C)$ is well defined (no $\infty - \infty$ form).

T13 Theorem. $\phi: H \to \mathcal{M}_0^2$ *defined by* (4.5) *is an isometry.*

PROOF. The proof is somewhat similar to that of I, T5. We will sketch it in the case $m = 1$ [notation $N_t(1) = N_t$, $\lambda(t, 1) = \lambda(t)$, $C_t(1) = C_t$, $M_t(1) = M_t$].

Step 1

E15 Exercise. Show that if $C_t = 1_A 1(u < t \leq v)$ where $0 \leq u \leq v \leq 1$, $A \in \mathscr{F}_u^N$, then $\phi(C)_t$ is indeed a square-integrable $(P, \mathscr{F}_t^N)$-martingale over $[0, 1]$. Moreover, for all $t \in [0, 1]$

$$E\left[\int_0^t |C_s|^2 \lambda(s)\, ds\right] = E[|\phi(C)_t|^2]$$

$$= \tfrac{1}{2}\left\{E\left[\int_0^t |C_s|^2 \lambda(s)\, ds\right] + E\left[\int_0^t |C_s|^2\, dN_s\right]\right\}. \quad (4.6)$$

Step 2. Now let $\mathscr{H}$ be the vector space of measurable processes C_t such that $\phi(C)_t = \int_0^t C_s\, dM_s$ is in $\mathscr{M}_0^2$ and (4.6) is verified. Let $\mathscr{S}$ be the set of subsets of $\Omega \times (0, 1]$ of the form $A \cap \{u < t \leq v\}$, $0 \leq u \leq v \leq 1$, $A \in \mathscr{F}_u^N$. Apply T4 of Appendix A1 to obtain that $\mathscr{H}$ contains all bounded $\mathscr{F}_t^N$-predictable processes C_t (a similar argument was used in the proof of I, T5).

Step 3. By step 2, ϕ is an isometry from H_b into $\mathscr{M}_0^2$. But H_b is dense in H, so that by the Hahn–Banach theorem this isometry can be extended (uniquely) into an isometry—still denoted ϕ—from H into $\mathscr{M}_0^2$. The problem is now to show that for any $C \in H$

$$\phi(C)_t = \int_0^t C_s\, dM_t. \quad (4.7)$$

(This is true for $C \in H_b$, but there is *a priori* no guarantee that it is true for all $C \in H$, since ϕ is not defined constructively as was its restriction to H_b; ϕ is only defined "globally" by means of an extension procedure.) Inspired by the proof of I, T5, the reader should not find the following exercise too difficult. □

E16 Exercise. Show that (4.7) holds for all $C \in H$.

The next result is the basic step in the proof of T12. It is called the *totality theorem for exponentials*. In order to state it, a few definitions are needed.

Let $L^2(\mathscr{G}_1, P)$ be the Hilbert space of (classes of equivalences of) square-integrable random variables of $(\Omega, \mathscr{G}_1, P)$, and let $L_0^2(\mathscr{G}_1, P)$ be the Hilbert subspace of $L^2(\mathscr{G}_1, P)$ consisting of the 0-mean random variables of $L^2(\mathscr{G}_1, P)$. Let $\mathscr{K}_0$ be the set consisting of the following random variables:

$$\exp\left\{\sum_{i=1}^m \int_0^1 f(s, i)\, dN_s(i) - \int_0^1 \left(1 - \sum_{i=1}^m e^{f(s, i)}\lambda(s, i)\right) ds\right\} - 1, \quad (4.8)$$

where the $f(t, i)$'s are *bounded* real-valued measurable functions (deterministic). In other notations, which we will use later on,

$$\mathscr{K}_0 = \{\overline{M}_1^f R_1^f - 1 \,|\, f(t, i) \text{ bounded deterministic}\} \quad (4.8')$$

or

$$\mathscr{K}_0 = \{M_1^f - 1 \,|\, f(t, i) \text{ bounded deterministic}\}, \quad (4.8'')$$

where

$$M_t^f = \exp\left\{\sum_{i=1}^m \int_0^t f(s, i)\, dN_s(i) - \int_0^t \left(1 - \sum_{i=1}^m e^{f(s, i)}\lambda(s, i)\right) ds\right\}, \quad (4.9)$$

$$\overline{M}_t^f = \exp\left\{\sum_{i=1}^m \int_0^t f(s, i)\, dN_s(i)\right\}, \quad (4.10)$$

and

$$R_t^f = \exp\left\{\int_0^t \left(1 - \sum_{i=1}^m e^{f(s, i)}\lambda(s, i)\, ds\right)\right\}. \quad (4.11)$$

E17 Exercise. Show that if the $f(t, i)$'s are bounded measurable deterministic functions, then

$$M_t^f = 1 + \sum_{i=1}^m \int_0^t M_{s-}^f (e^{f(s, i)} - 1)\, dM_s(i), \quad t \geq 0, \quad (4.12)$$

and $M_t^f - 1$ is a martingale of $\mathcal{M}_0^2$. In particular,

$$E[|M_1^f|^2] = 1 + \sum_{i=1}^m E\left[\int_0^1 |M_s^f(e^{f(s, i)} - 1)|^2 \lambda(s, i)\, ds\right]. \quad (4.13)$$

L14 Lemma [9]. $\mathcal{K}_0$ is total in $L_0^2(\mathcal{G}_1, P)$, that is to say, the linear combinations of random variables of $\mathcal{K}_0$ are dense in $L_0^2(\mathcal{G}_1, P)$.

PROOF. By the definition of totality, we have to prove that if an element Y of $L_0^2(\mathcal{G}_1, P)$ is orthogonal to $\mathcal{K}_0$ (that is to say, if $E[YX] = 0$ for all $X \in \mathcal{K}_0$), then necessarily $Y = 0$. Now, since $E[Y] = 0$, and since the R_t^f's of (4.11) are deterministic, it is enough to show that if $E[Y\overline{M}_1^f] = 0$ for all bounded deterministic functions $f(t, i)$, then $Y = 0$. □

E18 Exercise. Prove the last assertion.

We are now ready to proceed to the

PROOF OF T12. The set $\mathcal{K}_0$ is total in $L_0^2(\mathcal{G}_1, P)$, and M_1 is an element of $L_0^2(\mathcal{G}_1, P)$; it is therefore the $L_0^2(\mathcal{G}_1, P)$-limit of linear combinations of random variables in $\mathcal{K}_0$. By E17 and the isometry of T13, M_1 has the form

$$M_1 = \sum_{i=1}^m \int_0^1 H_s(i)\, dM_s(i), \quad (4.14)$$

where the $H_t(i)$'s are $\mathcal{G}_t$-predictable processes verifying (4.2). Now, by the definition of a martingale, $M_t = E[M_1 | \mathcal{G}_t]$, and therefore

$$M_t = \sum_{i=1}^m E\left[\int_0^1 H_s(i)\, dM_s(i) \Big| \mathcal{G}_t\right]. \quad (4.15)$$

By the isometry theorem T13, $\sum_{i=1}^{m} \int_0^t H_s(i) \, dM_s(i)$ as a $(P, \mathcal{G}_t)$-martingale, and therefore

$$\sum_{i=1}^{m} E\left[\int_0^1 H_s(i) \, dM_s(i) \mid \mathcal{G}_t\right] = \sum_{i=1}^{m} \int_0^t H_s(i) \, dM_s(i),$$

and this terminates the proof of T12. $\square$

T15 Theorem (Fundamental Isometry for Square-Integrable Point-Process Martingales). *Let $(N_t(1), \ldots, N_t(m))$ be an m-variate point process defined on $(\Omega, \mathcal{F}, P)$, with the $(P, \mathcal{G}_t)$-intensity $(\lambda_t(1), \ldots, \lambda_t(m))$, where $\mathcal{G}_t$ is the internal history $\bigvee_{i=1}^{m} \mathcal{F}_t^N(i)$. Let $\mathcal{M}_0^2$ be the Hilbert space of zero-mean square-integrable $(P, \mathcal{G}_t)$-martingales over $[0, 1]$ with the scalar product*

$$\langle m, m' \rangle_{\mathcal{M}_0^2} = E[m_1 m_1']. \quad (4.16)$$

Let H be the Hilbert space of $\mathcal{G}_t$-predictable processes $C_t = (C_t(1), \ldots, C_t(m))$ such that

$$\sum_{i=1}^{m} E\left[\int_0^1 |C_s(i)|^2 \lambda_s(i) \, ds\right] < \infty \quad (4.17)$$

with the scalar product

$$\langle C, C' \rangle_H = \sum_{i=1}^{m} E\left[\int_0^1 C_s(i) S_s'(i) \lambda_s(i) \, ds\right]. \quad (4.18)$$

Suppose that for all $1 \leq i \leq m$, $E[N_1(i)] < \infty$.[3] Then $\mathcal{M}_0^2$ and H are isometric with respect to the mapping $H \xrightarrow{\phi} \mathcal{M}_0^2$ defined by

$$\phi(C)_t = \sum_{i=1}^{m} \int_0^t C_s(i)(dN_s(i) - \lambda_s(i) \, ds). \quad (4.19)$$

Comment. Of course, H and $\mathcal{M}_0^2$ are sets of equivalence classes, as explained in the lines following the statement of T12.

PROOF. We do the case $m = 1$ for simplicity. By II, T8, ϕ takes H into $\mathcal{M}_0$, the (equivalence) set of $(P, \mathcal{G}_t)$-martingales. But because C_t is a $\mathcal{G}_t$-predictable process such that $E[\int_0^1 |C_s|^2 \lambda_s \, ds] < \infty$, $\phi(C)_t = \int_0^t C_s(dN_s - \lambda_s \, ds)$ must be in $\mathcal{M}_0^2$, as we now proceed to show.

Since $\phi(C)_t = \sum_{n \geq 1} C_{T_n} 1(T_n \leq t) - \int_0^t C_s \lambda_s \, ds$, it follows that

$$|\phi(C)_t|^2 \leq \sum_{n \geq 1} (1 \vee |C_{T_n}|^2) 1(T_n \leq t) + \int_0^t (1 \vee |C_s|^2) \lambda_s \, ds, \quad (4.20)$$

---

[3] This is a condition whih can be dispensed with at the price of a more involved proof.

and therefore $E[|\phi(C)_t|^2] < \infty$, since, by II, T8,

$$E\left[\sum_{n\geq 1}(1 \vee |C_{T_n}|^2)1(T_n \leq t)\right] = E\left[\int_0^t (1 \vee |C_s|^2)\lambda_s \, ds\right]$$
$$\leq E\left[\int_0^t \lambda_s \, ds\right] + E\left[\int_0^t C_s^2 \lambda_s \, ds\right] < \infty,$$

the last inequality being part of the hypothesis.

Thus the mapping ϕ is *into* $\mathcal{M}_0^2$. By T11, ϕ is *onto*. By the last equality in the proof of T11, ϕ preserves the norms. In summary, ϕ is isometric. □

The above proof for the representation of square-integrable poissonian martingales can be reproduced step by step to obtain the representation of square-integrable brownian martingales. After reading Appendix A3, the reader will have at his disposition all the basic definitions and tools that will allow him to devise such a proof.

5. Useful Extensions

Let $(N_t(1), \ldots, N_t(m))$ be an m-variate point process with the $(P, \mathcal{G}_t)$-intensity $(\lambda_t(1), \ldots, \lambda_t(m))$, where $\mathcal{G}_t$ is the internal history.

Let now M_t be a $(P, \mathcal{G}_t)$-martingale. If it is not right-continuous, T9 is not directly applicable. It would be if we were granted the existence of a right-continuous version, i.e. a process M_t' that is adapted to $\mathcal{G}_t$ and such that for all $t \geq 0$, $M_t' = M_t$, P-a.s. Indeed such a process M_t' would also be a $(P, \mathcal{G}_t)$-martingale, and being right-continuous, would admit an integral representation (3.1). Unfortunately, at this point the existence of a right-continuous version of M_t has been proven in general only for the case where P, $\mathcal{F}$, and $\mathcal{G}_t$ satisfy the following technical conditions (C):

(C.1) $\mathcal{F}$ is P-complete,
(C.2) $\mathcal{G}_t$ is right-continuous,
(C.3) $\mathcal{G}_0$ contains all the P-null sets of $\mathcal{G}_t$.

Those conditions are called by Dellacherie the "usual conditions" (see p. 52 of {47}).

L16 Lemma (Right-Continuous Modifications of Martingales). *Let M_t be a $(P, \mathcal{G}_t)$-martingale where P, $\mathcal{F}$, and $\mathcal{G}_t$ satisfy the usual conditions. There exists a right-continuous version M_t' of M_t. Moreover M_t' admits left-hand limits $M_{t-}' = \lim_{s\uparrow t, s<t} M_s$.*

This result is due to Meyer (see VI, T4 of {106}).

A modified form of T9 is available. First let us define $\bar{\mathscr{F}}$, $\bar{P}$, and $\bar{\mathscr{G}}_t$, "modifications" of $\mathscr{F}$, P, $\mathscr{G}_t$ in the following sense: $(\Omega, \bar{\mathscr{F}}, \bar{P})$ is the completion of $(\Omega, \mathscr{F}, P)$, and $\bar{\mathscr{G}}_t = \sigma(\mathscr{G}_t \cup \mathscr{N}_\infty)$, where $\mathscr{N}_\infty$ is the family of the $\bar{P}$-null events of $\mathscr{G}_\infty = \bigvee_{t \geq 0} \mathscr{G}_t$. Here $\mathscr{G}_t$ is the internal history of $(N_t(1), \ldots, N_t(m))$, or, more generally,

$$\mathscr{G}_t = \mathscr{F}_0 \vee \left(\bigvee_{i=1}^{m} \mathscr{F}_t^N(i) \right) \quad (5.1)$$

[i.e., one "adds" a "germ σ-field" $\mathscr{F}_0$ to the internal history; $\mathscr{F}_0$ can be thought of as the "prehistory" of the point process $(N_t(i), \ldots, N_t(m))$]. It is assumed that this point process admits the $(P, \mathscr{G}_t)$-intensity $(\lambda_t(i), \ldots, \lambda_t(m))$, and therefore the $(\bar{P}, \bar{\mathscr{G}}_t)$-intensity $(\lambda_t(1), \ldots, \lambda_t(m))$.

T17 Theorem (General Martingale Representation Theorem [5]). *Let M_t be a $(\bar{P}, \bar{\mathscr{G}}_t)$-martingale. Then there exists a $\mathscr{G}_t$-predictable process $(H_t(1), \ldots, H_t(m))$ such that*

$$M_t = M_0 + \sum_{i=1}^{m} \int_0^t H_s(i)(dN_s(i) - \lambda_s(i)\,ds) \quad \bar{P}\text{-a.s.} \quad (5.2)$$

and

$$\sum_{i=1}^{m} \int_0^t |H_s(i)| \lambda_s(i)\,ds < \infty \quad \bar{P}\text{-a.s.} \quad (5.3)$$

PROOF. First one observes that, $\mathscr{G}_t$ being right-continuous, $\bar{\mathscr{G}}_t$ is also right-continuous (Appendix A2, T35). Therefore $\bar{\mathscr{F}}$, $\bar{P}$, and $\bar{\mathscr{G}}_t$ satisfy the usual conditions. Therefore M_t can be taken, without loss of generality, right-continuous [by L16 and the fact that if the representation (5.2) is true for any modification of M'_t it is true for M_t].

The rest of the proof then follows the same line of argument as in the proof of T9, because the structure of the modified history $\bar{\mathscr{G}}_t$ is the same as that of $\mathscr{G}_t$ (i.e., any result written for $\mathscr{G}_t$ is true if it is true when $\bar{\mathscr{G}}_t$ is replaced by $\mathscr{G}_t$ in the statement; see A2). □

Notational Convention. In the sequel we will always assume that the triples $(\Omega, P, \mathscr{G}_t)$ have been properly modified as shown above, but we will keep the notation $(\Omega, P, \mathscr{G}_t)$ for the modified triple, instead of $(\bar{\Omega}, \bar{P}, \bar{\mathscr{G}}_t)$.

References

[1] Boel, R., Varaiya, P., and Wong, E. (1975) Martingales on jump processes; Part I: Representation results; Part II: Applications, SIAM J. Control **13**, pp. 999–1061.
[2] Chou, C. S. and Meyer, P. A. (1974) Sur la représentation des martingales comme intégrales stochastiques dans les processus ponctuels, in *Séminaire Proba VIII*, Lect. Notes in Math. **381**, Springer, Berlin.

[3] Courrège, Ph. and Priouret, P. (1965) Temps d'arrêt d'une fonction aléatoire, Publ. Inst. Stat. Univ. Paris, pp. 245–274.
[4] Davis, M. H. A. (1976) The representation of martingales of jump processes, SIAM J. Control **14**, 623–638.
[5] Jacod, J. (1975) Multivariate point processes: predictable projection, Radon–Nikodym derivatives, representation of martingales, Z. für W. **31**, pp. 235–253.
[6] Kunita, H. and Watanabe, S. (1967) On square integrable martingales, Nagoya Math. J. **30**, pp. 209–245.
[7] Lazaro, J. (1974) Sur les hélices du flot spécial sous une fonction, Z. für W. **30**, pp. 279–302.
[8] Papangelou, F. (1972) Integrability of expected increments of point processes and a related change of scale, Trans. Amer. Math. Soc. **165**, pp. 483–506.
[9] Yor, M. (1976) Représentation des martingales de carré intégrable relative aux processus de Wiener et de Poisson à n paramètres, Z. für W. **39**, pp. 121–129.

SOLUTIONS TO EXERCISES, CHAPTER III

E1. See Appendix A2, T23.

E2. See Appendix A2, T24.

E3. $R_k \equiv 0$; if $k < n$, $R_n \equiv \infty$; otherwise take R_n to be 0, for instance.

E4. See Appendix A2, T34.

E5. Let $A \in \mathscr{F}_{T_n}^N$; then

$$E[1_A 1(S_{n+1} \geq t - T_n) 1(t \geq T_n)] = E[1_A G^{(n+1)}([t - T_n, \infty]) 1(t \geq T_n)]. \quad (1)$$

The left-hand side of (1) is also equal to

$$E[1_A 1(N_t - N_{T_n} = 0) 1(t \geq T_n)]. \quad (2)$$

Now, for $t \geq T_n$,

$$1(N_t - N_{T_n} = 0) = 1 - \int_{(T_n, t]} 1(N_{s-} - N_{T_n} = 0)\, dN_s, \quad (3)$$

and therefore

$$E[1_A 1(N_t - N_{T_n} = 0) 1(t \geq T_n)]$$
$$= E\left[1_A \left(1 - \int_{(T_n, t]} 1(N_{s-} - N_{T_n} = 0)\, dN_s\right) 1(t \geq T_n)\right]. \quad (4)$$

The process C_t defined by

$$C_s = 1_A 1(N_{s-} - N_{T_n} = 0) 1(s > T_n) \quad (5)$$

is $\mathscr{F}_t^N$-predictable, so that, by the definition of λ_t as stochastic intensity,

$$E\left[1_A \int_{(T_n, t]} 1(N_{s-} - N_{T_n} = 0)\, dN_s\, 1(t \geq T_n)\right]$$
$$= E\left[1_A \int_{T_n}^t 1(N_s - N_{T_n} = 0) \lambda_s\, ds\, 1(t \geq T_n)\right]. \quad (6)$$

Also clearly the right-hand side of (6) is

$$E\left[1_A \int_{T_n}^t 1(N_s - N_{T_n} = 0)\lambda^{(n)}(s)\, ds\, 1(t \geq T_n)\right], \quad (7)$$

which in turn equals (after conditioning by $\mathscr{F}_{T_n}^N$)

$$E\left[1_A \int_{T_n}^t G^{(n+1)}([s - T_n, \infty])\lambda^{(n)}(s)\, ds\, 1(t \geq T_n)\right]. \quad (8)$$

Combining all the above equalities we obtain

$$E[1_A G^{(n+1)}((0, t - T_n))1(t \geq T_n)]$$
$$= E\left[1_A \int_{T_n}^t G^{(n+1)}([s - T_n, \infty])\lambda^{(n)}(s)\, ds\, 1(t \geq T_n)\right]. \quad (9)$$

This proves the existence of the density $g^{(n+1)}$ and (2.12) simultaneously.

E6. Sketch: Let T_n be the sequence of jumps of Q_t, and let $Z_n = \Delta Q_{T_n}$. To be computed:

$$P[S_{n+1} \in dt, Z_{n+1} = -1 | S_{n+1} \geq t - T_n, \mathscr{F}_{T_n}^N]. \quad (1)$$

But this quantity is null when $Q_{T_n} = 0$; thus it is equal to

$$P[S_{n+1} \in dt, Z_{n+1} = -1 | S_{n+1} \geq t - T_n, \mathscr{F}_{T_n}^N]1(Q_{T_n} > 0). \quad (2)$$

Let V_{n+1} be the first time of departure of a customer after T_n. Then, letting $\mathscr{G}_{T_n} = (T_0, T_1, Z_1, \ldots, T_n, Z_n)$,

$$P[S_{n+1} \in dt, Z_{n+1} = -1 | S_{n+1} \geq t - T_n, \mathscr{G}_{T_n}]1(Q_{T_n} > 0)$$
$$= P[V_{n+1} - \theta_{T_n} \in dt | V_{n+1} - \theta_{T_n} \geq t - \theta_{T_n}]1(Q_{T_n} > 0). \quad (3)$$

But $V_{n+1} - \theta_{T_n}$ has the density $f(x)$; therefore

$$P[S_{n+1} \in dt, Z_{n+1} = -1 | S_{n+1} \geq t - T_n, \mathscr{F}_{T_n}^N] = G^{(n+1)}(dt, -1)$$
$$= \frac{F(dt - \theta_{T_n})}{1 - \int_0^{t - \theta_{T_n}} f(u)\, du} 1(Q_{T_n} > 0). \quad (4)$$

E7. By E5, for all $n \geq 0$,

$$G_1^{(n+1)} = G_2^{(n+1)}, \quad (1)$$

that is to say,

$$P_1[S_{n+1} \in A | T_0, T_1, \ldots, T_n] = P_2[S_{n+1} \in A | T_0, T_1, \ldots, T_n], \quad A \in \mathscr{B}_+. \quad (2)$$

Hence $P_1 \equiv P_2$ on $\sigma(T_0, T_1, \ldots) \equiv \mathscr{F}_\infty^N$.

E8. $M_{T_{n+1}} 1(T_{n+1} < \infty)$ is $\mathscr{G}_{T_{n+1}}$-measurable But $\mathscr{G}_{T_{n+1}} = \sigma(T_0, Z_0, \ldots, T_n, Z_n, T_{n+1}, Z_{n+1})$; hence, by Appendix A1, T6,

$$M_{T_{n+1}} 1(T_{n+1} < \infty)$$
$$= g^{(n)}(T_0, Z_0, \ldots, T_n, Z_n, T_{n+1}, Z_{n+1})1(T_{n+1} < \infty) \quad (1)$$

Solutions to Exercises, Chapter III

for some measurable $g^{(n)}: (R^{n+2} \times E^{n+2}, \mathscr{B}^{n+2} \otimes E^{n+2}) \to (R, \mathscr{B})$. Now take

$$f^{(n)}(t, i, \omega) = g^{(n)}(T_0(\omega), Z_0(\omega), \ldots, T_n(\omega), Z_n(\omega), T_n(\omega) + t, i). \quad (2)$$

E9. Let $B \in \mathscr{G}_t$. By T5, on $\{T_n < \infty\}$,

$$B \cap \{T_n \le t \le T_{n+1}\} = C \cap \{T_n \le t < T_{n+1}\}, \quad (1)$$

where $C \in \mathscr{G}_{T_n}$. The representation (3.6) follows from (1) and the approximation theorem (Appendix A1, T3).

E10. Case $m = 1$ for simplicity. By T9,

$$M_t = M_0 + \int_0^t H_s(dN_s - \lambda_s \, ds). \quad (1)$$

By the continuity assumption,

$$H_{T_n} = \Delta M_{T_n} = 0 \quad P\text{-a.s., on } \{T_n < \infty\}. \quad (2)$$

That is to say,

$$H_t(\omega) \equiv 0 \quad P(d\omega) \, dN_t(\omega)\text{-a.e.}, \quad (3)$$

or, equivalently, since H_t is $\mathscr{F}_t^N$-predictable and $P(d\omega) \, dN_t(\omega) = P(d\omega)\lambda_t(\omega) \, dt$ on $\mathscr{P}(\mathscr{F}_t^N)$,

$$H_t(\omega) \equiv 0 \quad P(d\omega)\lambda_t(\omega) \, dt\text{-a.e.} \quad (4)$$

By Fubini [Remark (2) after Appendix A1, T13], there exists a P-null event $\mathscr{N}$ such that, for ω outside $\mathscr{N}$,

$$H_t(\omega) = 0 \quad \lambda_t(\omega) \, dt\text{-a.e.}, \quad (5)$$

and therefore

$$\int_0^t H_s \lambda_s \, ds = 0 \quad P\text{-a.s.}, \quad t \ge 0. \quad (6)$$

E11. Apply II, T8(β); one has to check

$$E\left[\int_0^t 1(s \le U_n \wedge V_m) | 2M_{s-} + H_s | |H_s| \lambda_s \, ds\right] < \infty. \quad (1)$$

By definition of U_n and V_m,

$$2M_{t-} 1(t \le U_n) \le 2n \quad (2)$$

and

$$\int_0^{t \wedge V_m} H_s^2 \lambda_s \, ds \le m. \quad (3)$$

Also

$$\int_0^{t \wedge V_m \wedge U_n} |H_s| \lambda_s \, ds \le \int_0^{t \wedge V_m \wedge U_n} (H_s^2 + 1)\lambda_s \, ds \le m + n. \quad (4)$$

Hence

$$\int_0^t 1(s \le U_n \wedge V_m) | 2M_{s-} + H_s | \lambda_s \, ds \le 2n(m + n) + m < \infty. \quad (5)$$

E12. M_n is $\mathscr{F}_n^X$-measurable; hence by Appendix A1, T6

$$M_n = h(n, X_0, \ldots, X_n) \tag{1}$$

for a deterministic function from $N_+ \times \{0, 1\}^{n+1}$ into R. By the definition of a martingale, for $n \geq 1$,

$$M_{n-1} = E[M_n | \mathscr{F}_{n-1}^X] = E[h(n, X_0, \ldots, X_n) | \mathscr{F}_{n-1}^X]. \tag{2}$$

Now, since X_n takes the value 0 or 1,

$$h(n, X_0, \ldots, X_n) = h(n, X_0, \ldots, X_{n-1}, 1)X_n$$
$$+ h(n, X_0, \ldots, X_{n-1}, 0)(1 - X_n). \tag{3}$$

Hence, from (2) and the definition of intensity,

$$M_{n-1} = h(n, X_0, \ldots, X_{n-1}, 1)\lambda_n + h(n, X_0, \ldots, X_{n-1}, 0)(1 - \lambda_n). \tag{4}$$

Therefore

$$M_n - M_{n-1} = h(n, X_0, \ldots, X_{n-1}, 1)(X_n - \lambda_n)$$
$$- h(n, X_0, \ldots, X_{n-1}, 0)([(1 - X_n) - (1 - \lambda_n)]$$
$$= (h(n, X_0, \ldots, X_{n-1}, 1) - h(n, X_0, \ldots, X_{n-1}, 0))(X_n - \lambda_n). \tag{5}$$

Hence the representation (3.19) with

$$H_n = h(n, X_0, \ldots, X_{n-1}, 1) - h(n, X_0, \ldots, X_{n-1}, 0). \tag{6}$$

It remains to prove (3.20). For this observe that

$$E[|H_n(X_n - \lambda_n)|] = E[|M_n - M_{n-1}|] < \infty \tag{7}$$

and that

$$H_n(X_n - \lambda_n) = H_n(1 - \lambda_n)X_n + H_n(-\lambda_n)(1 - X_n), \tag{8}$$

so that

$$|H_n(X_n - \lambda_n)| = |H_n|((1 - \lambda_n)X_n + \lambda_n(1 - X_n)). \tag{9}$$

Now, since H_n and λ_n are $\mathscr{F}_{n-1}^X$-measurable, we have

$$E[|H_n|(1 - \lambda_n)X_n] = E[|H_n|(1 - \lambda_n)\lambda_n], \tag{10}$$

$$E[|H_n|\lambda_n(1 - X_n)] = E[|H_n|\lambda_n(1 - \lambda_n)], \tag{11}$$

$$E[|H_n(X_n - \lambda_n)|] = 2E[|H_n|\lambda_n(1 - \lambda_n)], \tag{12}$$

and (3.20) follows from (7).

E13. By the definition of intensity (II, D7),

$$E\left[\int_0^1 |C_s(i)|^2 \, dN_s(i)\right] = E\left[\int_0^1 |C_s(i)|^2 \lambda(s, i) \, ds\right], \tag{1}$$

since $|C_t(i)|^2$ is a nonnegative $\mathscr{G}_t$-predictable process. Hence the result.

Solutions to Exercises, Chapter III   81

E14. Using the inequality $|x| \leq x^2 + 1$, which is valid for all real numbers x, we have, for all $t \in [0, 1]$,

$$\left| \int_0^t C_s(i) \, dN_s(i) \right| \leq \int_0^1 |C_s(i)| \, dN_s(i)$$

$$\leq \int_0^1 (|C_s(i)|^2 + 1) \, dN_s(i) < \infty, \qquad (1)$$

where the last inequality follows from $E[\int_0^1 |C_s(i)|^2 \, dN_s(i)] = E[\int_0^1 |C_s(i)|^2 \times \lambda(s, i) \, ds] < \infty$ (proven in E13), and $\int_0^1 dN_s(i) = N_1(i) < \infty$ [the $N_t(i)$'s are nonexplosive]. Similarly,

$$\left| \int_0^t C_s(i) \lambda(s, i) \, ds \right| \leq \int_0^1 |C_s(i)| \lambda(s, i) \, ds$$

$$\leq \int_0^1 (|C_s(i)|^2 + 1) \lambda(s, i) \, ds < \infty. \qquad (2)$$

E15. This is routine computation. For (4.6), for instance:

(i) case $t \leq a$: $\Phi(C)_t = 0$ and $\int_0^t C_s^2 \lambda(s) \, ds = 0$;
(ii) case $t \geq a$: $\Phi(C)_t = M_{t \wedge b} - M_a$ and $\int_0^t C_s^2 \lambda(s) \, ds = \int_0^{t \wedge b} \lambda(s) \, ds$;

hence the result, since $N_{t \wedge b} - N_a$ is a Poisson random variable with mean (and variance) $\int_a^{t \wedge b} \lambda(s) \, ds$.

E16. See (I, T5).

E17. The proof of (4.12) is the same as that of II, (2.8). We do the rest of the proof in the case $m = 1$. Let S_n be, for each $n \geq 1$, the $\mathscr{G}_t$-stopping time defined by

$$S_n = \begin{cases} \inf\{t \,|\, M_{t-}^f \geq n\} & \text{if } \{\cdots\} \neq \varnothing, \\ +\infty & \text{otherwise.} \end{cases} \qquad (1)$$

By T12,

$$M_{t \wedge S_n}^f = 1 + \int_0^t M_{s-}^f 1(s \leq S_n)(e^{f(s)} - 1) \, dM_s \qquad (2)$$

is a square-integrable martingale, and

$$E[(M_{t \wedge S_n}^b - 1)^2] = E\left[\int_0^{t \wedge S_n} |M_s^f|^2 (e^{f(s)} - 1)^2 \lambda(s) \, ds \right]. \qquad (3)$$

But $E[(M_{t \wedge S_n}^f - 1)^2] = E[|M_{t \wedge S_n}^f|^2] - 1$ (why ?), and $E[|M_{t \wedge S_n}^f|^2] = E[|M_t^f|^2 \times 1(t \leq S_n) + |M_{S_n}^f|^2 1(t > S_n)] \geq E[|M_t^f|^2 1(t \leq S_n)]$, so that

$$E[|M_t^f|^2 1(t \leq S_n)] \leq 1 + \int_0^t E[|M_s^f|^2 1(s \leq S_n)](e^{f(s)} - 1)^2 \lambda(s) \, ds. \qquad (4)$$

By Gronwall's lemma,

$$E[|M_t^f|^2 1(t \leq S_n)] \leq \exp\left\{ \int_0^t (e^{f(s)} - 1)^2 \lambda(s) \, ds \right\} \qquad (5)$$

and therefore, by monotone convergence,

$$E[|M_t^f|^2] \leq \exp\left\{\int_0^t (e^{f(s)} - 1)^2 \lambda(s)\, ds\right\}$$

$$\leq \exp\left\{\int_0^1 (e^{f(s)} - 1)^2 \lambda(s)\, ds\right\} < \infty \qquad t \in [0, 1]. \qquad (6)$$

In particular,

$$E\left[\int_0^1 |M_t^f|^2 \lambda(s)\, ds\right] < \infty, \qquad (7)$$

so that by T12 and (4.12), M_t^f is a square-integrable $\mathcal{G}_t$-martingale.

E18. We take the case $m = 1$ for simplicity.

The set of random variables of the form

$$g(N_{t_1} - N_{t_0}, \ldots, N_{t_n} - N_{t_{n-1}}), \qquad (1)$$

where $g: N_+^n \to R$ is bounded, is dense in $L^2(\mathcal{G}_1, P)$. It suffices therefore to show that the linear combinations of random variables of the form

$$\exp\left(\sum_{i=1}^n u_i(N_{t_i} - N_{t_{i-1}})\right) \qquad (2)$$

approximate the random variable (1) in the $L^2(\mathcal{G}_1, P)$ sense. This is true since the function $g: N_+^n \to R$ can be approximated pointwise and boundedly by the linear combinations of exponentials

$$e_{u_2,\ldots,u_n}: N_+^{n+1} \to R$$

$$e_{u_2,\ldots,u_n}(l_1, \ldots, l_n) = \exp\left(\sum_{i=1}^n u_i l_i\right). \qquad (3)$$

CHAPTER IV

Filtering

1. The Theory of Innovations
2. State Estimate for Queues and Markov Chains
3. Continuous States and Nontrivial Prehistory
References
Solutions to Exercises, Chapter IV

1. The Theory of Innovations

Recursive Filtering

Filtering is another word for *bayesian estimation* of the state at time t of a given dynamical stochastic system, based on the available incomplete information at the same time t. If the observation at time t consists of the trajectory of the state up to time $t - a$, where a is a strictly positive number, the problem is one of *prediction*. Such problems have been dealt with in great detail in the framework of second-order stationary processes by various authors: Kolmogorov {93}, Wiener {154}, Wold {155}, Cramer {38}. Their solutions were given in terms of frequency spectra, whereas the treatment of the filtering problem with point-process observations which will be described in this chapter follows Kalman's time-domain approach {84}. Besides the fact that time-domain models are more versatile, there is another advantage with them to be found in the recursive form of the solutions. What this means will be explained in a few lines.

The general setting of filtering theory is the following. On a probability space $(\Omega, \mathscr{F}, P)$ are given two histories:

(i) the *global history* $\mathscr{F}_t$,
(ii) the *observed history* $\mathcal{O}_t$, such that $\mathcal{O}_t \subset \mathscr{F}_t, t \geq 0$.

Frequently, $\mathscr{F}_t$ and $\mathscr{O}_t$ admit the representation

$$\mathscr{F}_t = \mathscr{F}_t^Y \vee \mathscr{F}_t^X,$$
$$\mathscr{O}_t = \mathscr{F}_t^X, \qquad (*)$$

where X_t is the *observation* process, E-valued, and Y_t is the *state* process, U-valued.

The goal is to find expressions for the process $E[h_t|\mathscr{O}_t]$, where h_t is some integrable process adapted to $\mathscr{F}_t$. We will be uniquely concerned in this book with a state–observation structure of type $(*)$ and of finding expressions for

$$\hat{h}(Y, t) = E[h(Y_t)|\mathscr{F}_t^X],$$

where h is a measurable function from $(U, \mathscr{U})$ into $(R, \mathscr{B})$ such that $h(Y_t)$ is integrable. These expressions should preferably be in *recursive form*. In nonmathematical terms, we want to find a function or device f such that for all t

$$\hat{h}(Y, t + dt) = f(t, \hat{h}(Y, t), dX_t). \qquad (*)$$

In other words, the estimate of $h(Y_{t+dt})$ given $\mathscr{F}_{t+dt}^X$ should depend only upon the estimate of $h(Y_t)$ given $\mathscr{F}_t^X$ and the increment $dX_t = X_{t+dt} - X_t$.

As we said, the latter sentence does not have much mathematical meaning, and Equation $(*)$ is merely a string of symbols. We will not try to formalize the concept of recursiveness in filtering theory, but various examples will show what really stands behind this apparently mysterious concept.

The practical implication of recursiveness is the implementability of estimation in real time: the device f in $(*)$ can be thought of, in the computer-science vocabulary, as *hardware* or—and this is the same for our purpose—as a *program* that does not depend upon the data; $\hat{h}(Y, t)$ is all that has to be stored in *memory*, and dX_t is the *new data* that is fed into the program f. Recursiveness is there to save *memory-space* since instead of storing the whole past $(X_s, s \in [0, t])$ of the observation, it is only necessary to store $\hat{h}(Y, t)$. Statisticians are familiar with this concept: they would say that $(\hat{h}(Y, t), dX_t)$ is a *sufficient statistic* for the problem of estimating $h(Y_{t+dt})$ on the basis of $\mathscr{F}_{t+dt}^X$.

Before proceeding to the technical details of obtaining recursive filters, the reader is entitled to ask the following question: if h_t is the partially observed process of interest to us, why are we taking $E[h_t|\mathscr{F}_t^X]$ as the "best" estimate of h_t at time t? As is well known, such an estimate is optimal for the *quadratic criterion*: more precisely, if h_t is square-integrable, then

$$E[|h_t - E[h_t|\mathscr{F}_t^X]|^2] \le E[|h_t - X|^2]$$

for any square-integrable random variable X that is $\mathscr{F}_t^X$-measurable. Therefore, looking for such estimates seems to imply that we have chosen the quadratic criterion, and to base a whole theory on a criterion which has not received general consensus is a questionable choice.

1. The Theory of Innovations

Although it is true that the quadratic criterion is not a universal one, and in fact quite inappropriate for many purposes, the estimates $E[h_t|\mathscr{F}_t^X]$ have their own *raison d'être*. For instance, it will be seen in the next chapter that such estimates play a crucial role in proving or disproving the poissonian character of various flows in queueing networks. Also, in Chapter VI on likelihood ratios, especially in the theorem of separation of detection and filtering (VI, T12), such estimates appear naturally, independently of any criterion.

We now close this heuristic discussion, only adding that recursiveness is in general not achieved or even achievable. However, the martingale methods of filtering are quite well adapted to this purpose, as will soon become apparent.

Innovations Method

The reader familiar with the innovations theory of Kalman filters (see Kailath [7])[1] knows that the main ingredients therein are: (i) the innovating representation of the state process and the projection of this representation on the observed history, (ii) the representation of the *martingales with respect to the observed history*, which leads to filtering formulas in terms of the innovations gain and of the innovating part of the observation, (iii) the *martingale calculus*, which provides a means of identifying the innovations gain. In continuous time, the same ingredients do exist (at least for point-process observations or Wiener-type observations) and the same program can be carried out. Some technical details have to be worked out, but they are not really difficult, and once familiarity with the mathematical subtleties is acquired, the application of the innovations method becomes almost automatic.

The Innovating Structure of the Filter

State Equations. In a number of situations, the process $Z_t = h(Y_t)$ to be filtered satisfies an equation of the type

$$Z_t = Z_0 + \int_0^t f_s\, ds + m_t, \tag{1.1}$$

where f_t is an $\mathscr{F}_t$-progressive process satisfying $\int_0^t |f_s|\, ds < \infty$, P-a.s., $t \geq 0$, and m_t is an $\mathscr{F}_t$-local martingale of mean 0. In most cases of practical interest, one can show directly the existence of the representation (1.1), and this is done with the help of martingale calculus. Such a representation is

---

[1] We do not assume previous knowledge of Kalman filters. The reader may skip a few lines and go to the next subsection.

called a semi-martingale representation of Z_t. A few examples will illustrate the basic mechanisms for obtaining it.

EXAMPLE 1 (Signal in Noise). (The necessary background is contained in A3.) Here, Y_t consists of the sum of an integrated signal and of an integrated white noise. More precisely,

$$Y_t = Y_0 + \int_0^t S_s \, ds + W_t, \tag{1.2}$$

where S_t is a measurable process adapted to $\mathscr{F}_t$ and such that $\int_0^t |S_s| \, ds < \infty$, P-a.s., $t \geq 0$, and W_t is an $\mathscr{F}_t$-Wiener process. Clearly, $Z_t = Y_t$ is of the form (1.1).

Now let h be a twice continuously differentiable function, and let $Z_t = h(Y_t)$. By application of Ito's differentiation rule (A3, T11),

$$Z_t = h(Y_t) = Z_0 + \int_0^t \left(\frac{\partial h}{\partial y}(Y_s) S_s + \frac{1}{2} \frac{\partial^2 h}{\partial^2 y}(Y_s) \right) ds$$

$$+ \int_0^t \frac{\partial h}{\partial y}(Y_s) \, dW_s, \tag{1.3}$$

and this again is a representation of type (1.1) with

$$f_t = \frac{\partial h}{\partial y}(Y_t) S_t + \frac{1}{2} \frac{\partial^2 h}{\partial y^2}(Y_t), \quad m_t = \int_0^t \frac{\partial h}{\partial y}(Y_s) \, dW_s.$$

EXAMPLE 2 (Queueing Process). Consider a single-channel queueing process Q_t with arrival process A_t and departure process D_t. Suppose that Q_t admits the $\mathscr{F}_t$-parameters λ_t and μ_t. Let Z_t be the process

$$Z_t = \exp\{iuQ_t\} \tag{1.4}$$

for some real number u (a more complete notation would be Z_t^u instead of Z_t).

E1 Exercise. Show that if we define the processes m_t^+ and m_t^- by

$$m_t^+ = \int_0^t Z_{s-}(dA_s - \lambda_s \, ds),$$

$$m_t^- = \int_0^t Z_{s-}(dD_t - \mu_s 1(Q_s > 0)) \, ds,$$

then a representation of the type (1.1) is available, namely

$$Z_t = Z_0 + \int_0^t [(e^{iu} - 1)\lambda_s + (e^{-iu} - 1)\mu_s 1(Q_s > 0)] Z_s \, ds + m_t, \tag{1.5}$$

where $m_t = (e^{iu} - 1)m_t^+ + (e^{-iu} - 1)m_t^-$ is an $\mathscr{F}_t$-local martingale.

1. The Theory of Innovations

EXAMPLE 3 (Markov Processes With a Transition Semigroup; see A1, §8)). Let Y_t be a U-valued homogeneous $\mathscr{F}_t$-Markov process with the $\mathscr{F}_t$-transition semigroup $(P_t, t \geq 0)$; P_t is a mapping from $b(\mathscr{U})$, the set of bounded measurable functions from $(U, \mathscr{U})$ into (R, B), into itself, such that

$$P_t 1 = 1, \quad t \geq 0; \qquad P_0 = I \text{ (identity)}$$
$$P_t P_s = P_{t+s}, \qquad t \geq 0, \quad s \geq 0.$$

The $\mathscr{F}_t$-Markov property of Y_t reads:

$$E[f(Y_t)|\mathscr{F}_s] = P_{t-s} f(Y_s), \qquad 0 \leq s \leq t.$$

E2 Exercise. Let $f \in b(\mathscr{U})$, and T be some strictly positive real time; show that

$$m_t = P_T f(Y_0) - P_{T-t} f(Y_t) \text{ is an } \mathscr{F}_t\text{-martingale over } [0, T].$$

Thus, defining Z_t by

$$Z_t = P_{T-t} f(Y_t), \tag{1.6}$$

we have found a semi-martingale representation of Z_t with the particular feature that $f_t \equiv 0$:

$$Z_t = Z_0 + m_t. \tag{1.7}$$

EXAMPLE 4 (Markov Processes With a Generator). The process Y_t is defined as above, and it is now assumed that the semigroup $(P_t, t \geq 0)$ has an infinitesimal generator $\mathscr{L}$ of domain $\mathscr{D}(\mathscr{L})$. Then for any $f \in \mathscr{D}(\mathscr{L})$, Dynkin's formula yields

$$f(X_t) = f(X_0) + \int_0^t \mathscr{L} f(X_s) \, ds + m_t, \tag{1.8}$$

where m_t is an $\mathscr{F}_t$-martingale. This is a semi-martingale representation of $Z_t = f(X_t)$.

Projection of the State on the Observation. The first step in the innovations method consists in projecting the state equation (1.1) on the observed history $\mathscr{O}_t$.

T1 Theorem (Projection of the State [5, 6, 9]). *Let Z_t be an integrable real-valued process with the semi-martingale representation:*

$$Z_t = Z_0 + \int_0^t f_s \, ds + m_t, \tag{1.9}$$

where

(i) f_t is a $\mathscr{F}_t$-progressive process such that

$$E\left[\int_0^t |f_s| \, ds\right] < \infty, \qquad t \leq 0,$$

(ii) m_t is a 0-mean $\mathscr{F}_t$-martingale.

Let $\mathcal{O}_t$ be some history such that $\mathcal{O}_0 = (\Omega, \varnothing)$, $\mathcal{O}_t \subset \mathscr{F}_t$, $t \geq 0$. Then

$$E[Z_t | \mathcal{O}_t] = E[Z_0] + \int_0^t \hat{f}_s \, ds + \hat{m}_t, \tag{1.10}$$

where

(a) $\hat{m}_t$ is a 0-mean $\mathcal{O}_t$ martingale,
(b) $\hat{f}_t$ is an $\mathcal{O}_t$-progressive process defined by

$$E\left[\int_0^t C_s f_s \, ds\right] = E\left[\int_0^t C_s \hat{f}_s \, ds\right], \tag{1.11}$$

for all nonnegative bounded $\mathcal{O}_t$-progressive processes C_t.

Remarks

(α) If there exists for all $t \geq 0$ a version of $E[f_t | \mathcal{O}_t]$ (denoted $\tilde{f}_t$) such that the mapping $(t, \omega) \to \tilde{f}_t(\omega)$ is $\mathcal{O}_t$-progressively measurable, then the choice $\hat{f}_t(\omega) = \tilde{f}_t(\omega)$ satisfies the requirements. Indeed, by Fubini,

$$E\left[\int_0^t C_s f_s \, ds\right] = \int_0^t E[C_s f_s] \, ds = \int_0^t E[C_s \tilde{f}_s] \, ds = E\left[\int_0^t C_s \tilde{f}_s \, ds\right].$$

(β) The existence of $\hat{f}_t$ is always granted by the Radon–Nikodym derivative theorem: $\hat{f}_t(\omega)$ is the Radon–Nikodym derivative of measure μ_2 with respect to measure μ_1, the two measures being defined on $((0, \infty) \times \Omega$, prog $\mathcal{O}_t)$ by

$$\mu_2(dt \times d\omega) = f_t(\omega) \, dt \, P(d\omega),$$

$$\mu_1(dt \times d\omega) = dt \, P(d\omega).$$

Also, two versions of $\hat{f}_t(\omega)$ differ only on a set of $P(d\omega) \, dt$-measure 0.

(γ) In Remark (α) we have seen that if one can find an $\mathcal{O}_t$-progressively measurable version of $E[f_t | \mathcal{O}_t]$, then $\hat{f}_t$ can be chosen to be such a version. Such a version of $E[f_t | \mathcal{O}_t]$ will always exist in practice; however, in the general statement of Theorem T1 one must resort to this rather abstract definition of $\hat{f}_t$ because it is not possible to define $\int_0^t E[f_s | \mathcal{O}_s] \, ds$ if nothing is known concerning the measurability in t of $E[f_t | \mathcal{O}_t]$.

PROOF OF THEOREM T1. After conditioning, (1.9) becomes

$$E[Z_t | \mathcal{O}_t] = E[Z_0 | \mathcal{O}_t] + E\left[\int_0^t f_s \, ds \,\Big|\, \mathcal{O}_t\right] + E[m_t | \mathcal{O}_t].$$

Defining

$$\hat{m}_t = E[Z_0 | \mathcal{O}_t] - E[Z_0] + E\left[\int_0^t f_s \, ds \,\Big|\, \mathcal{O}_t\right] - \int_0^t \hat{f}_s \, ds + E[m_t | \mathcal{O}_t],$$

one obtains (1.10). It remains to show that $\hat{m}_t$ is an $\mathcal{O}_t$-martingale, mean 0.

1. The Theory of Innovations

Since this is clearly true for $E[Z_0|\mathcal{O}_t] - E[Z_0]$ and $E[m_t|\mathcal{O}_t]$, it suffices to prove that $E[\int_0^t f_s\,ds|\mathcal{O}_t] - \int_0^t \hat{f}_s\,ds$ is an $\mathcal{O}_t$-martingale, or, equivalently, that for all $0 \leq s \leq t$ and all $A \in \mathcal{O}_s$

$$E\left[1_A\left\{E\left[\int_0^t f_u\,du\Big|\mathcal{O}_t\right] - E\left[\int_0^s f_u\,du\Big|\mathcal{O}_s\right]\right\}\right] = E\left[1_A\int_s^t \hat{f}_u\,du\right]. \quad (*)$$

The left-hand side of the above potential equality can be rewritten as

$$E\left[1_A\left\{\int_0^t f_u\,du - \int_0^s f_u\,du\right\}\right] = E\left[1_A\int_s^t f_u\,du\right] = E\left[\int_0^t C_u f_u\,du\right],$$

where $C_u(\omega) = 1_A(\omega)1_{(s,t]}(u)$. The right-hand side of $(*)$ is

$$E\left[1_A\int_s^t \hat{f}_u\,du\right] = E\left[\int_0^t C_u \hat{f}_u\,du\right];$$

$(*)$ then follows by the definition of $\hat{f}_t$ and the fact that C_t is a nonnegative bounded $\mathcal{O}_t$-progressive process, since $A \in \mathcal{O}_s$.

Point-Process Observation and Bounded-Variation State: Theory. The observed history $\mathcal{O}_t$ is the internal history $\mathcal{G}_t$ of an integrable k-variate point process $(N_t(1), \ldots, N_t(k))$ with $\mathcal{F}_t$-intensity $(\lambda_t(1), \ldots, \lambda_t(k))$, where $\mathcal{F}_t$ is the global history recording events outside $\mathcal{G}_t$. By (II, T13), this k-variate point process also admits a $\mathcal{G}_t$-intensity $(\hat{\lambda}_t(1), \ldots, \hat{\lambda}_t(k))$ which can be assumed predictable.

A state process Z_t is given which satisfies all the requirements of Theorem T1. We suppose moreover that

(H1) Z_t is bounded.
(H2) m_t has paths P-a.s. *of bounded variation on finite intervals,* and $E[\int_0^t |dm_s|] < \infty, t \geq 0$.

The conventions stated at the end of Chapter III relative to the "usual conditions" are supposed to hold. As explained in Chapter III, they do not entail a loss of generality.

Assumption (H1) is crucially used in the proof of T2, but it is not a hindrance in applications. Assumption (H2) can be dispensed with, as will be seen in detail in the final version of the filtering result (T8).

The following notation will be useful.

$$\hat{Z}_t = E[Z_t|\mathcal{F}_t^N],$$

$$M_t(i) = N_t(i) - \int_0^t \lambda_s(i)\,ds, \qquad M_t = (M_t(1), \ldots, M_t(k)),$$

$$\hat{M}_t(i) = N_t(i) - \int_0^t \hat{\lambda}_s(i)\,ds, \qquad \hat{M}_t = (\hat{M}_t(1), \ldots, \hat{M}_t(k)).$$

By the representation theorem (III, T17) there exists a $\mathscr{G}_t$-predictable k-vector process $K_t = (K_t(1), \ldots, K_t(k))$ such that for all $t \geq 0$, all $1 \leq i \leq k$

$$\int_0^t K_s(i) \hat{\lambda}_s(i)\, ds < \infty \quad P\text{-a.s.},$$

and such that the representation (1.10) takes the more precise form

$$\hat{Z}_t = E[Z_0] + \int_0^t \hat{f}_s\, ds + \sum_{i=1}^k \int_0^t K_s(i)\, d\hat{M}_s(i). \tag{1.12}$$

Equation (1.12) is called the *filter* of Z_t, $\hat{M}_t$ the $\mathscr{G}_t$-*innovations process*,[2] and K_t the *innovations gain*.

Computation of the Innovations Gain

In this subsection, we are not going to solve the filtering problem. All that is possible in the generality in which the model is described is to find an abstract expression for the innovations gain K_t. Examples will be treated later on that show its usefulness for effective computation.

A few preliminaries are needed: in particular we will introduce some $\mathscr{G}_t$-predictable processes that will appear in the expression of the innovations gain. Namely, for each $1 \leq i \leq k$, we denote by processes $\Psi_{1,t}(i)$, $\Psi_{2,t}(i)$, and $\Psi_{3,t}(i)$ $\mathscr{G}_t$-predictable processes which satisfy

$$E\left[\int_0^t C_s Z_s \lambda_s(i)\, ds\right] = E\left[\int_0^t C_s \Psi_{1,s}(i) \hat{\lambda}_s(i)\, ds\right],$$

$$E\left[\int_0^t C_s Z_s \hat{\lambda}_s(i)\, ds\right] = E\left[\int_0^t C_s \Psi_{2,s}(i) \hat{\lambda}_s(i)\, ds\right], \tag{1.13}$$

$$E\left[\sum_{0 < s \leq t} C_s \Delta m_s \Delta N_s(i)\right] = E\left[\int_0^t C_s \Psi_{3,s}(i)\, ds\right]$$

for all $\mathscr{G}_t$-predictable nonnegative bounded processes C_t and all $t \geq 0$.

These processes are defined as Radon–Nikodym derivatives. For instance, $\Psi_{1,t}(i)(\omega)$ is the Radon–Nikodym derivative of the measure $Z_t(\omega) \lambda_t(i)(\omega)\, dt\, P(d\omega)$ with respect to the measure $\hat{\lambda}_t(i)(\omega)\, dt\, P(d\omega)$, both measures on $((0, \infty) \times \Omega, \mathscr{P}(\mathscr{G}_t))$. The first measure is a signed measure, σ-finite (since Z_t is bounded), and is absolutely continuous with respect to

---

[2] A heuristic justification of the word *innovations* is as follows: dN_t is what one observes in $[t, t+dt)$, whereas $\hat{\lambda}_t\, dt = E[dN_t | \mathscr{F}_t^N]$ is what one expected to happen in $[t, t+dt)$ on the basis of the previous observation $\mathscr{F}_t^N$. Thus $dN_t - \hat{\lambda}_t\, dt$ is what is really new. Wold introduced the innovations concept. Although he did it in the context of second-order stationary processes and in terms of Hilbert-space projectors (see {39}), the idea is essentially the same. Kailath [7] recognized this analogy and presented a derivation of Kalman's filtering formulas in a way that clarified the role of the innovating structure and the usefulness of martingale representations.

1. The Theory of Innovations

the second [use the fact that $\hat{\lambda}_t(i)(\omega)\, dt\, P(d\omega) = \lambda_t(i)(\omega)\, dt\, P(d\omega)$ on $((0, \infty) \times \Omega, \mathscr{P}(\mathscr{G}_t))$].[3] The existence of the Radon–Nikodym derivatives $\Psi_{1,t}(i)(\omega)$ is therefore granted. Moreover, as Radon–Nikodym derivatives of measures on $\mathscr{P}(\mathscr{G}_t)$, they are $\mathscr{P}(\mathscr{G}_t)$-measurable, that is to say, $\Psi_{1,t}(i)$ is a $\mathscr{G}_t$-predictable process. Also $\Psi_{1,t}(i)(\omega)$ is defined up to a set $\mathscr{N}_1(i)$ in $\mathscr{P}(\mathscr{G}_t)$ of $\hat{\lambda}_t(i)(\omega)\, dt\, P(d\omega)$-measure 0. The existence and $\hat{\lambda}_t(i)(\omega)\, dt\, P(d\omega)$-uniqueness of $\Psi_{2,t}(i)(\omega)$ is proven in the same way. As for $\Psi_{3,t}(i)(\omega)$, it is obtained as the Radon–Nikodym derivative of the measure $dm_t(\omega)\, dN_t(i)(\omega)\, P(d\omega)$ with respect to $\hat{\lambda}_t(i)(\omega)\, dt\, P(d\omega)$, both measures on $((0, \infty) \times \Omega, \mathscr{P}(\mathscr{G}_t))$. The first measure is a signed measure, σ-finite (since Z_t and hence $|\Delta m_t|$ is bounded), and is absolutely continuous with respect to the second, because $\hat{\lambda}_t(i)(\omega)\, dt\, P(d\omega) = dN_t(i)(\omega)P(d\omega)$ on $((0, \infty) \times \Omega, \mathscr{P}(\mathscr{G}_t))$. The existence and $\hat{\lambda}_t(i)(\omega)\, dt\, P(d\omega)$-uniqueness of $\Psi_{3,t}(i)(\omega)$ then follows from these remarks exactly in the same way as for $\Psi_{1,t}(i)$ and $\Psi_{2,t}(i)$.

We are now in a position to state and prove the main result of this section:

T2 Theorem (Expression of the Innovations Gain [6, 9, 4, 1, 2]). *Under the conditions prevailing in this section,*

$$\hat{Z}_t = E[Z_0] + \int_0^t \hat{f}_s\, ds + \sum_{i=1}^k \int_0^t K_s(i)(dN_s(i) - \hat{\lambda}(i)\, ds) \quad P\text{-a.s.}, \quad t \geq 0, \tag{1.14}$$

where for all $1 \leq i \leq k$

$$K_t(i)(\omega) = \Psi_{1,t}(i)(\omega) - \Psi_{2,t}(i)(\omega) + \Psi_{3,t}(i)(\omega), \tag{1.15}$$

the $\Psi(i)$'s being defined by the equations (1.13).

Remark. One always has

$$\Psi_{2,t}(i) = \hat{Z}_{t-}. \tag{1.16}$$

Indeed, in the second equality of (1.13), $E[\int_0^t C_s Z_s \hat{\lambda}_s(i)\, ds] = E[\int_0^t C_s \hat{Z}_s \hat{\lambda}_s(i)\, ds] = E[\int_0^t C_s \hat{Z}_{s-} \hat{\lambda}_s(i)\, ds]$. Also, if Z_t and $N_t(i)$ have no common jumps, $\Psi_{3,t}(i) = 0$.

Remark. The Innovations Gain is Unique. For $1 \leq i \leq k, j = 1, 2, 3$, we have seen that two versions of $\Psi_{j,t}(i)(\omega)$ differ on a set $\mathscr{N}_j(i)$ or $\mathscr{P}(\mathscr{G}_t)$ of $v(i)$ $(dt \times d\omega)$-measure 0, where

$$v(i)(dt \times d\omega) = \hat{\lambda}_t(i)(\omega)\, dt\, P(d\omega) \text{ or } \Delta N_t(i)(\omega)\, P(d\omega).$$

---

[3] By definition of the $\mathscr{F}_t$- and $\mathscr{G}_t$-intensities, for all nonnegative $\mathscr{G}_t$-predictable processes C_t

$$E\left[\int_0^t C_s\, dN_s(i)\right] = E\left[\int_0^t C_s \lambda_s(i)\, ds\right] = E\left[\int_0^t C_s \hat{\lambda}_s(i)\, ds\right],$$

that is to say, $dN_t(i)(\omega)\, P(d\omega) \equiv \lambda_t(i)(\omega)\, dt\, P(d\omega) \equiv \hat{\lambda}_t(i)(\omega)\, dt\, P(d\omega)$ on $((0, \infty) \times \Omega, \mathscr{P}(\mathscr{G}_t))$.

The corresponding $K_t(i)(\omega)$'s are given by (1.15) and (1.13), and therefore may differ on a $\mathscr{P}(\mathscr{G}_t)$-set included in $\mathscr{N}(i) = \bigcup_{j=1}^{3} \mathscr{N}_j(i)$. By Fubini's theorem (Appendix A1, T13), the section $\mathscr{N}(i)(\omega)$ of $\mathscr{N}(i)$ by ω is P-a.s. of measure 0 for the measures $\hat{\lambda}_t(i)(\omega)\,dt$ and $\Delta N_t(i)(\omega)$ on $(R_+, \mathscr{B}_+)$. Since $K_t(i)$ enters (1.14) only through $K_s(i)(\omega)\hat{\lambda}_s(i)(\omega)\,ds$ and $K_t(i)(\omega)\Delta N_t(i)(\omega)$, two different versions yield the same filter.

PROOF OF THEOREM T2. We will present the proof in the case $k = 1$, in order to avoid upper indices and transposition signs.

Since the existence of a $\mathscr{G}_t$-predictable gain K_t is granted, it is enough to show that (1.15) is a necessary form for it. To do this, we first express that if U_t is a process of the form

$$U_t = \int_0^t H_s \, d\hat{M}_s, \tag{1.17}$$

where H_t is a $\mathscr{G}_t$-predictable process such that U_t is a *bounded* $\mathscr{G}_t$-martingale, satisfying moreover $E[\int_0^t |H_s K_s|\hat{\lambda}_s \, ds] < \infty$, $t \geq 0$, then

$$E[Z_t U_t] = E[\hat{Z}_t U_t]. \tag{1.18}$$

Before we start computing both sides of (1.18), let us observe that since U_t is bounded, then H_{T_n} is bounded for all $n \geq 0$; in particular

$$E\left[\int_0^t |H_s|\hat{\lambda}_s \, ds\right] = E\left[\int_0^t |H_s|\lambda_s \, ds\right]$$

$$= E\left[\int_0^t |H_s|\, dN_s\right] < \infty, \quad t \geq 0. \qquad (*)$$

Step 1. By the Stieltjes–Lebesgue rule of integration by parts, (Appendix A4, T2)

$$Z_t U_t = \int_0^t Z_{s-} \, dU_s + \int_0^t U_s \, dZ_s$$

$$= \int_0^t Z_{s-} H_s \, d\hat{M}_s + \int_0^t U_s(dm_s + f_s \, ds)$$

and therefore, after rearrangement

$$Z_t U_t = \int_0^t Z_{s-} H_s \, dM_s + \int_0^t Z_{s-} H_s(\lambda_s - \hat{\lambda}_s) \, ds$$

$$+ \int_0^t U_s f_s \, ds + \int_0^t U_{s-} \, dm_s + \sum_{s \leq t} H_s \Delta m_s \Delta N_s.$$

From I, T5 and II, T8, and taking into account the boundedness of Z_t and U_t and inequality $(*)$, it follows that $\int_0^t Z_{s-} H_s \, dM_s$ and $\int_0^t U_{s-} \, dm_s$ are 0-mean

1. The Theory of Innovations   93

$\mathscr{F}_t$-martingales. Therefore, by the definition of the Ψ processes,

$$E[Z_t U_t] = E\left[\int_0^t U_s f_s \, ds\right] + E\left[\int_0^t H_s(\Psi_{1,s} - \Psi_{2,s} + \Psi_{3,s})\hat{\lambda}_s \, ds\right]. \quad (1.19)$$

Step 2. Another application of the Stieltjes–Lebesgue product formula yields

$$\hat{Z}_t U_t = \int_0^t \hat{Z}_{s-} \, dU_s + \int_0^t U_s \, d\hat{Z}_s = \int_0^t \hat{Z}_{s-} H_s \, d\hat{M}_s + \int_0^t U_s(K_s \, d\hat{M}_s + \hat{f}_s \, ds)$$

or

$$\hat{Z}_t U_t = \int_0^t \hat{Z}_{s-} H_s \, d\hat{M}_s + \int_0^t U_s \hat{f}_s \, ds + \int_0^t U_s K_s \, d\hat{M}_s$$

$$= \int_0^t \hat{Z}_{s-} H_s \, d\hat{M}_s + \int_0^t U_s \hat{f}_s \, ds + \int_0^t U_{s-} K_s \, d\hat{M}_s + \sum_{s \leq t} H_s K_s \Delta N_s.$$

In view of I, T5, II, T8, and (∗) and since U_t and $\hat{Z}_t$ are bounded, we have that $\int_0^t \hat{Z}_{s-} H_s \, d\hat{M}_s$ and $\int_0^t U_{s-} K_s \, d\hat{M}_s$ are martingales. Also $E[\Sigma H_s K_s \Delta N_s] = E[\int_0^t H_s K_s \hat{\lambda}_s \, ds]$, where we make use of the assumption $E[\int_0^t |H_s K_s| \hat{\lambda}_s \, ds] < \infty$. Therefore

$$E[\hat{Z}_t U_t] = E\left[\int_0^t U_s \hat{f}_s \, ds\right] + E\left[\int_0^t H_s K_s \hat{\lambda}_s \, ds\right]. \quad (1.20)$$

Step 3. By the definition of $\hat{f}_t$, $E[\int_0^t U_s f_s \, ds] = E[\int_0^t U_s \hat{f}_s \, ds]$, so that the equality resulting from (1.18), (1.19), and (1.20) reduces to

$$E\left[\int_0^t H_s K_s \hat{\lambda}_s \, ds\right] = E\left[\int_0^t H_s(\Psi_{1,s} - \Psi_{2,s} + \Psi_{3,s})\hat{\lambda}_s \, ds\right] \quad (1.21)$$

for all $t \geq 0$ and all H_t satisfying the requirements stated at the beginning of the proof. For instance, if C_t is any nonnegative bounded $\mathscr{G}_t$-predictable process, and if we define the $\mathscr{G}_t$-stopping times S_n by

$$S_n \begin{cases} \inf\{t \mid N_t \geq n \text{ or } \int_0^t (1 + |K_s|)\hat{\lambda}_s \, ds \geq n\} & \text{if } \{\cdots\} \neq 0, \\ +\infty & \text{otherwise,} \end{cases} \quad (1.22)$$

then $H_t = C_t 1(t \leq S_n)$ satisfies the requirements, and

$$E\left[\int_0^t C_s K_s 1(s \leq S_n) \hat{\lambda}_s \, ds\right]$$

$$= E\left[\int_0^t C_s(\Psi_{1,s} - \Psi_{2,s} + \Psi_{3,s}) 1(s \leq S_n) \hat{\lambda}_s \, ds\right]. \quad (1.23)$$

The latter equality being true for all nonnegative bounded $\mathscr{G}_t$-predictable processes C_t, it follows that

$$K_t(\omega)1(t \le S_n(\omega))$$
$$= (\Psi_{1,t}(\omega) - \Psi_{2,t}(\omega) + \Psi_{3,t}(\omega))1(t \le S_n(\omega)) \quad v(dt \times d\omega)\text{-a.e.}, \quad (1.24)$$

where $v(dt \times d\omega) = \hat{\lambda}_t(\omega) \, dt \, P(d\omega)$ or $dN_t(\omega) \, P(d\omega)$. Now $S_n(\omega) \uparrow \infty$, P-a.s. (recall that $\int_0^t \hat{\lambda}_s \, ds < \infty$ and $\int_0^t |K_s| \hat{\lambda}_s \, ds < \infty$, $t \ge 0$). Therefore

$$K_t(\omega) = \Psi_{1,t}(\omega) - \Psi_{2,t}(\omega) + \Psi_{3,t}(\omega) \quad v(dt \times d\omega)\text{-a.e.}, \quad (1.25)$$

and this terminates the proof. □

A useful extension of the filtering formulas of T2 consists in replacing the deterministic time t by a finite $\mathscr{G}_t$-stopping time τ.

E3 Exercise. Prove the following

T3 Theorem. *Let the conditions of* T2 *prevail. Then for any finite stopping time* τ, $E[Z_\tau | \mathscr{G}_\tau]$ *admits the version* $\hat{Z}_\tau$, *where* $\hat{Z}_t$ *is defined by* (1.14) *and* (1.15).

A Simple Example: The Disruption Problem

We have to detect the random time of change of the intensity of a point process, on the basis of the observation of the point process itself. More precisely: Let $F(t) = \int_0^t f(u) \, du$ be the cumulative distribution of a nonnegative random variable and let N_t and Z_t be two processes on $(\Omega, \mathscr{F}, P)$. Define $\mathscr{F}_t = \mathscr{F}_t^Z \vee \mathscr{F}_t^N$. Suppose that Z_t is an increasing right-continuous process taking its values in $\{0, 1\}$ and such that, if we let τ be the *disruption time* at which Z_t jumps from 0 to 1,

$$P[\tau \le t | \mathscr{F}_t] = F(t). \quad (1.26)$$

Then, by III, T7 for instance,

$$Z_t - \int_0^{t \wedge \tau} h(s) \, ds \text{ is an } \mathscr{F}_t\text{-martingale}, \quad (1.27)$$

where

$$h(t) = \begin{cases} \dfrac{f(t)}{1 - F(t)} & \text{if } F(t) < 1, \\ 0 & \text{otherwise.} \end{cases} \quad (1.28)$$

Now, since $1(\tau \le t) = Z_t$, Equation (1.27) can be written

$$Z_t = \int_0^t h(s)(1 - Z_s) \, ds + m_t, \quad (1.29)$$

1. The Theory of Innovations

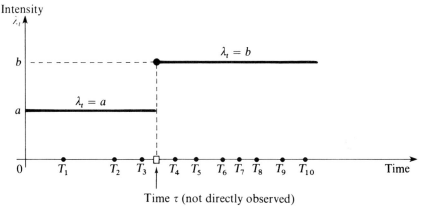

Figure 1 The disruption.

where m_t is a $(P, \mathscr{F}_t)$-martingale. Thus we have obtained a semi-martingale representation of Z_t.

The observation N_t is a point process without jumps in common with Z_t, and its $\mathscr{F}_t$-intensity is

$$\lambda_t = a + (b - a)Z_t, \tag{1.30}$$

that is to say: before τ, $\lambda_t = a$, and after τ, $\lambda_t = b$ (see Figure 1). The problem is that of computing

$$\hat{Z}_t = E[Z_t | \mathscr{F}_t^N] = P[\tau \le t | \mathscr{F}_t^N].$$

The process $\hat{Z}_t$ has a version of the form (1.10), that is to say, a right-continuous version. Therefore $\hat{f}_t = h(t)(1 - \hat{Z}_t)$. Also $\hat{\lambda}_t = a + (b-a)\hat{Z}_{t-}$, and

$$\hat{Z}_t = \int_0^t h(s)(1 - \hat{Z}_s)\,ds + \int_0^t K_s(dN_s - \{a + (b-a)\hat{Z}_s\}\,ds). \tag{1.31}$$

It remains to compute K_t, using T2. Clearly $\Psi_{3,t} = 0$, since $\Delta m_s \Delta N_s = 0$ (Z_t and N_t have no common jump). $\Psi_{1,t}$ is characterized by

$$E\left[\int_0^t C_s Z_s(a + (b-a)Z_s)\,ds\right] = E\left[\int_0^t C_s \Psi_{1,s}(a + (b-a)\hat{Z}_{s-})\,ds\right]$$

for all nonnegative bounded $\mathscr{F}_t^N$-predictable processes C_t, and $\Psi_{2,t}$ is given by

$$\Psi_{2,t} = \hat{Z}_{t-}. \tag{1.32}$$

The equality defining $\Psi_{1,t}$ can be rewritten, in view of the identity $Z_t^2 = Z_t$, as

$$E\left[\int_0^t C_s b Z_s\,ds\right] = E\left[\int_0^t C_s \Psi_{1,s}(a + (b-a)\hat{Z}_{s-})\,ds\right],$$

and therefore

$$\Psi_{1,t} = \frac{b\hat{Z}_{t-}}{a + (b-a)\hat{Z}_{t-}}. \qquad (1.33)$$

In summary [4]:

$$\hat{Z}_t = \int_0^t h(s)(1 - \hat{Z}_s)\,ds$$

$$+ \int_0^t \frac{(b-a)(\hat{Z}_{t-} - \hat{Z}_{t-}^2)}{a + (b-a)\hat{Z}_{t-}} (dN_t - (a + (b-a)\hat{Z}_{t-})\,dt), \qquad (1.34)$$

or

$$\hat{Z}_t = \int_0^t [h(s) + (b-a)\hat{Z}_s](1 - \hat{Z}_s)\,ds$$

$$+ \sum_{n \geq 1} \frac{(b-a)\hat{Z}_{T_n-}(1 - \hat{Z}_{T_n-})}{a + (b-a)\hat{Z}_{T_n-}} 1(T_n \leq t). \qquad (1.35)$$

The above expression is recursive. Suppose for instance that τ is distributed according to an exponential of mean $1/\lambda$. Then $h(t) = \lambda$ and equation (1.35) reads

Between the jumps $(t \in [T_n, T_{n+1}))$:

$$\hat{Z}_t = \hat{Z}_{T_n} + \int_{T_n}^t (\lambda + (b-a)\hat{Z}_s)(1 - \hat{Z}_s)\,ds; \qquad (1.36)$$

at the jumps:

$$\hat{Z}_{T_n} = \frac{b\hat{Z}_{T_n-}}{a + (b-a)\hat{Z}_{T_n-}}. \qquad (1.37)$$

From (1.36) we see that for $t \in [T_n, T_{n+1})$

$$\frac{d}{dt}\hat{Z}_t = (\lambda + (b-a)\hat{Z}_t)(1 - \hat{Z}_t) > 0$$

(we have assumed that $b - a > 0$), and moreover, after easy computations,

$$\frac{d^2}{dt^2}\hat{Z}_t > 0.$$

The trajectory of $\hat{Z}_t$ between jumps is therefore increasing and concave. From (1.37) we see that $\hat{Z}_{T_n} > \hat{Z}_{T_n-}$ (making use of the fact that $\hat{Z}_{T_n-} \in [0, 1)$). Finally, a trajectory of $\hat{Z}_t$ looks like Figure 2.

In the above model for the disruption problem, one could allow $F(t)$ to depend upon the past $(N_s, s \in [0, t])$ by making $F(t)$ an increasing process such that $F(0) = 0$, $F(\infty) = 1$, and $F(t) = \int_0^t f(u)\,du$ for some nonnegative

1. The Theory of Innovations

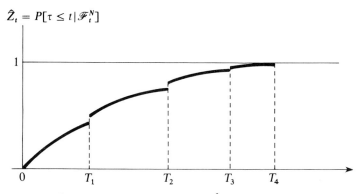

Figure 2 A typical trajectory of $\hat{Z}_t$ in the case $b > a$.

$\mathscr{F}_t^N$-progressive process $f(t)$. Exactly the same equations would have been obtained.

We now turn to a generalization of the disruption problem: we will estimate the state of a Markov chain driving a conditional Poisson process.

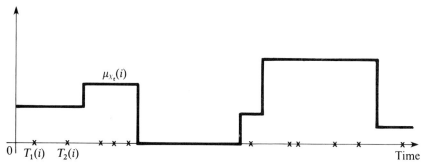

Figure 3 Modulation of a point process by a Markov chain.

Recursive Estimation of a Markov Chain Driving a Doubly Stochastic Poisson Process

Let $(N_t(1), \ldots, N_t(k))$ be a k-variate point process and X_t be an S-valued process (where $S = \{1, 2, \ldots\}$), both defined on $(\Omega, \mathscr{F}, P)$. Let $\mathscr{F}_t$ be the history

$$\mathscr{F}_t = \mathscr{G}_t \vee \mathscr{F}_\infty^X \tag{1.38}$$

where $\mathscr{G}_t$ is the internal history of the k-variate point process:

$$\mathscr{G}_t = \bigvee_{i=1}^{k} \mathscr{F}_t^N(i). \tag{1.39}$$

We make the following assumptions. First

$$\lambda_t(i) = \mu_{X_t}(i) \tag{1.40}$$

and therefore, $N_t(i)$ is a conditional Poisson process by (II, T4). Secondly, X_t is a stable and conservative $\mathscr{F}_t$-Markov chain with infinitesimal parameters q_{ij} (see Appendix A1, §8). Hence, the process $N_t(l, m)$ counting the transitions $l \to m$ has the $\mathscr{F}_t$-intensity $1(X_t = l)q_{lm}$. It is assumed moreover that $X_t, N_t(1), \ldots, N_t(k)$ have no common jumps. We want to find recursive equations for

$$\hat{Z}_t(l) = P[X_t = l | \mathscr{G}_t]. \tag{1.41}$$

In order to state the result, we introduce the following notation:

$$\hat{Z}_t = \begin{pmatrix} \hat{Z}_t(1) \\ \hat{Z}_t(2) \\ \hat{Z}_t(3) \\ \vdots \end{pmatrix}, \quad \mu(i) = \begin{pmatrix} \mu_1(i) \\ \mu_2(i) \\ \mu_3(i) \\ \vdots \end{pmatrix}, \quad P(0) = \begin{pmatrix} P[X_0 = 1] \\ P[X_0 = 2] \\ P[X_0 = 3] \\ \vdots \end{pmatrix}$$

$$Q = \begin{pmatrix} -q_1 & q_{12} & q_{13} & \cdots \\ q_{21} & -q_2 & q_{23} & \cdots \\ q_{31} & q_{32} & -q_3 & \\ \vdots & \vdots & \vdots & \ddots \end{pmatrix}, \quad M(i) = \begin{pmatrix} \mu_1(i) & & 0 \\ & \mu_2(i) & \\ & & \mu_3(i) \\ 0 & & & \ddots \end{pmatrix}$$

$$(1.42)$$

R4 Result. *The filter $\hat{Z}_t$ is given by*

$$\hat{Z}_t = p(0) + \int_0^t Q'\hat{Z}_s \, ds$$

$$+ \sum_{i=1}^k \int_0^t \left(-\hat{Z}_{s-} + \frac{M(i)\hat{Z}_{s-}}{\mu(i)'\hat{Z}_{s-}} \right) (dN_s(i) - \mu(i)'\hat{Z}_s \, ds). \tag{1.43}$$

PROOF. We do the case $k = 1$ for simplicity. Let J_t be the process that counts all transitions of the chain X_t:

$$J_t = \sum_{i \in S} \sum_{j \in S - \{i\}} N_t(i, j), \tag{1.44}$$

and define

$$Z_t(l) = 1(X_t = l). \tag{1.45}$$

We seek a semimartingale decomposition for $Z_t(l)$; since it is a step process jumping if and only if J_t jumps, $Z_t(l)$ admits the decomposition

$$Z_t(l) = Z_0(l) + \int_0^t (Z_s(l) - Z_{s-}(l)) \, dJ_s.$$

1. The Theory of Innovations

After a little thought, we see that

$$Z_s(l)\, dJ_s = Z_s(l) \sum_{i \in S} \sum_{j \in S - \{i\}} dN_s(i, j) = \sum_{i \in S - \{l\}} Z_{s-}(i)\, dN_s(i, l)$$

and

$$Z_{s-}(l)\, dJ_s = Z_{s-}(l) \sum_{i \in S} \sum_{j \in S - \{i\}} dN_s(i, j)$$

$$= Z_{s-}(l) \sum_{j \in S - \{l\}} dN_s(l, j).$$

Hence

$$Z_t(l) = Z_0(l) + \sum_{i \in S - \{l\}} \int_0^t Z_{s-}(i)\, dN_s(i, l)$$

$$- \sum_{j \in S - \{l\}} \int_0^t Z_{s-}(l)\, dN_s(l, j). \tag{1.46}$$

The $\mathscr{F}_t$-intensity of $N_t(l, m)$ is $Z_t(l) q_{lm}$, and therefore the $\mathscr{F}_t$-intensity of $\sum_{j \in S - \{l\}} dN_s(l, j)$ is $Z_t(l) \sum_{j \in S - \{l\}} q_{lj} = Z_t(l) q_l$; therefore we obtain from (1.46) after compensation

$$Z_t(l) = Z_0(l) + \int_0^t \left(\sum_{i \in S - \{l\}} Z_s(i) q_{il} - Z_s(l) q_l \right) ds + m_t(l), \tag{1.47}$$

where $m_t(l)$ is an $\mathscr{F}_t$-martingale. By the general theory (T2),

$$\hat{Z}_t(l) = P[X_0 = l] + \int_0^t \left(\sum_{i \in S - \{l\}} \hat{Z}_s(i) q_{il} - \hat{Z}_s(l) q_l \right) ds$$

$$+ \int_0^t K_s(l) \left(dN_s - \sum_{l \in S} \mu_l \hat{Z}_s(l)\, ds \right), \tag{1.48}$$

where $K_t(l)$ is the innovations gain to be computed, and $\sum_{l \in S} \mu_l \hat{Z}_s(l) = E[\mu_{X_s} | \mathscr{G}_s]$ is the $\mathscr{G}_t$-intensity of N_t.

The gain K has the form $K = \Psi_1 - \Psi_2 - \Psi_3$. Here $\Psi_3 \equiv 0$, since $Z_t(l)$ and N_t have no common jumps. Also, as always, $\Psi_{2,t}(l) = Z_{t-}(l)$. As for Ψ_1, it is computed from

$$E\left[\int_0^t C_s Z_s(l) \mu_{X_s}\, ds \right] = E\left[\int_0^t C_s \Psi_{1,s}(l) \sum_{l \in S} \mu_l \hat{Z}_s(l)\, ds \right].$$

Observing that $\mu_{X_s} Z_s(l) = \mu_l Z_s(l)$ and $\hat{Z}_s(l)\, ds = \hat{Z}_{s-}(l)\, ds$, we obtain

$$\Psi_{1,t}(l) = \frac{\mu_l \hat{Z}_{t-}(l)}{\sum_{l \in S} \mu_l \hat{Z}_{t-}(l)}.$$

Therefore

$$K_t(l) = -Z_{t-}(l) + \frac{\mu_l \hat{Z}_{t-}(l)}{\sum_{l \in S} \mu_l \hat{Z}_{t-}(l)}, \tag{1.49}$$

and the result is proven. □

We now proceed to solve filtering problems in the situation where the state and the observation have common jumps (hence the Ψ_3 term in $K = \Psi_1 - \Psi_2 + \Psi_3$ does not vanish).

2. State Estimates for Queues and Markov Chains

Filtering with Respect to the Output of a Simple Queue

In this subsection, we apply the main results of §1 to situations involving queueing processes. Let Q_t be a queueing process with the representation *in standard form*

$$Q_t = Q_0 + A_t - D_t,$$

and suppose that it admits the $\mathscr{F}_t$-parameters λ_t and μ_t. Assume moreover that

$$E[Q_0] < \infty, \qquad E[A_t] < \infty, \quad t \geq 0 \tag{2.1}$$

(the above two inequalities also imply

$$E[D_t] < \infty, \qquad t \geq 0, \tag{2.1'}$$

since Q_t is a nonnegative process). Our goal is to find recursive expressions for the conditional expectations $P[Q_t = n | \mathscr{F}_t^D]$ for all $t \geq 0$ and all $n \geq 0$. Such expressions play a central role in the output theory for markovian networks to be developed in Chapter V.

The theory of §1 is applicable to the present situation if for each $n \geq 0$ one chooses the state process to be

$$Z_t(n) = 1(Q_t = n), \tag{2.2}$$

and if one observes the departure process D_t. The first step of the procedure leading to recursive equations consists in finding for $Z_t(n)$ a semimartingale equation of type (1.1).

E4 Exercise. Define for each $n \geq 0$

$$M_t^A(n) = \int_0^t (Z_{s-}(n-1)1(n>0) - Z_{s-}(n))(dA_s - \lambda_s\, ds),$$

$$M_t^D(n) = \int_0^t (Z_{s-}(n+1) - Z_{s-}(n)1(n>0))(dD_s - \mu_s 1(Q_s > 0)\, ds), \tag{2.3}$$

and

$$m_t(n) = M_t^A(n) + M_t^D(n). \tag{2.4}$$

2. State Estimates for Queues and Markov Chains

Show that we have for $Z_t(n)$ the sought semimartingale equation:

$$Z_t(n) = Z_0(n) + \int_0^t [(Z_s(n-1)1(n>0) - Z_s(n))\lambda_s \\ + (Z_s(n+1) - Z_s(n)1(n>0))\mu_s] \, ds + m_t(n). \quad (2.5)$$

Equation (2.5) is of the form (1.1) and satisfies the requirements of T1 and T2. Let

$$f_t(n) = (Z_t(n-1)1(n>0) - Z_t(n))\lambda_t \\ + (Z_t(n+1) - Z_t(n)1(n>0))\mu_t. \quad (2.6)$$

The equation for the filter is

$$\hat{Z}_t(n) = P[Q_0 = n] + \int_0^t \hat{f}_s(n) \, ds + \int_0^t K_s(n)(dD_s - \hat{h}_s \, ds), \quad (2.7)$$

where $\hat{h}_t$ is the $\mathscr{F}_t^D$-intensity of D_t:

$$\hat{h}_t = E[\mu_t 1(Q_t > 0) | \mathscr{F}_t^D], \quad (2.8)$$

a predictable version (see II, T13). By T2, the innovations gain has the form

$$K_t(n) = \Psi_{1,t}(n) - \Psi_{2,t}(n) + \Psi_{3,t}(n), \quad (2.9)$$

where the $\Psi_{1,t}(n)$ and $\Psi_{3,t}(n)$ are $\mathscr{F}_t^D$-predictable processes such that for all nonnegative $\mathscr{F}_t^D$-predictable processes C_t

$$E\left[\int_0^t C_s Z_s(n)\mu_s(1 - Z_s(0)) \, ds\right] = E\left[\int_0^t C_s \Psi_{1,s}(n)\hat{h}_s \, ds\right],$$

$$E\left[\sum_{0 < s \le t} C_s \Delta m_s(n) \Delta D_t\right] = E\left[\int_0^t C_s \Psi_{3,s}(n)\hat{h}_s \, ds\right]. \quad (2.10)$$

[As always, $\Psi_{2,t}(n) = \hat{Z}_{t-}(n)$.] In order to carry out the computations to their end, we must make assumptions on λ_t and μ_t. Essentially, we do not allow the parameters to depend upon the whole past of the queueing process Q_t; instead, we require that they depend only upon the whole past of the departure process D_t and the *present state* of Q_t. More precisely:

Assumption (A). Let $(t, \omega) \to \lambda(t, \omega, n)$ and $(t, \omega) \to \mu(t, \omega, n)$, $n \ge 0$, be two families of $\mathscr{F}_t^D$-predictable processes, nonnegative, and suppose that

$$\lambda_t(\omega) \equiv \lambda(t, \omega, Q_t(\omega)), \\ \mu_t(\omega) \equiv \mu(t, \omega, Q_t(\omega)). \quad (2.11)$$

Under such an assumption,

$$\lambda_t = \sum_{n \ge 0} \lambda(t, n) Z_t(n),$$

$$\mu_t = \sum_{n \ge 1} \mu(t, n) Z_t(n), \quad (2.12)$$

and:

$$f_t(n) = Z_t(n-1)1(n>0)\lambda(t, n-1)$$
$$- Z_t(n)(\lambda(t, n) + \mu(t, n)1(n>0)) + Z_t(n+1)\mu(t, n+1).$$

Hence, since the $\lambda(t, n)$'s and $\mu(t, n)$'s are $\mathscr{F}_t^D$-progressive (being $\mathscr{F}_t^D$-predictable),

$$\hat{f}_t(n) = \hat{Z}_t(n-1)1(n>0)\lambda(t, n-1)$$
$$- \hat{Z}_t(n)(\lambda(t, n) + \mu(t, n)1(n>0)) + \hat{Z}_t(n+1)\mu(t, n+1). \quad (2.13)$$

Also

$$\hat{h}_t = \sum_{n>0} \mu(t, n)\hat{Z}_{t-}(n). \quad (2.14)$$

Now, using (2.12) and (2.14), it is easy to find from (2.10) that

$$\Psi_{1,t}(n) = \frac{\hat{Z}_{t-}(n)1(n>0)\mu(t, n)}{\sum_{j>0} \mu(t, j)\hat{Z}_{t-}(j)}.$$

It remains to compute $\Psi_{3,t}(n)$. Using (2.4) and (2.3), one observes that

$$\Delta m_t(n)\, \Delta D_t = (Z_{t-}(n+1) - Z_{t-}(n)1(n>0))\, \Delta D_t,$$

and therefore by the integration theorem (II, T8), for all nonnegative $\mathscr{F}_t^D$-predictable processes C_t

$$E\left[\sum_{0<s\leq t} C_s\, \Delta m_s(n)\, \Delta D_s\right]$$
$$= E\left[\int_0^t C_s(Z_s(n+1) - Z_s(n)1(n>0))\mu_s 1(Q_s>0)\, ds\right].$$

But, by (2.12), the right-hand side of the above equality is equal to

$$E\left[\int_0^t C_s(Z_s(n+1)\mu(s, n+1) - Z_s(n)\mu(s, n)1(n>0))\, ds\right],$$

and therefore the last equality of (2.10) can be written, under Assumption (A),

$$E\left[\int_0^t C_s(\hat{Z}_s(s+1)\mu(s, n+1) - \hat{Z}_s(n)\mu(s, n)1(n>0))\, ds\right]$$
$$= E\left[\int_0^t C_s\Psi_{3,s}(n) \sum_{j>0} \mu(s, j)\hat{Z}_{s-}(j)\, ds\right].$$

Hence

$$\Psi_{3,t}(n) = \frac{\hat{Z}_{t-}(n+1)\mu(t, n+1) - \hat{Z}_{t-}(n)\mu(t, n)1(n>0)}{\sum_{j>0} \mu(t, j)\hat{Z}_{t-}(j)}.$$

By combining the above results, the gain $K_t(n)$ is found to be

$$K_t(n) = \frac{\mu(t, n+1)\hat{Z}_{t-}(n+1)}{\sum_{j>0} \mu(t, j)\hat{Z}_{t-}(j)} - \hat{Z}_{t-}(n), \qquad (2.15)$$

and finally the filter for $Z_t(n)$ is, under Assumption (A),

$$\hat{Z}_t(n) = P[Q_0 = n]' + \int_0^t (\hat{Z}_s(n-1)1(n>0)\lambda(s, n-1)$$
$$- \hat{Z}_s(n)(\lambda(s, n) + \mu(s, n)1(n>0)) + \hat{Z}_s(n+1)\mu(s, n+1))\, ds$$
$$+ \int_0^t \left\{\mu(s, n+1)\hat{Z}_{s-}(n+1) - \hat{Z}_{s-}(n)\left(\sum_{j>0} \mu(s, j)\hat{Z}_{s-}(j)\right)\right\}$$
$$\times \left(\frac{dD_s}{\sum_{j>0} \mu(s, j)\hat{Z}_{s-}(j)} - ds\right). \qquad (2.16)$$

We summarize the above results, giving a slight extension of them along the lines of T3.

R5 Result [2]. *Let Q_t be a queueing process represented in standard form by*

$$Q_t = Q_0 + A_t - D_t$$

and admitting the $\mathscr{F}_t$-parameters λ_t and μ_t. It is supposed that

$$E[Q_0] < \infty, \qquad E[A_t] < \infty, \quad t \geq 0,$$

and that

$$\lambda_t(\omega) \equiv \lambda(t, \omega, Q_t(\omega)),$$
$$\mu_t(\omega) \equiv \mu(t, \omega, Q_t(\omega)), \qquad (2.17)$$

where, for each $n \geq 0$, $(t, \omega) \to \lambda(t, \omega, n)$, $(t, \omega) \to \mu(t, \omega, n)$ are nonnegative $\mathscr{F}_t^D$-predictable mappings. Then, for any finite $\mathscr{F}_t^D$-stopping time τ and any $n \geq 0$, the conditional probability $P[Q_\tau = n | \mathscr{F}_\tau^D]$ admits a version $\hat{Z}_\tau(n)$, where $\hat{Z}_t(n)$ is given by (2.16).

Comment (Predictable and Innovating Parts). Suppose that $\lambda(t, n)$ and $\mu(t, n)$ are deterministic; then $p(t, n) = P(Q_t = n)$ satisfies the forward Kolmogorov equation:

$$p(t, n) = p(0, n) + \int_0^t \mathscr{K}(p, s, n)\, ds, \qquad (2.18)$$

where:

$$\mathscr{K}(p, s, n) = p(s, n-1)1(n>0)\lambda(s, n-1) - p(s, n)(\lambda(s, n)$$
$$+ \mu(s, n)1(n>0)) + p(s, n+1)\mu(s, n+1). \qquad (2.19)$$

The filter for $\hat{Z}_t(n) = P[Q_t = n | \mathcal{F}_t^D]$ has the form

$$\hat{Z}_t(n) = \hat{Z}_0(n) + \int_0^t \mathcal{K}(\hat{Z}, s, n)\, ds + \int_0^t K_s(n)(dD_s - \hat{\mu}_s\, ds). \qquad (2.20)$$

Thus, if not for the innovations term $\int_0^t K_s(n)(dD_s - \hat{\mu}_s\, ds)$, $\hat{Z}_t(n)$ would evolve as $p(t, n)$. It is therefore natural to call the part

$$\hat{Z}_0(n) + \int_0^t \mathcal{K}(\hat{Z}, s, n)\, ds$$

the "predictable" part (where "predictable" is to be understood in its worldly sense); the part $\int_0^t K_s(n)(dD_s - \hat{\mu}_s\, ds)$ then appears to be "innovating" (with respect to the forward Kolmogorov equations). The same comment applies to R4; indeed if we suppress the innovating term in (1.44), what remains is $\hat{Z}_t = p(0) + \int_0^t Q' \hat{Z}_s\, ds$, the forward Kolmogorov equation for the Markov chain X_t.

E5 Exercise. In a control situation based on the observation of the output only, the $\mathcal{F}_t$-parameter μ_t can be controlled: for instance $\mu_t = \mu u_t$, where u_t is the number of servers ($u_t = 0, 1, 2, \ldots, K$), this number being chosen by the controller on the basis of the observation of the output. Thus u_t, and hence μ_t, must be adapted to $\mathcal{F}_t^D$. Write the filtering equations in this case, with the arrival parameter λ_t of the form $\lambda_t = \lambda(Q_t)$.

Filtering in an $M/M/1$ Feedback Queue

Here is another example of the application of the innovations theory of filtering; it is slightly more computational than the preceding one in that one must take into account the fact that some of the observed point processes have common jumps.

Let Q_t be an $M/M/1$ feedback queue of the type described in II, §5. Recall that $Q_t = Q_0 + A_t - D_t = Q_0 + E_t - S_t$, where A_t and D_t have no common jumps, whereas E_t and S_t have in common the jumps of the feedback stream F_t: $\Delta E_t \Delta S_t = \Delta F_t$.

Letting $\mathcal{F}_t = \mathcal{F}_t^Q \vee \mathcal{F}_t^F$, we know that F_t, E_t, and S_t admit the $\mathcal{F}_t$-intensities $\mu p 1(Q_t > 0)$, $\lambda + \mu p 1(Q_t > 0)$, and $\mu 1(Q_t > 0)$ respectively. We seek the filter of Q_t with respect to N_t for $N_t = A_t, F_t, E_t$, and S_t successively. Define, as above, $Z_t(n) = 1(Q_t = n)$ and $\hat{Z}_t^N(n) = E[Z_t(n) | \mathcal{F}_t^N]$, where N takes the "values" F, E, and S successively. As in §1, starting from the semi-martingale equation (2.5) with $\lambda_t \equiv \lambda$, $\mu_t \equiv \mu(1 - p)$, one finds that

$$\hat{Z}_t^N(n) = \hat{Z}_0^N(n) + \int_0^t (\hat{Z}_s^N(n - 1) 1(n > 0)\lambda - (\lambda + \mu(1 - p) 1(n > 0))\hat{Z}_s^N(n)$$

$$+ \mu(1 - p)\hat{Z}_s^N(n + 1))\, ds + \int_0^t K_s^N(n)(dN_s - \hat{h}_s^N\, ds), \qquad (2.20')$$

2. State Estimates for Queues and Markov Chains

where $\hat{h}_t^N$ is an $\mathscr{F}_t^N$-intensity of N_t (any version for that purpose), and $K_t^N(n)$ is a gain to be computed. Since the $\mathscr{F}_t$-intensities of the N_t in terms of $Z_t(n)$ are known, the $\hat{h}_t^N$ are readily available: the right-continuous versions are

$$\hat{h}_t^F = \mu p(1 - \hat{Z}_t^F(0)),$$
$$\hat{h}_t^E = \lambda + \mu p(1 - \hat{Z}_t^E(0)) \qquad (2.21)$$
$$\hat{h}_t^S = \mu(1 - \hat{Z}_t^S(0)).$$

The predictable versions are obtained by replacing $\hat{Z}_t^N(0)$ with $\hat{Z}_{t-}^N(0)$ in the above expressions.

E6 Exercise. Prove the following:

R6 Result [3]. *The gains relative to the $M/M/1$ feedback queue are*

$$K_t^F(n) = -\hat{Z}_{t-}^F(n) + \frac{\hat{Z}_{t-}^F(n)1(n > 0)}{1 - \hat{Z}_{t-}^F(0)},$$

$$K_t^E(n) = -\hat{Z}_{t-}^E(n) + \frac{\mu p 1(n > 0)\hat{Z}_{t-}^E(n) + \lambda 1(n > 0)\hat{Z}_{t-}^E(n-1)}{\lambda + \mu p(1 - Z_{t-}^E(0))}, \qquad (2.22)$$

$$K_t^S(n) = -\hat{Z}_{t-}^S(n) + \frac{(1 - p)\hat{Z}_{t-}^T(n+1) + p\hat{Z}_{t-}^S(n)1(n > 0)}{1 - \hat{Z}_{t-}^S(0)}.$$

It is instructive at this point to show that gain expressions such as (2.22) have an intuitive interpretation. For instance, let us look at the way $K_t^F(n)$ intervenes in the filter $\hat{Z}_t^F(n)$ at a jump t of F_t; we find that, by (2.20) *(begin heuristics)*

$$\hat{Z}_t^F(n) - \hat{Z}_{t-}^F(n) = K_t^F(n), \qquad (2.23)$$

and therefore, by the expression for $K_t^F(n)$ given in (2.22),

$$\hat{Z}_t^F(n) = \frac{\hat{Z}_{t-}^F(n)1(n > 0)}{1 - \hat{Z}_{t-}^F(0)}. \qquad (2.24)$$

If $n = 0$, (2.24) yields

$$\hat{Z}_t^F(0) = 0, \qquad (2.25)$$

which is quite natural, since at a jump t of F_t, we know that Q_t must be strictly greater than 0. For $n \geq 1$, (2.24) reads

$$\hat{Z}_t^F(n) = \frac{\hat{Z}_{t-}^F(n)}{1 - \hat{Z}_{t-}^F(0)}, \qquad (2.26)$$

and this reads in turn

$$P[Q_t = n | \Delta F_t = 1, \mathscr{F}_{t-}^F] = \frac{P[Q_{t-} = n | \mathscr{F}_{t-}^F]}{P[\Delta F_t = 1 | \mathscr{F}_{t-}^F]}, \qquad (2.27)$$

which is nothing but Bayes's formula if we observe that $P[Q_{t-} = n | \mathscr{F}^F_{t-}, \Delta F_t = 1] = P[Q_t = n | \mathscr{F}^F_{t-}, \Delta F_t = 1]$, since at a jumps of F_t, $Q_t = Q_{t-}$ (end heuristics).

All the examples given at the beginning of the present paragraph are particular instances of the more general situation where the state of a Markov chain is to be estimated on the basis of the observation of selected transitions.

Observing Part of a Markov Chain

Let X_t be a standard homogeneous Markov chain taking its values in $S = \{1, 2, \ldots\}$, stable and conservative, i.e., for all $i \in S$

$$q_i = \sum_{j \in S} q_{ij} < \infty.^4 \qquad (2.28)$$

Let $N_t(i, j)$ be the number of transitions from i to j in the interval $(0, t]$. As was seen in I, §2, $N_t(i, j)$ has the $\mathscr{F}^X_t$-intensity $1(X_t = i)q_{ij}$.

Now, let D be a mapping from S into $\mathscr{P}(S)$, the family of subsets of S: $D(\gamma)$ is, for all $\gamma \in S$, a subset of S. Let N_t be the flow corresponding to any transition from a state $\gamma \in C$ to a state $\delta \in D(\gamma)$, i.e.

$$N_t = \sum_{\gamma \in C} \sum_{\delta \in D(\gamma)} N_t(\gamma, \delta).^4 \qquad (2.29)$$

We will say that N_t is the *flow from C to D* (beware: D is a mapping, not a subset).

For each $1 \leq l \leq k$, let $N_t(l)$ be the flow from C_l to D_l, and suppose that if $h \neq l$, $N_t(h)$ and $N_t(l)$ have no common jumps. Let

$$\mathscr{G}_t = \bigvee_{i=1}^m \mathscr{F}^N_t(i). \qquad (2.30)$$

R7 Result. *Define* $\hat{Z}_t(i) = P[X_t = i | \mathscr{G}_t]$. *Then*

$$\hat{Z}_t(i) = P[X_0 = i] + \int_0^t \left(\sum_{j \in S} \hat{Z}_s(j) q_{ji} - \hat{Z}_s(i) q_i \right) ds$$

$$+ \sum_{l=1}^k \int_0^t K_s(l, i) \left(dN_s(l) - \sum_{\gamma \in C_l} \sum_{\delta \in D_l(\gamma)} q_{\gamma\delta} \hat{Z}_s(\gamma) \, ds \right), \qquad (2.31)$$

where

$$K_t(l, i) = -\hat{Z}_{t-}(i) + \frac{\sum_{\gamma \in C_l} 1(i \in D_l(\gamma)) \hat{Z}_{t-}(\gamma) q_{\gamma i}}{\sum_{\gamma \in C_l} \hat{Z}_{t-}(\gamma) \sum_{\delta \in D_l(\gamma)} q_{\gamma\delta}}. \qquad (2.32)$$

---

[4] The notational convention for this subsection is $q_{ii} = 0$ and $N_t(i, i) \equiv 0$.

PROOF. We do the case $m = 1$. Just as in the proof of R4, we reach the equation

$$\hat{Z}_t(i) = P[X_0 = i] + \int_0^t \left(\sum_{j \in S} \hat{Z}_s(j) q_{ji} - \hat{Z}_s(i) q_i \right) ds + \int_0^t K_s(i)(dN_s - \hat{\lambda}_s\, ds), \qquad (2.33)$$

where

$$\hat{\lambda}_t = \sum_{\gamma \in C} \hat{Z}_{t-}(\gamma) \sum_{\delta \in D(\gamma)} q_{\gamma\delta} \qquad (2.34)$$

is the $\mathscr{G}_t$-intensity of N_t, and $K_t(i)$ is a gain to be computed according to T2: $K_t(i) = \Psi_{1,t}(i) - \Psi_{2,t}(i) + \Psi_{3,t}(i)$. As usual, $\Psi_{2,t}(i) = \hat{Z}_{t-}(i)$. As to $\Psi_{1,t}(i)$, it is obtained from

$$E\left[\int_0^t C_s \Psi_{1,s}(i) \hat{\lambda}_s\, ds \right] = E\left[\int_0^t C_s Z_s(i) \lambda_s\, ds \right],$$

where

$$\lambda_t = \sum_{\gamma \in C} \sum_{\delta \in D(\gamma)} q_{\gamma\delta} Z_t(\gamma) \qquad (2.35)$$

is the $\mathscr{F}_t^X$-intensity of N_t [recall the definition $Z_t(\gamma) = 1(X_t = \gamma)$]. Now familiar routine computations give

$$\Psi_{1,t}(i) = \frac{1(i \in C)\hat{Z}_{t-}(i) \sum_{\delta \in D(i)} q_{i\gamma}}{\hat{\lambda}_t}.$$

The term $\Psi_{3,t}(i)$ is obtained from

$$E\left[\int_0^t C_s \Psi_{3,s}(i) \hat{\lambda}_s\, ds \right] = E\left[\int_0^t C_s \Delta Z_s(i)\, dN_s \right],$$

and we obtain, after easy computations,

$$\Psi_{3,t}(i) = \frac{\sum_{\gamma \in C} 1(i \in D(\gamma) - \{\gamma\})\hat{Z}_{t-}(\gamma) q_{\gamma i}}{\hat{\lambda}_t} - \Psi_{1,t}(i)$$

[just observe that

$$\Delta Z_t(i)\, dN_t = \sum_{\gamma \in C} 1(i \in D(\gamma))Z_{t-}(\gamma)\, dN_t(\gamma, i) - 1(i \in C)Z_{t-}(i) \sum_{\delta \in D(i)} dN_t(i, \delta)].$$

Hence the announced result. □

3. Continuous States and Nontrivial Prehistory

Continuous States

In Theorem T2 we have assumed that the martingale m_t of the representation

$$Z_t = Z_0 + \int_0^t f_s\, ds + m_t$$

has paths of bounded variation over finite intervals. This hypothesis is not essential, as we will see; however, in the proof of T2 we have explicitly used it, since we have applied the rule of Stieltjes calculus $d(A_t B_t) = A_t dB_t + B_{t-} dA_t$. In the case where m_t has paths of unbounded variation, a similar rule of integration by parts is available and is proven in Appendix A2, T19. It is due to Dellacherie {47} and will play the same role in the derivation of the extension of T2 as the Stieltjes rule. The extension of T2 to martingales m_t of *unbounded variation* is important, since *all continuous martingales are such*. This general result of martingale theory (see Meyer {106}) is beyond the scope of this introductory text; however, the reader is already acquainted with it if he has studied Wiener processes (which are indeed continuous and with paths of unbounded variation (A3, T3)).

Although the proof of the extension of T3 to a *continuous*-state process Z_t requires more sophisticated tools of martingale theory such as Dellacherie's result (Appendix A2, T19), the result takes a simpler form: it reads exactly like T2 with $\Psi_{3,t}(i) = 0$ [formally: $\Delta m_t \neq 0$; hence $\Delta m_t \Delta N_t(i) \equiv 0$; hence $\Psi_{3,t}(i) \equiv 0$ by (1.13)].

Nontrivial Observation at the Origin of Time

In some applications the initial observed σ-field $\mathcal{O}_0$ contains information, say $\mathcal{O}_0 = \sigma(Z_0)$ for instance (the initial state is observed). The extension of T2 to this case is not difficult at all: indeed, the proof of T2 was based on the representation of $\mathcal{G}_t$-martingales as stochastic integrals, and a similar representation holds for $\mathcal{O}_0 \vee \mathcal{G}_t$-martingales (III, T17).

Let us now state the general filtering theorem:

T8 Theorem [1]. *Let there be given on a probability space* $(\Omega, \mathcal{F}, P)$ *two histories $\mathcal{F}_t$ and $\mathcal{O}_t$ such that*

$$\mathcal{F}_t \supset \mathcal{O}_t, \quad t \geq 0,$$

$$\mathcal{O}_t = \mathcal{O}_0 \vee \mathcal{G}_t, \quad t \geq 0, \quad \text{where } \mathcal{G}_t = \bigvee_{i=1}^{k} \mathcal{F}_t^N(i), \tag{3.1}$$

where $N_t = (N_t(1), \ldots, N_t(k))$ is a nonexplosive k-variate point process with the $\mathcal{F}_t$-intensity $\lambda_t = (\lambda_t(1), \ldots, \lambda_t(k))$ and the $\mathcal{O}_t$-intensity $\hat{\lambda}_t = (\hat{\lambda}_t(1), \ldots, \hat{\lambda}_t(k))$. Let Z_t, the state process, be adapted to $\mathcal{F}_t$ and of the form

$$Z_t = Z_0 + \int_0^t f_s \, ds + m_t, \tag{3.2}$$

where f_t is an $\mathcal{F}_t$-progressive process such that

$$E\left[\int_0^t |f_s| \, ds\right] < \infty, \quad t > 0. \tag{3.3}$$

3. Continuous States and Nontrivial Prehistory

The process m_t is an $\mathscr{F}_t$-martingale of the form

$$m_t = m_t^c + m_t^d, \tag{3.4}$$

where m_t^c is a continuous $\mathscr{F}_t$-martingale and m_t^d is an $\mathscr{F}_t$-martingale of integrable bounded variation, both martingales having mean 0. Assume that $Z_t - m_t^c$ is bounded (do not assume however that m_t^c is bounded). Let $\hat{Z}_t$ be defined by

$$\hat{Z}_t = E[Z_t | \mathcal{O}_t]. \tag{3.5}$$

Then

$$\hat{Z}_t = E[Z_0 | \mathcal{O}_0] + \int_0^t \hat{f}_s \, ds + \sum_{i=1}^k \int_0^t K_s(i)(dN_s(i) - \hat{\lambda}_s(i) \, ds), \tag{3.6}$$

where $\hat{f}_t$ is an $\mathcal{O}_t$-progressive process satisfying

$$E\left[\int_0^t C_s f_s \, ds\right] = E\left[\int_0^t C_s \hat{f}_s \, ds\right] \tag{3.7}$$

for all bounded $\mathcal{O}_t$-progressive processes C_t, and the $K_t(i)$ are $\mathcal{O}_t$-predictable processes defined by

$$K_t(i) = \Psi_{1,t}(i) - \Psi_{2,t}(i) + \Psi_{3,t}(i), \tag{3.8}$$

the Ψ's being $\mathcal{O}_t$-predictable processes satisfying, for all bounded $\mathcal{O}_t$-predictable processes C_t and all $t \geq 0$,

$$E\left[\int_0^t C_s Z_s \lambda_s(i) \, ds\right] = E\left[\int_0^t C_s \Psi_{1,s}(i) \hat{\lambda}_s(i) \, ds\right],$$

$$E\left[\int_0^t C_s Z_s \hat{\lambda}_s(i) \, ds\right] = E\left[\int_0^t C_s \Psi_{2,s}(i) \hat{\lambda}_s(i) \, ds\right], \tag{3.9}$$

$$E\left[\int_0^t C_s \Delta m_s \, dN_s(i)\right] = E\left[\int_0^t C_s \Psi_{3,s}(i) \hat{\lambda}_s(i) \, ds\right].$$

Comment. In particular, if Z_t is continuous,

$$K_t(i) = \Psi_{1,t}(i) - \Psi_{2,t}(i). \tag{3.10}$$

Also note that

$$\Psi_{2,t}(i) = \hat{Z}_{t-}. \tag{3.11}$$

PROOF OF T8

Step 1: The case $m_t^c \equiv 0$, $\mathcal{O}_0 \neq \{\Omega, \emptyset\}$. The proof does not present any difficulty and is identical to that of T2 [indeed, the representation theorem is valid for a history of type (3.1)].

Step 2: The case $Z_t = m_t^c$, $\mathcal{O}_0 \neq \{\Omega, \emptyset\}$. We restrict ourselves to $k = 1$, with the correlative change of notation $K_t(1) = K_t$, $\hat{\lambda}_s(1) = \lambda_s$, etc. Let U_t

be a bounded $\mathcal{O}_t$-martingale of the form $U_t = \int_0^t H_s \, d\hat{M}_s$, where $\hat{M}_t = N_t - \int_0^t \hat{\lambda}_s \, ds$ and H_t is a bounded $\mathcal{O}_t$-predictable process. The process $\hat{Z}_t = E[m_t^c | \mathcal{O}_t]$ is an $\mathcal{O}_t$-martingale, and therefore, by II, T17,

$$\hat{Z}_t = \int_0^t K_s \, d\hat{M}_s, \tag{3.12}$$

where K_t is an $\mathcal{O}_t$-predictable process such that

$$\int_0^t K_s \hat{\lambda}_s \, ds < \infty, \qquad t \geq 0. \tag{3.13}$$

As in the proof of T2, we will find a necessary form for K_t by writing

$$E[Z_t U_t] = E[\hat{Z}_t U_t] \tag{3.14}$$

for all the above martingales U_t such that, in the representation $U_t = \int_0^t H_s \, d\hat{M}_s$, H_t satisfies

$$E\left[\int_0^t |H_s||K_s|\hat{\lambda}_s \, ds\right] + E\left[\int_0^t |H_s|(\lambda_s + \hat{\lambda}_s) \, ds\right]$$
$$+ E\left[\int_0^t |H_s||Z_s|(\lambda_s + \hat{\lambda}_s) \, ds\right] < \infty. \tag{3.15}$$

The computation of $E[\hat{Z}_t U_t]$ has already been carried out in the proof of T2:

$$E[\hat{Z}_t U_t] = \left[\int_0^t H_s K_s \hat{\lambda}_s \, ds\right]. \tag{3.16}$$

As to $E[Z_t U_t]$, we will write

$$E[Z_t U_t] = E\left[Z_t \int_0^t H_s \, dM_s\right] + E\left[Z_t \int_0^t H_s(\lambda_s - \hat{\lambda}_s) \, ds\right].$$

By Appendix A2, T19,

$$E\left[Z_t \int_0^t H_s \, dM_s\right] = E\left[\int_0^t Z_s H_s \, dM_s\right].$$

By II, T8, using the condition (3.15) and the continuity of Z_t (which makes it a $\mathscr{F}_t$-predictable process), we see that $\int_0^t Z_s H_s \, dM_s$ is a zero-mean $\mathscr{F}_t$-martingale, and therefore $E[\int_0^t Z_s H_s \, dM_s] = 0$. Thus

$$E[Z_t U_t] = E\left[Z_t \int_0^t H_s(\lambda_s - \hat{\lambda}_s) \, ds\right].$$

By Appendix A2, T19 again,

$$E\left[Z_t \int_0^t H_s(\lambda_s - \hat{\lambda}_s) \, ds\right] = E\left[\int_0^t H_s Z_s(\lambda_s - \hat{\lambda}_s) \, ds\right].$$

3. Continuous States and Nontrivial Prehistory

Hence
$$E[Z_t U_t] = E\left[\int_0^t H_s Z_s(\lambda_s - \hat{\lambda}_s)\, ds\right]. \tag{3.17}$$

Therefore, combining (3.14), (3.16), and (3.17),
$$E\left[\int_0^t H_s K_s \hat{\lambda}_s\, ds\right] = E\left[\int_0^t H_s Z_s(\lambda_s - \hat{\lambda}_s)\, ds\right],$$
or, in view of the definition of Ψ_1 and Ψ_2,
$$E\left[\int_0^t H_s K_s \hat{\lambda}_s\, ds\right] = E\left[\int_0^t H_s Z_s(\Psi_{1,s} - \Psi_{2,s})\hat{\lambda}_s\, ds\right]. \tag{3.18}$$

The rest of the proof follows almost the same lines as the end of the proof of T2. More precisely, we select H_t of the form
$$H_t = C_t 1(t \leq S_n), \tag{3.19}$$
where S_n is the first time t at which any of the following events occurs:
$$t \geq T_n, \quad \int_0^t |K_s|\hat{\lambda}_s\, ds \geq n, \quad \int_0^t (\lambda_s + \hat{\lambda}_s)\, ds \geq n, \quad Z_t \geq n \tag{3.20}$$
(taking $S_n = \infty$ if none of these events occur for $t < \infty$) and C_t is any bounded $\mathcal{O}_t$-predictable process. Clearly the corresponding U_t and H_t satisfy the technical assumptions which allowed us to prove (3.18). Hence
$$E\left[\int_0^t C_s 1(s \leq S_n) K_s \hat{\lambda}_s\, ds\right]$$
$$= E\left[\int_0^t C_s 1(s \leq S_n)(\Psi_{1,s} - \Psi_{2,s})\hat{\lambda}_s\, ds\right]. \tag{3.21}$$

C_t being an arbitrary bounded $\mathcal{O}_t$-predictable process, it follows that
$$K_t(\omega) = \Psi_{t,1}(\omega) - \Psi_{t,2}(\omega) \quad P(d\omega)\, dN_t(\omega)\text{-a.e.} \quad \text{on } t \leq S_n(\omega). \tag{3.22}$$

Now S_n goes to infinity with n P-almost surely (here we use the fact that Z_t is a finite continuous process). Hence
$$K_t(\omega) = \Psi_{1,t}(\omega) - \Psi_{2,t}(\omega) \quad P(d\omega)\, dN_t(\omega)\text{-a.e.} \tag{3.23}$$

Step 3. The general case follows from Step 1 and Step 2 by linearity of the filter [i.e., if $Z_t = Z_t(1) + Z_t(2)$, then $\hat{Z}_t = \hat{Z}_t(1) + \hat{Z}_t(2)$]. □

White Noise and Point-Process Observation

Here is an illustration of T8 relevant to communications theory (see Snyder [8] for engineering motivation).

Let Y_t be a real-valued process of the form
$$Y_t = Y_0 + \int_0^t \phi_s\, ds + W_t, \tag{3.24}$$

where W_t is an $\mathscr{F}_t$-Wiener process and ϕ_t is an $\mathscr{F}_t$-progressive process such that $\int_0^t |\phi_s|\, ds < \infty$, P-a.s. Let now N_t be a point process with the $\mathscr{F}_t$-intensity $\lambda_t = \lambda(Y_t)$, where λ is a nonnegative borelian function from R into itself. In other words N_t is a doubly stochastic point process driven by Y_t.

The point process N_t can be thought of as being modulated by process Y_t, the modulation being implemented by the function λ. Also Y_t can be interpreted as the integral of a signal ϕ_t corrupted by white noise (see A3). The problem is to derive a recursive expression for the conditional statistics of Y_t given $\mathscr{F}_t^N$. The required background in the theory of Wiener-driven stochastic systems is given in A3.

E7 Exercise. For any $u \in R$, define $Z_t^u = \exp(iu Y_t)$. Show that

$$Z_t^u = Z_0^u + \int_0^t Z_s^u \left(iu\phi_s - \frac{u^2}{2}\right) ds + iu \oint_0^t Z_s^u \, dW_s \qquad (3.25)$$

where $\oint$ denotes Ito integration.

E8 Exercise [8]. Show that $\hat{Z}_t^u = E[Z_t^u | \mathscr{F}_t^N] = E[\exp(iu Y_t)|\mathscr{F}_t^N]$ satisfies

$$\hat{Z}_t^u = E[Z_0^u] + iu \int_0^t \widehat{Z_s^u \phi_s}\, ds - \frac{u^2}{2} \int_0^t \hat{Z}_s^u \, ds$$
$$+ \int_0^t \widehat{(Z_s^u - \hat{Z}_s^u)(\lambda_s - \hat{\lambda}_s)} \left(\frac{dN_s}{\hat{\lambda}_s} - ds\right) \qquad (3.26)$$

with the right interpretation of the symbol $\hat{\ }$ as "predictable projection."

Innovations Gain, Conditional Variance, and Noise-Signal Correlation

We can give a "physical" interpretation of the innovations gain in terms that are quite similar to those of the theory of Kalman filters, following Van Schuppen [9] and Snyder [8]. This interpretation is rather formal and of no practical value. We include it for historical reasons, because the pioneers of filtering theory for point processes came from the engineering community where Kalman filtering is a popular technique.

In the case of univariate point processes, we have seen that

$$K_t = \Psi_{1,t} - \Psi_{2,t} + \Psi_{3,t}. \qquad (3.27)$$

The term $\Psi_{1,t} - \Psi_{2,t} = \Psi_t$ is defined by

$$E\left[\int_0^t C_s \Psi_s \hat{\lambda}_s \, ds\right] = E\left[\int_0^t C_s Z_s (\lambda_s - \hat{\lambda}_s)\, ds\right] \qquad (3.28)$$

3. Continuous States and Nontrivial Prehistory

for all $\mathscr{F}_t^N$-predictable bounded processes C_t. Now since $E[\int_0^t \hat{Z}_s(\lambda_s - \hat{\lambda}_s)\,ds] = 0$, the above equality is equivalent to

$$E\left[\int_0^t C_s \Psi_s \hat{\lambda}_s\,ds\right] = E\left[\int_0^t (Z_s - \hat{Z}_s)(\lambda_s - \hat{\lambda}_s)\,ds\right]. \tag{3.29}$$

In other words, $\Psi_t(\omega)\hat{\lambda}_t(\omega)$ is the Radon–Nikodym derivative of the restriction of $(Z_t(\omega) - \hat{Z}_t(\omega))(\lambda_t(\omega) - \hat{\lambda}_t(\omega))\,dt\,P(d\omega)$ with respect to $dt\,P(d\omega)$, both restrictions being relative to the σ-field $\mathscr{P}(\mathscr{F}_t^N)$. We denote this by

$$\Psi_t \hat{\lambda}_t = \widehat{(Z_t - \hat{Z}_t)(\lambda_t - \hat{\lambda}_t)}. \tag{3.30}$$

The process in (3.30) is the predictable conditional correlation between the error estimates of the state and of the intensity.

Let us now examine the term $\Psi_{3,t}$. By definition

$$E\left[\int_0^t C_s \Psi_{3,s} \hat{\lambda}_s\,ds\right] = E\left[\sum_{s \le t} C_s \Delta Z_s \Delta N_s\right] \tag{3.31}$$

for all $\mathscr{F}_t^N$-predictable bounded processes C_t. We therefore write, in a *purely formal* way,

$$\Psi_{3,t}\hat{\lambda}_t = \widehat{\Delta Z_t \Delta N_t} = \widehat{\Delta m_t \Delta M_t}. \tag{3.32}$$

This process takes into account the coupling between the state noise m_t and the observation noise M_t. It can also be interpreted as the state–observation conditional correlation.

With all the above notation, the filter becomes

$$\hat{Z}_t = E[Z_0] + \int_0^t \hat{f}_s\,ds + \int_0^t \left[\widehat{(Z_s - \hat{Z}_s)(\lambda_s - \hat{\lambda}_s)} + \widehat{\Delta m_s \Delta N_s}\right]\left(\frac{dN_s}{\hat{\lambda}_s} - ds\right), \tag{3.33}$$

or, in a form appealing to electrical engineers,

$$\hat{Z}_t = E[Z_0] + \int_0^t \hat{f}_s\,ds + \int_0^t \left(\frac{\widehat{\tilde{Z}_s \tilde{\lambda}_s}}{\widehat{\Delta N_s \Delta N_s}} + \frac{\widehat{\Delta m_s \Delta N_s}}{\widehat{\Delta N_s \Delta N_s}}\right) d\hat{M}_s, \tag{3.34}$$

where $\tilde{Z}_t = Z_t - \hat{Z}_t$, $\tilde{\lambda}_t = \lambda_t - \hat{\lambda}_t$ are the *filtering errors* for the state Z_t and the intensity λ_t, and $\hat{M}_t$ is the *innovations process*; $\widehat{\tilde{Z}_t \tilde{\lambda}_t}$ is the *conditional error cross-correlation between state Z_t and modulation*; $\widehat{\Delta N_t \Delta N_t}$ is the *observation conditional autocorrelation*.

E10 Exercise. Show that we indeed have

$$\hat{\lambda}_t = \widehat{\Delta N_t \Delta N_t}. \tag{3.35}$$

We now conclude this chapter, but more will be said on filtering: in Chapter V we will see a number of applications of the innovations theory of filtering to networks of queues, and in Chapter VI we will give another method of filtering: the method of the *probability of reference*, which yields filtering formulas of a different form.

Additional Exercises

E11 Exercise. Let X_t be a telegraph signal, that is to say, X_t is a Markov chain with two states 0 and 1, and infinitesimal parameters q_{01} and q_{10}; X_t influences in a doubly stochastic manner a point process N_t by modulating its intensity $\lambda_t = \mu_{X_t}$. Let $\hat{Z}_t = P[X_t = 1 | \mathcal{F}_t^N]$. Show that

$$\hat{Z}_t = P[X_0 = 0] + \int_0^t [q_{01} + \hat{Z}_s((\mu_0 - \mu_1) - (q_{01} + q_{10}))$$

$$+ \hat{Z}_s^2 (\mu_1 - \mu_0)] \, ds + \sum_{n \geq 1} \frac{\mu_1 \hat{Z}_{\tau_n-}}{\mu_0 + \hat{Z}_{\tau_n-}(\mu_1 - \mu_0)} 1(\tau_n \leq t), \quad (3.36)$$

where τ_n is the sequence of jumps of X_t.

E12 Exercise. Let X_t be the solution of the stochastic differential equation

$$X_t = X_0 + \int_0^t a(s) X_s \, ds + W_t, \quad (3.37)$$

where $a(t)$ is a continuous deterministic function. Suppose X_t modulates in a doubly stochastic way a point process N_t through its intensity $\lambda_t = |X_t|$. Find an expression for $\hat{X}_t = E[X_t | \mathcal{F}_t^N]$.

E13 Exercise (Recognizing the Points). Let $N_t(1)$ and $N_t(2)$ be two point processes with common history $\mathcal{F}_t$ and the $\mathcal{F}_t$-intensities $\lambda_1(N_t(1), N_t(2))$ and $\lambda_2(N_t(1), N_t(2))$ respectively, where the mappings $(n_1, n_2) \to \lambda_i(n_1, n_2)$ are nonnegative for $i = 1, 2$. Let $N_t = N_t(1) + N_t(2)$. Compute for all $n_1 \geq 0$, $n_2 \geq 0$

$$\hat{Z}_t(n_1, n_2) = P[N_t(1) = n_1, N_t(2) = n_2 | \mathcal{F}_t^N]. \quad (3.38)$$

Show that if

$$\lambda_1(n_1, n_2) + \lambda_2(n_1, n_2) = \mu(n_1 + n_2), \quad (3.39)$$

then $\hat{Z}_t(n_1, n_2)$ is constant between jumps of N_t, and that, denoting by T_k the kth jump of N_t,

$$\hat{Z}_{T_k}(n_1, n_2) = \hat{Z}_{T_k-}(n_1, n_2) \frac{\lambda_1(n_1 - 1, n_2) 1(n_1 > 0)}{\mu(k)}$$

$$+ \hat{Z}_{T_k-}(n_1, n_2) \frac{\lambda_2(n_1, n_2 - 1) 1(n_2 > 0)}{\mu(k)} \quad (3.40)$$

for all $n_1 \geq 0$, $n_2 \geq 0$ such that $n_1 + n_2 = k$. Otherwise, if $n_1 + n_2 \neq k$, then $\hat{Z}_{T_k}(n_1, n_2) = 0$.

References

[1] Brémaud, P. (1975) La méthode des semi-martingales en filtrage lorsque l'observation est un processus ponctuel, in *Séminaire Proba X*, Université de Strasbourg, Lect. Notes in Math. **511**, Springer, Berlin, pp. 1–18.
[2] Brémaud, P. (1975) Estimation de l'état d'une file d'attente et du temps de panne d'une machine par la méthode des semi-martingales, *Adv. Appl. Probab.* **7**, pp. 845–863.
[3] Brémaud, P. (1978) Streams of a $M/M/1$ feedback queue in equilibrium, Zeitschrift für Wahrscheinlichkeitstheorie **45**, 1, pp. 21–33.
[4] Davis, M. H. A., Kailath, T., and Segall, A. (1975) Nonlinear filtering with counting observations, IEEE Transactions **IT-21**, pp. 143–150.
[5] Fujisaki, M., Kallianpur, G., and Kunita, H. (1972) Stochastic differential equations for the non-linear filtering problem, Osaka J. Math. **9**, pp. 19–40.
[6] Grigelionis, B. (1973) On nonlinear filtering theory and absolute continuity of probability measures corresponding to stochastic processes, *2nd Proc. Japan–USSR Symposium on Probability Theory*, Lect. Notes in Math., **330**, Springer, Berlin, pp. 80–94.
[7] Kailath, T. (1968) An innovations approach to least squares estimation, Part I: Linear filtering in additive white noise, IEEE Transactions **AC-13**, pp. 646–655.
[8] Snyder, P. (1972) Filtering and detection for doubly stochastic Poisson processes, IEEE Transactions **IT-18**, pp. 97–102.
[9] Van Schuppen, J. (1977) Filtering, prediction and smoothing for counting process observations, a martingale approach, SIAM J. Appl. Math. **32**, pp. 552–570.
[10] Wong, E. (1971) *Stochastic Processes in Information and Dynamical Systems*, McGraw-Hill, New York.

SOLUTIONS TO EXERCISES, CHAPTER IV

E1. Since $t \to Z_t = \exp\{iuQ_t\}$ is a step function,

$$Z_t = Z_0 + \sum_{0 < s \le t} \Delta Z_s. \tag{1}$$

Now $\Delta Z_t = \Delta \exp\{iuQ_t\} = \exp\{iuQ_{t-}\} - \exp\{iuQ_{t-}\}$ is a null quantity except at the jumps of Q_t. The jumps of Q_t fall into two categories: the upward jumps ($Q_t = Q_{t-} + 1$), which occur at an arrival ($\Delta A_t = 1$), and the downward jumps ($Q_t = Q_{t-} - 1$), which occur at a departure ($\Delta D_t = -1$). Thus

$$\Delta Z_t = \Delta A_t (\exp\{iu(Q_{t-} + 1)\} - \exp\{iuQ_{t-}\})$$
$$+ \Delta D_t (\exp\{iu(Q_{t-} - 1)\} - \exp\{iuQ_t\})$$
$$= (e^{iu} - 1)Z_{t-} \Delta A_t + (e^{-iu} - 1)Z_{t-} \Delta D_t. \tag{2}$$

Hence

$$Z_t = Z_0 + (e^{iu} - 1)\int_0^t Z_{s-}\, dA_s + (e^{-iu} - 1)\int_0^t Z_{s-}\, dD_s, \tag{3}$$

or

$$Z_t = Z_0 + (e^{iu} - 1)\int_0^t Z_{s-}\lambda_s\, ds + (e^{-iu} - 1)\int_0^t Z_{s-}\mu_s 1(Q_s > 0)\, ds + m_t, \tag{4}$$

where m_t is defined in the statement of the exercise.

E2. By the definition of a transition semigroup,

$$P_s f(X_t) = E[f(X_{t+s})|\sigma(X_t)] = E[f(X_{t+s})|\mathscr{F}_t)] \tag{1}$$

(where the last equality is an avatar of the $\mathscr{F}_t$-markovian property of X_t). Thus

$$m_t = P_T f(X_0) - P_{T-t} f(X_t) = E[f(X_T)|\mathscr{F}_0] - E[f(X_T)|\mathscr{F}_t], \tag{2}$$

and therefore m_t is an $\mathscr{F}_t$-martingale, mean 0.

E3. For all $t \geq 0$, $X_{t \wedge \tau}$ admits the representation

$$X_{t \wedge \tau} = X_0 + \int_0^t 1(s \leq \tau) f_s \, ds + \int_0^t 1(s \leq \tau) \, dm_s, \tag{1}$$

which is of the form required for the application of the theory. The observation history is now taken to be $\mathscr{O}_t = \mathscr{F}_{t \wedge \tau}^N = \sigma(N_{s \wedge \tau}, s \in [0, t])$ (see III, T3). But $N_{t \wedge \tau}$ admits the $\mathscr{F}_t^N$-intensity $\hat{\lambda}_t 1(t \leq \tau)$, as can be readily checked (taking into account the hypothesis that τ is an $\mathscr{F}_t^N$-stopping time), and therefore $N_{t \wedge \tau}$ admits the $\mathscr{O}_t$-intensity $\hat{\lambda}_t 1(t \leq \tau)$ (see II, T14). Also observe that the $\mathscr{O}_t$-projection of $1(t \leq \tau) f_t$ in the sense of T1, Equation 1.11, is $1(t \leq \tau) \hat{f}_t$. Hence

$$E[X_{t \wedge \tau}|\mathscr{F}_{t \wedge \tau}^N] = E[X_0] + \int_0^t 1(s \leq \tau) \hat{f}_s \, ds + \int_0^t K_s^\tau (dN_{s \wedge \tau} - \hat{\lambda}_s 1(s \leq \tau) \, ds), \tag{2}$$

or

$$E[X_{t \wedge \tau}|\mathscr{F}_{t \wedge \tau}^N] = E[X_0] + \int_0^{t \wedge \tau} \hat{f}_s \, ds + \int_0^{t \wedge \tau} K_s^\tau (dN_s - \hat{\lambda}_s \, ds), \tag{3}$$

where K_s^τ is some $\mathscr{F}_{t \wedge \tau}^N$-predictable process to be computed from (1.15) and (1.13) [with X_t, λ_t, $\hat{\lambda}_t$, N_t replaced respectively by $X_{t \wedge \tau}$, $\hat{\lambda}_t 1(t \leq \tau)$, $\hat{\lambda}_t 1(t \leq \tau)$, $N_{t \wedge \tau}$]. This computation leads to

$$K_t^\tau = K_t 1(t \leq \tau). \tag{4}$$

Therefore the case where τ is of the form $\tau \wedge t$ is proven. For passage to the general case, first observe that

$$E[X_{t \wedge \tau}|\mathscr{F}_{t \wedge t}^N] = E[X_\tau 1(\tau \leq t)|\mathscr{F}_{t \wedge t}^N] + E[X_t 1(t < \tau)|\mathscr{F}_{t \wedge t}^N]. \tag{5}$$

Now, by Appendix A2, T7, $\{t < \tau\} = \{\tau \wedge t < \tau\}$ is $\mathscr{F}_{t \wedge t}^N$-measurable, so that

$$E[X_t 1(t < \tau)|\mathscr{F}_{t \wedge t}^N] = E[X_t|\mathscr{F}_{t \wedge t}^N] 1(t < \tau). \tag{6}$$

But X_t is bounded, and therefore $E[X_t|\mathscr{F}_{t \wedge t}^N]$ is bounded. Consequently, τ being finite,

$$\lim_{t \uparrow \infty} E[X_t 1(t < \tau)|\mathscr{F}_{t \wedge t}^N] = 0. \tag{7}$$

By definition of conditional expectation, if $A \in \mathscr{F}_\tau^N$,

$$E[1_A X_\tau 1(\tau \leq t)] = E[1_A E[X_\tau 1(\tau \leq t)|\mathscr{F}_\tau^N]]. \tag{8}$$

By Appendix A2, T7, $A \cap \{\tau \leq t\} = A \cap \{\tau \leq \tau \wedge t\}$ belongs to $\mathscr{F}_{t \wedge t}^N$, so that

$$E[1_A X_\tau 1(\tau \leq t)] = E[1_A E[X_\tau|\mathscr{F}_{t \wedge t}^N] 1(\tau \leq t)]$$
$$= E[1_A E[X_\tau 1(\tau \leq t)|\mathscr{F}_{t \wedge t}^N]] \tag{9}$$

Solutions to Exercises, Chapter IV

(use the fact that $\{\tau \leq t\} = \{\tau \leq \tau \wedge t\} \in \mathscr{F}^N_{\tau \wedge t}$). A being arbitrary in $\mathscr{F}^N_\tau$, and $E[X_\tau 1(\tau \leq t)|\mathscr{F}^N_{\tau \wedge t}]$ being $\mathscr{F}^N_\tau$-measurable (by Appendix A2, T8, since $\tau \wedge t \leq \tau$), it follows from (8) and (9) that

$$E[X_\tau 1(\tau \leq t)|\mathscr{F}^N_{\tau \wedge t}] = E[X_\tau|\mathscr{F}^N_\tau]1(\tau \leq t). \tag{10}$$

Since τ is finite, it follows from (10) that

$$\lim_{t \uparrow \infty} E[X_\tau 1(\tau \leq t)|\mathscr{F}^N_{\tau \wedge t}] = E[X_\tau|\mathscr{F}^N_\tau]. \tag{11}$$

Hence, from (5), (7), and (11),

$$\lim_{t \uparrow \infty} E[X_{\tau \wedge t}|\mathscr{F}^N_{\tau \wedge t}] = E[X_\tau|\mathscr{F}^N_\tau]. \tag{12}$$

The proof is terminated by recalling that

$$E[X_{\tau \wedge t}|\mathscr{F}^N_{\tau \wedge t}] = \hat{Z}_{\tau \wedge t} \tag{13}$$

and observing that

$$\lim_{t \uparrow \infty} \hat{Z}_{\tau \wedge t} = \hat{Z}_\tau. \tag{14}$$

E4. We proceed in essentially the same manner as in E1:

$$Z_t(n) = Z_0(n) + \sum_{0 < s \leq t} (Z_s(n) - Z_{s-}(n))$$

$$= \sum_{0 < s \leq t} (Z_s(n) - Z_{s-}(n)) \Delta A_s + \sum_{0 < s \leq t} (Z_s(n) - Z_{s-}(n)) \Delta D_s$$

$$= \int_0^t (Z_{s-}(n - 1)1(n > 0) - Z_{s-}(n)) \, dA_s$$

$$+ \int_0^t (Z_{s-}(n + 1) - Z_{s-}(n)1(n > 0)) \, dD_s. \tag{1}$$

Therefore

$$Z_t(n) = Z_0(n) + \int_0^t (Z_s(n - 1)1(n > 0) - Z_s(n))\lambda_s \, ds$$

$$+ \int_0^t (Z_s(n + 1) - Z_s(n)1(n > 0))\mu_s 1(Q_s > 0) \, ds$$

$$+ M_t^A(n) + M_t^D(n). \tag{2}$$

Observing that $Z_s(n)1(Q_s > 0) = Z_s(n)1(n > 0)$, we see that

$$\int_0^t (Z_s(n - 1)1(n > 0) - Z_s(n))\lambda_s \, ds$$

$$+ \int_0^t (Z_s(n + 1) - Z_s(n)1(n > 0))\mu_s 1(Q_s > 0) \, ds$$

$$= \int_0^t (Z_s(n - 1)\lambda_s 1(n > 0) - Z_s(n)(\lambda_s + \mu_s 1(n > 0))$$

$$+ Z_s(n + 1)\mu_s) \, ds. \tag{3}$$

E5. Solution:

$$\hat{Z}_t(n) = P[Q_0 = n] + \int_0^t \{\hat{Z}_s(n-1)\lambda(n-1)1(n>0) - \hat{Z}_s(n)(\lambda + \mu u_s 1(n>0))$$
$$+ \hat{Z}_s(n+1)\mu u_s\} ds$$
$$+ \int_0^t \left(\frac{\hat{Z}_{s-}(n+1)}{1 - \hat{Z}_{s-}(0)} - \hat{Z}_{s-}(n)\right)(dD_s - \mu u_s(1 - Z_s(0)) ds).$$

[Observe that $\sum_{j>0} \mu(s, j)\hat{Z}_{s-}(j) = \sum_{j>0} \mu u_s \hat{Z}_{s-}(j) = \mu u_s(1 - \hat{Z}_{s-}(0))$.]

E6. We apply the general theory, and compute K_t^N according to (1.15) and (1.13). It is known that

$$\Psi_{2,t}^N(n) = \hat{Z}_{t-}^N(n). \tag{1}$$

Also (same type of calculations as for the proof of T2)

$$\Psi_{1,t}^N(n)\hat{h}_t^N = \frac{Z_t(n)h_t^N dt dP}{dt dP}, \tag{2}$$

where h_t^N is the $\mathscr{F}_t^N$-intensity of N_t. Now since

$$h_t^F = \mu p 1(Q_t > 0),$$
$$h_t^E = \lambda + \mu p 1(Q_t > 0), \tag{3}$$
$$h_t^S = \mu 1(Q_t > 0),$$

it follows from (1) that

$$\Psi_{1,t}^F(n) = \frac{\hat{Z}_{t-}^F(n)1(n>0)}{1 - \hat{Z}_{t-}^F(0)},$$

$$\Psi_{1,t}^E(n) = \frac{\hat{Z}_{t-}^E(n)(\lambda + \mu p 1(n>0))}{\lambda + \mu p(1 - \hat{Z}_{t-}^E(0))}, \tag{4}$$

$$\Psi_{1,t}^S(n) = \frac{\hat{Z}_{t-}^S(n)1(n>0)}{1 - \hat{Z}_{t-}^S(0)}.$$

In order to compute the last term $\Psi_{3,t}^N(n)$ using (1.13), one must evaluate $\sum_{0 < s \le t} C_s \Delta m_s(n) \Delta N_s$ for $N = F, E, S$ successively. Since

$$\Delta m_s(n) = (Z_{s-}(n+1) - Z_{s-}(n)1(n>0)) \Delta D_s$$
$$+ (Z_{s-}(n-1)1(n>0) - Z_{s-}(n)) \Delta A_s$$

and

$$\Delta F_s \Delta D_s = 0, \qquad \Delta F_s \Delta A_s = 0,$$
$$\Delta E_s \Delta D_s = 0, \qquad \Delta E_s \Delta A_s = \Delta A_s, \tag{5}$$
$$\Delta S_s \Delta D_s = \Delta D_s, \qquad \Delta S_s \Delta A_s = 0,$$

Solutions to Exercises, Chapter IV

we obtain

$$C_s \Delta m_s(n) \Delta F_s = 0,$$
$$C_s \Delta m_s(n) \Delta E_s = (Z_{s-}(n-1)1(n>0) - Z_{s-}(n)) \Delta A_s, \quad (6)$$
$$C_s \Delta m_s(n) \Delta S_s = (Z_{s-}(n+1) - Z_{s-}(n)1(n>0)) \Delta D_s.$$

Hence, using the definitions of the $\mathscr{F}_t^N$-intensity of $N_t (N = F, E, S)$,

$$E\left[\sum_{s \leq t} C_s \Delta m_s(n) \Delta F_s\right] = 0,$$

$$E\left[\sum_{s \leq t} C_s \Delta m_s(n) \Delta E_s\right] = E\left[\int_0^t C_s(Z_s(n-1)1(n>0) - Z_s(n))\lambda \, ds\right], \quad (7)$$

$$E\left[\sum_{s \leq t} C_s \Delta m_s(n) \Delta S_s\right] = E\left[\int_0^t C_s(Z_s(n+1) - Z_s(n)1(n>0)) \right.$$
$$\left. \times \mu(1-p)1(Q_s > 0) \, ds\right],$$

where the equalities must hold for all $t \geq 0$ and all nonnegative $\mathscr{F}_t^N$-predictable processes $C_t (N = F, E, S$ successively). Each of the right-hand sides of the equalities in (7) should be equated to

$$E\left[\int_0^t C_s \Psi_{3,s}^N(n) \hat{h}_s^N \, ds\right], \quad N = F, E, S \text{ successively}. \quad (8)$$

The rest of the proof is purely computational.

E7. This is just Ito's differentiation rule (Appendix A3, T11) applied to $F(x) = e^{iux}$.

E8. We have to show that the gain corresponding to Z_t^u is

$$K_t = \frac{\widehat{(Z_t^u - \hat{Z}_t^u)(\lambda_t - \hat{\lambda}_t)}}{\hat{\lambda}_t} \quad dN_t(\omega) \, P(d\omega)\text{-a.e.,} \quad (1)$$

the $\widehat{}$ referring to predictable projections in the following sense:

$$E\left[\int_0^t \widehat{(Z_s^u - \hat{Z}_s^u)(\lambda_s - \hat{\lambda}_s)} C_s \, ds\right] = E\left[\int_0^t (Z_s^u - \hat{Z}_s^u)(\lambda_s - \hat{\lambda}_s) C_s \, ds\right] \quad (2)$$

for all nonnegative $\mathscr{F}_t^N$-predictable processes C_t. Now

$$E\left[\int_0^t \hat{Z}_s^u(\lambda_s - \hat{\lambda}_s) C_s \, ds\right] = 0; \quad (3)$$

this follows from the fact that $\hat{\lambda}_t$ is the $\mathscr{F}_t^N$-predictable version of the $\mathscr{F}_t^N$-intensity of N_t, and therefore

$$E\left[\int_0^t \lambda_s C_s \hat{Z}_s^u \, ds\right] = E\left[\int_0^t \hat{\lambda}_s C_s \hat{Z}_s^u \, ds\right] \quad (4)$$

for all nonnegative predictable processes C_t; recall that $\hat{Z}^u_{t-}$ is $\mathscr{F}^N_t$-predictable. Therefore we find

$$E\left[\int_0^t (Z^u_s - \hat{Z}^u_s)(\lambda_s - \hat{\lambda}_s)C_s\, ds\right] = E\left[\int_0^t Z^u_s(\lambda_s - \hat{\lambda}_s)C_s\, ds\right], \quad (5)$$

and

$$(Z^u_t - \hat{Z}^u_t)(\lambda_t - \hat{\lambda}_t) = (\Psi_{1,t} - \Psi_{2,t})\hat{\lambda}_t. \quad (6)$$

But $\Psi_{3,t} \equiv 0$ [comment (α) after the statement of T4], and therefore the equality (1) is proven.

E10. This is just a matter of interpretation of the formal symbol $\widehat{\Delta N_t\, \Delta N_t}$. We do this along the lines of (3.31) and (3.32), i.e., $\widehat{\Delta N_t\, \Delta N_t}$ is the $\mathscr{F}^N_t$-predictable process Ψ_t satisfying

$$E\left[\int_0^t C_s \Psi_s\, ds\right] = E\left[\sum_{0 < s \le t} C_s \Delta N_s\, \Delta N_s\right] \quad (1)$$

for all $\mathscr{F}^N_t$-predictable processes C_t which are bounded; and clearly, since $\Delta N_t\, \Delta N_t = \Delta N_t$, we have $\Psi_t = \hat{\lambda}_t$, by the definition of intensity.

E11. A particular case of R4.

E12. Solution:

$$\hat{X}_t = E[X_0] + \int_0^t a(s)\hat{X}_s\, ds$$

$$+ \int_0^t \frac{[\widehat{X^{+2}_{s-}} - (\widehat{X^+_{s-}})^2] - [\widehat{X^{-2}_{s-}} - (\widehat{X^-_{s-}})^2]}{\hat{X}^+_{s-} + \hat{X}^-_{s-}} (dN_s - (\widehat{X^+_s} + \widehat{X^-_s})\, ds). \quad (1)$$

E13. The observed point process N_t has the $\mathscr{F}_t$-intensity $\lambda(N_t(1), N_t(2)) = \lambda_1(N_t(1), N_t(2)) + \lambda_2(N_t(1), N_t(2))$ and the $\mathscr{F}^N_t$-predictable intensity $\hat{\lambda}_t = \hat{\lambda}_t(1) + \hat{\lambda}_t(2)$, where

$$\hat{\lambda}_t(i) = \sum_{n_1, n_2} \lambda_i(n_1, n_2)\hat{Z}_{t-}(n_1, n_2). \quad (1)$$

Let $Z_t(n_1, n_2) = 1(N_t(1) = n_1, N_t(2) = n_2)$. Then

$$Z_t(n_1, n_2) = Z_0(n_1, n_2) + \int_0^t \{Z_s(n_1 - 1, n_2)1(n_1 > 0)\lambda_1(n_1 - 1, n_2)$$

$$+ Z_s(n_1, n_2 - 1)1(n_2 > 0)\lambda_2(n_1, n_2 - 1)$$

$$- Z_s(n_1, n_2)\lambda(n_1, n_2)\}\, ds$$

$$+ \int_0^t (Z_{s-}(n_1 - 1, n_2)1(n_1 > 0) - Z_{s-}(n_1, n_2))$$

$$\times (dN_s(1) - \lambda_1(N_s(1), N_s(2))\, ds)$$

$$+ \int_0^t (Z_{s-}(n_1, n_2 - 1)1(n_2 > 0) - Z_{s-}(n_1, n_2))$$

$$\times (dN_s(2) - \lambda_2(N_s(1), N_s(2))\, ds) \quad (2)$$

Solutions to Exercises, Chapter IV

and

$$\hat{Z}_t(n_1, n_2) = \hat{Z}_0(n_1, n_2) + \int_0^t \{\hat{Z}_s(n_1 - 1, n_2)1(n_1 > 0)\lambda_1(n_1 - 1, n_2)$$
$$+ \hat{Z}_s(n_1, n_2 - 1)1(n_2 > 0)\lambda_2(n_1, n_2 - 1)$$
$$- \hat{Z}_s(n_1, n_2)\lambda(n_1, n_2)\} \, ds$$
$$+ \int_0^t K_s(n_1, n_2)(dN_s - \hat{\lambda}_s \, ds). \quad (3)$$

One finds

$$K_t(n_1, n_2) = -\hat{Z}_{t-}(n_1, n_2)$$

$$+ \frac{\hat{Z}_{t-}(n_1 - 1, n_2)1(n_1 > 0)\lambda_1(n_1 - 1, n_2) + \hat{Z}_{t-}(n_1, n_2 - 1)1(n_2 > 0)\lambda_2(n_1, n_2 - 1)}{\hat{\lambda}_t}.$$
$$(4)$$

In particular, if $\lambda_1(n_1, n_2) + \lambda_2(n_1, n_2) = \mu(n_1 + n_2)$, then $\hat{\lambda}_t = \mu(N_{t-})$ and therefore the Lebesgue integral in (2) vanishes. What remains is

$$\hat{Z}_t(n_1, n_2) = \hat{Z}_0(n_1, n_2) + \int_0^t K_s(n_1, n_2)(dN_s - \mu(N_s) \, ds), \quad (5)$$

from which (3.40) follows. For the last assertion, just observe that

$$\hat{Z}_{T_k}(n_1, n_2) = P[N_{T_k}(1) = n_1, N_{T_k}(2) = n_2 | \mathscr{F}_{T_k}^N], \quad N_{T_k}(1) + N_{T_k}(2) = N_{T_k} = k.$$
$$(6)$$

CHAPTER V

Flows in Markovian Networks of Queues

1. Single Station: The Historical Results and the Filtering Method
2. Jackson's Networks
3. Burke's Output Theorem for Networks
4. Cascades and Loops in Jackson's Networks
5. Independence and Poissonian Flows in Markov Chains
References
Solutions to Exercises, Chapter V

1. Single Station: The Historical Results and the Filtering Method

Output Theorem for Birth-and-Death Processes Without Feedback or Feedforward

Consider a queueing process in standard form:

$$Q_t = Q_0 + A_t - D_t \qquad (1.1)$$

which satisfies the following integrability conditions:

$$E[Q_0] < \infty; \qquad E[A_t] < \infty, \quad t \geq 0, \qquad (1.2)$$

and admits the $\mathscr{F}_t^Q$-parameters λ_{Q_t} and μ_{Q_t}, where the λ_n's and μ_n's are nonnegative real numbers; it is supposed that

$$\mu_n > 0, \quad \lambda_n > 0, \qquad n \geq 0. \qquad (1.3)$$

1. Single Station: The Historical Results and the Filtering Method

This is a necessary and sufficient condition for the irreducibility of the chain. The following fact is recalled in A1, §8: under the condition

$$\sum_{n=0}^{\infty} \pi_n < \infty, \tag{1.4}$$

where

$$\pi_0 = 1; \quad \pi_n = \frac{\lambda_0 \lambda_1 \cdots \lambda_{n-1}}{\mu_1 \mu_2 \cdots \mu_n}, \quad n \geq 1, \tag{1.5}$$

there exists a unique *equilibrium distribution*:

$$p(n) = \frac{\pi_n}{\sum_{i=0}^{\infty} \pi_i}, \quad n \geq 0, \tag{1.6}$$

that is to say, if one starts from the equilibrium distribution p, then at any future time, Q_t will also be distributed according to p. In 1956, Burke [8] stated and proved a result which triggered a considerable amount of research in queueing theory.

Burke's Output Theorem. Suppose Q_t is a $M/M/s$ queueing process: Poisson arrivals, exponential service times independent and identically distributed, and s servers. It is of the general type described at the beginning of the section, with the following specifications:

$$\lambda_n \equiv \lambda, \tag{1.7}$$

$$\mu_n = \mu \inf(n, s). \tag{1.8}$$

Suppose also that one operates in light traffic:

$$\rho = \lambda/\mu s < 1 \tag{1.9}$$

(the quantity ρ is called the *traffic intensity*). The inequality (1.9) is equivalent, under (1.7) and (1.8), to (1.4), and therefore guarantees the existence of an equilibrium distribution in this particular case. In the above situation, if the queue is in equilibrium, then the output process D_t is poissonian, with the same intensity as the input, and moreover

$$P[Q_t = n | D_t] = p(n), \quad n \geq 0, \quad t \geq 0, \tag{1.10}$$

that is to say, the state Q_t at time t is independent of the number of departures during $(0, t]$. This result, which is the original result of Burke, can be extended using the filtering formula of IV, R5.

T1 Theorem ([8, 22]; see also [4]). *Under the hypotheses (1.2), (1.3), and (1.4), the following four statements are equivalent:*

(i) $P[Q_0 = n] = p(n), n \geq 0$, and $\lambda_n \equiv \lambda, n \geq 0$.
(ii) $P[Q_\tau = n] = p(n), n \geq 0$, for all finite $\mathscr{F}_t^D$-stopping times τ.
(iii) $P[Q_t = n | \mathscr{F}_t^D] = p(n), t \geq 0$, P-a.s.
(iv) $P[Q_\tau = n | \mathscr{F}_\tau^D] = p(n), n \geq 0$, P-a.s. for all finite $\mathscr{F}_t^D$-stopping times τ.

Moreover, any of the four above conditions implies that the output process D_t is a Poisson process with the intensity $\lambda = \sum_{i=1}^{\infty} \mu_i p(i)$.

Comments. Statement (iv) is (apparently) stronger than (iii): it says in particular that if we place ourselves at a departure, we do not get additional information as to the distribution of the queue; the implication (iii) $\Rightarrow$ (i) says in particular that one cannot hope to extend Burke's result in the direction where the input parameters would vary with n. Facing this negative statement, one could still expect some improvement, namely that in $P[Q_t = n] = p(n)$, $t \geq 0, n \geq 0$ (which does not depend upon the condition $\lambda_n \equiv$ constant), the deterministic time t could be replaced by a $\mathscr{F}_t^D$-stopping time. The implication (ii) $\Rightarrow$ (i) destroys such hopes.

Before proceeding to the proof of T1 via filtering, let us give the elegant proof of Reich for the poissonian character of the output process when $\lambda_n \equiv \lambda$.

REICH'S PROOF [22]. Consider a birth-and-death process Q_t defined on $(-\infty, +\infty)$ and at equilibrium; it is then *reversible*, since the detailed-balance equalities of Appendix A1, §8 hold: $p(n)\lambda_n = p(n+1)\mu_{n+1}, n \geq 0$. In consequence, the law of the sequence of downward jumps of $(Q_t, t \geq 0)$ is identical to the law of the sequence of downward jumps of $(Q_{-t}, t \geq 0)$. But the downward jumps of $(Q_{-t}, t \geq 0)$ are the upward jumps of $(Q_t, t \leq 0)$. If $\lambda_n \equiv \lambda$, the sequence of upward jumps of $(Q_t, t \leq 0)$ is poissonian with intensity λ, and therefore, the sequence of downward jumps of $(Q_t, t \geq 0)$ is poissonian with intensity λ. □

We will now give an alternative proof of T1 based on the filtering formulas of Chapter IV. Later, we will state various generalizations of T1, without proofs because the proofs are all similar to the following one:

PROOF OF T1. Under the conditions prevailing in this section, IV, (2.16) takes the form

$$\hat{Z}_t(n) = P(Q_0 = n) + \int_0^t \mathscr{K}(\hat{Z}_s, n) \, ds + \int_0^t K(\hat{Z}_{s-1}, n) \, d\hat{I}_s \quad (1.11)$$

where

$$K(x, n) = x(n-1)1(n > 0)\lambda_{n-1} - x(n)(\lambda_n + \mu_n 1(n > 0)) + x(n+1)\mu_{n+1},$$

$$\mathscr{K}(x, n) = -x(n) + \frac{\mu_{n+1} x(n+1)}{\sum_{i=1}^{\infty} \mu_i x(i)}, \quad (1.12)$$

$$\hat{I}_t = D_t - \int_0^t \left(\sum_{i=1}^{\infty} \mu_i \hat{Z}_s(i) \right) ds.$$

1. Single Station: The Historical Results and the Filtering Method

Equivalently, if we denote by τ_k the kth jump time of D_t,

$$\hat{Z}_t(n) = \hat{Z}_{\tau_k}(n) + \int_{\tau_k}^t \left(\mathcal{K}(\hat{Z}_s, n) - \left(\sum_{i=1}^\infty \mu_i \hat{Z}_s(i) \right) K(\hat{Z}_s, n) \right) ds,$$

$$t \in [\tau_k, \tau_{k+1}), \quad (1.13)$$

$$\hat{Z}_{\tau_k}(n) = \frac{\mu_{n+1} \hat{Z}_{\tau_k-}(n+1)}{\sum_{i=1}^\infty \mu_i \hat{Z}_{\tau_k-}(i)}.$$

(i) $\Rightarrow$ (iii): Consider the deterministic system of equations in the unknown $x_t = (x_t(n), n \geq 0)$:

$$x_t(n) = x_{t_k}(n) + \int_{t_k}^t \left(\mathcal{K}(x_s, n) - \left(\sum_{i=1}^\infty \mu_i x_s(i) \right) K(x_s, n) \right) ds, \quad t \in [t_k, t_{k+1}),$$

$$x_{t_k}(n) = \frac{\mu_{n+1} x_{t_k-}(n+1)}{\sum_{i=1}^\infty \mu_i x_{t_k-}(i)}, \quad (1.14)$$

for all $n \geq 0$, all $k \geq 0$, and all $t \in [t_k, t_{k+1})$, where $(t_k, k \geq 0)$ is an ordered sequence of $[0, \infty]$, without accumulation points at finite distance, and such that $t_0 = 0$. The system (1.14) admits, as can be readily checked, a stationary solution $x_t = p = (p(n), n \geq 0)$ where $p(n)$ is given by (1.5) and (1.6), with $\lambda_n \equiv \lambda$. But the right-continuous version $Z_t(n)$ of $P[Q_t = n | \mathscr{F}_t^D]$ has, outside a P-null set, trajectories that satisfy (1.14) where for each trajectory ω, $(t_k, k \geq 0)$ is replaced by $(\tau_k(\omega), k \geq 0)$. Therefore in order to prove (iii), it is sufficient to show that $x_t = p$ is the *unique* solution of (1.14) starting from the initial condition $x_0 = p$. A little thought convinces us that it is enough to prove uniqueness of the stationary solution $x_t = p$ for the system

$$x_t(n) = p(n) + \int_0^t \left(\mathcal{K}(x_s, n) - \left(\sum_{i=1}^\infty \mu_i x_s(i) \right) K(x_s, n) \right) ds, \quad (1.15)$$

since the second equation of (1.14), which provides the initial conditions at $t = t_k$, does not actually change the initial conditions: if $x_{t_k-} = p$, then $x_{t_k} = p$. But the system (1.15) is a differential equation on the Banach space B of sequences $x = (x(n), n \geq 0)$ endowed with the norm $\|x\|_B = \sup_{n>0} |x(n)| + \sum_{n>0} \mu_n |x(n)|$. One can check that this equation is locally lipschitzian around the solution $x_t = p$, and this suffices to prove uniqueness of the solution starting at $x_0 = p$.

(iii) $\Rightarrow$ (iv): Immediate from IV, T3, which states that $P[Q_\tau = n | \mathscr{F}_\tau^D] = \hat{Z}_\tau(n)$ for all finite $\mathscr{F}_t^D$-stopping times τ.

(iv) $\Rightarrow$ (ii): Obvious.

(iv) $\rightarrow$ (i): By hypothesis (iv), $P[Q_t = n | \mathscr{F}_t^D]$, as a process, admits a right-continuous version p_n. It also admits another right-continuous version $\hat{Z}_t(n)$ given by the filtering equations (1.13). Therefore $\hat{Z}_t(n)$ and p_n are indistinguishable processes, i.e., outside a P-null set, $\hat{Z}_t(\omega) = p$, $t \geq 0$. Thus, we have

$$\hat{Z}_{\tau_1 -} = \hat{Z}_{\tau_1} = p \quad P\text{-a.s.} \tag{1.16}$$

From (1.13) it follows that, P-a.s., for all $n \geq 0$,

$$\hat{Z}_{\tau_1}(n) = \frac{\mu_{n+1} \hat{Z}_{\tau_1 -}(n+1)}{\sum_{i=1}^{\infty} \mu_i \hat{Z}_{\tau_1 -}(i)},$$

and therefore, on account of (1.6), for all $n \geq 0$,

$$p(n) = \frac{\mu_{n+1} p(n+1)}{\sum_{i=1}^{\infty} \mu_i p(i)}. \tag{1.17}$$

Now, by assumption, the queue is in equilibrium, which imples that the stationary Kolmogorov equations are verified:

$$p(n+1)\mu_{n+1} = p(n)\lambda_n. \tag{1.18}$$

Hence $p(n)\lambda_n = p(n)(\sum_{i=1}^{\infty} \mu_i p(i))$. The announced result follows, since $p(n) > 0$.

(ii) → (iii): For each $n \geq 0$, the process $Z_t(n) = 1(Q_t = n)$ satisfies the semimartingale equation [IV, (2.5)]:

$$Z_t(n) = 1(Q_0 = n) + \int_0^t \mathcal{K}(Z_s, n) \, ds + m_t(n). \tag{1.19}$$

For any bounded $\mathcal{F}_t^Q$-stopping time τ, it follows by optional stopping (I, T2) applied to $m_t(n)$ that

$$E[Z_\tau(n)] = P[Q_\tau = n] = P[Q_0 = n] + E\left[\int_0^\tau \mathcal{K}(Z_s, n) \, ds\right]. \tag{1.20}$$

Now if τ is a bounded $\mathcal{F}_t^p$-stopping time, condition (ii) and the above equality yield

$$E\left[\int_0^\tau \mathcal{K}(Z_s, n) \, ds\right] = 0. \tag{1.21}$$

But $1(t \leq \tau)$ is adapted to $\mathcal{F}_t^p$. Therefore, from (1.21),

$$E\left[\int_0^\tau \mathcal{K}(\hat{Z}_{s-}, n) \, ds\right] = 0. \tag{1.22}$$

Since the mappings $(t, \omega) \to 1(t \leq \tau(\omega))$ generate the $\mathcal{F}_t^p$-predictable σ-field on $(0, \infty) \times \Omega$ when τ spans the set of bounded $\mathcal{F}_t^p$-stopping times (I, E9), it follows from (1.21) and the monotone class theorem (Appendix A1, T4) that $\mathcal{K}(\hat{Z}_{t-}, n, \omega) = 0$, $P(d\omega) \, dt$-a.e; in other words $(Z_{t-}(n, \omega), n \geq 0)$ satisfies the stationary backward Kolmogorov equations. Now there is a unique solution to these equations which is a probability distribution; hence $\hat{Z}_{t-}(\omega) = p$, $P(d\omega)$ dt-a.e., and therefore $\hat{Z}_t = p$, $t \geq 0$, P-a.s., as announced.

It now remains to prove the poissonian character of the output process. The $\mathscr{F}_t^Q$-intensity of D_t being $\mu_{Q_t} 1(Q_t > 0) = \sum_{i>0} \mu_i 1(Q_t = i)$, it follows that $\sum_{i>0} \mu_i \hat{Z}_t(i)$ is an $\mathscr{F}_t^D$-intensity of D_t. Since $\hat{Z}_t(i) = p(i)$, $i \geq 0$, this $\mathscr{F}_t^D$-intensity is a constant, equal to $\sum_{i>0} \mu_i p_i$, and therefore, by Watanabe's theorem (II, T5), D_t is a Poisson process. □

The Output of Feedforward Queues in Equilibrium

The output theorem of Burke has a variant discovered by Hadidi [12] and concerning the situation where part of the poissonian input stream is granted instantaneous service; such customers shunt the service facility, as is shown in Figure 1.

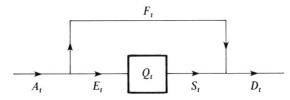

Figure 1   Feedforward queue.

Let $\mathscr{F}_t$ be a history common to all the flows, and suppose that the $\mathscr{F}_t$-intensities of these flows are given by the following table:

Exogenous input	A_t	λ
Service-facility input	E_t	$\lambda 1(Q_t \leq K - 1)$
Service-facility output	S_t	$\mu 1(Q_t > 0)$
Final output	D_t	$\mu 1(Q_t > 0) + \lambda 1(Q_t = K)$
Feedforward flow	F_t	$\lambda 1(Q_t = K)$

where K is a positive integer, and λ and μ are nonnegative real numbers. In other words: if $Q_0 < K$, the queue behaves as an $M/M/1$ queue with parameters λ and μ as long as $Q_t < K$. When it reaches K, the exogenous input A_t is diverted into F_t, this until Q_t returns to a value strictly less than K.

E1 Exercise [10]. Find the equilibrium distribution of Q_t assuming $\rho = \lambda/\mu < 1$. Is Q_t independent from $\mathscr{F}_t^S$ at equilibrium? Show that at equilibrium D_t is poissonian.

$M/M/1$ Feedback Queues in Equilibrium: Analysis of the Streams

Consider a feedback queue of the type described in II, §5. Figures 2 and 3 recall the definitions of the various streams. In Figure 3, note that $P[X_n = 1] = 1 - P[X_n = 0] = p$. Also the decision of recycling a customer at time τ_n is independent of the past of Q_t and F_t at time τ_n-.

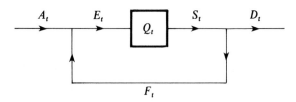

Figure 2   Feedback queue.

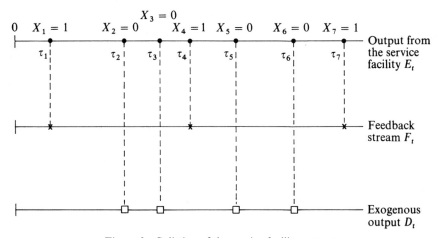

Figure 3   Splitting of the service-facility output.

The state process Q_t is a birth-and-death process with parameters $\lambda_n = \lambda$, $\mu_n = \mu(1 - p)$, where $\lambda > 0$, $\mu > 0$, and $0 < p < 1$. Under the light traffic condition

$$\rho = \frac{\lambda}{\mu(1 - p)} < 1, \qquad (1.23)$$

there exists an equilibrium distribution:

$$p(n) = \rho^n(1 - \rho), \qquad n \geq 0. \qquad (1.24)$$

Under the hypothesis (1.23), if the queue is at equilibrium, i.e.

$$P[Q_t = n] = p(n), \qquad n \geq 0, \qquad (1.25)$$

then we know by T1 that the output D_t is poissonian with rate λ. We now address the question of the poissonian character of the feedback flow F_t and of the flows E_t and S_t.

R2 Result [10, 5]. *In an $M/M/1$ feedback queue in equilibrium which is nontrivial (i.e $\lambda > 0, \mu > 0, 0 < p < 1$) the only poissonian flows are the exogenous input and the exogenous output.*

1. Single Station: The Historical Results and the Filtering Method

This is an apparently paradoxical result, especially for the feedback flow F_t: indeed, in the case where $p = \frac{1}{2}$ for instance, F_t and D_t are obtained from S_t by exactly the same random selection procedure, and if D_t is homogeneously poissonian, why should not F_t be homogeneously Poissonian, "by symmetry"? In fact, the roles of D_t an F_t are not symmetric: D_t is a flow that leaves the system, whereas a customer being recycled at time t might influence the intensity of the feedback stream later on. The mathematical proof of the nonpoissonian character of F_t, E_t, and S_t is based on the filtering formulas of IV, §2.

PROOF OF R2. *The feedback stream F_t is not homogeneously poissonian.* Suppose it were. It would then have an $\mathscr{F}_t^F$-intensity that is a deterministic constant, say ξ. But $\mu p(1 - \hat{Z}_{s-}^F(0))$ is also an $\mathscr{F}_t^F$-intensity for F_t. Therefore, by II, T12, $\mu p(1\hat{Z}_{t-}^F(0, \omega)) = \xi$, $\xi\, dt\, P(d\omega)$-a.e., or, since ξ must be strictly positive, $\mu p(1 - \hat{Z}_{t-}^F(0, \omega) = \xi$, $dt\, P(d\omega)$-a.e., which implies, by the left continuity of $t \to \hat{Z}_{t-}^F(0)$ and of $t \to \xi$,

$$\hat{Z}_t^F(0) = \frac{\mu p - \xi}{\mu p}, \qquad t \geq 0, \quad n \geq 0, \qquad P\text{-a.s.} \tag{1.26}$$

Let τ_1 be the first jump time of F_t. Then

$$\hat{Z}_{\tau_1}^F(0, \omega) = \frac{\mu p - \xi}{\mu p}. \tag{1.27}$$

By IV, (2.20), $\hat{Z}_{\tau_1}^F(0, \omega) = \hat{Z}_{\tau_1-}^F(0, \omega) + K_{\tau_1}^F(0, \omega)$. By IV, (2.22), which gives the explicit form of the gain K_t^F, $K_{\tau_1}^F(0, \omega) = -\hat{Z}_{\tau_1-}^F(0, \omega)$. The last two equalities imply

$$\hat{Z}_{\tau_1}^F(0, \omega) = 0. \tag{1.28}$$

Therefore, by (1.27) and (1.28), we have $(\mu p - \xi)/\mu p = 0$, which implies, by (1.26), that $\hat{Z}_t^F(0) = 0, t \geq 0$, P-a.s. Taking expectations, we see that $P[Q_t = 0] = 0$, a contradiction, since $P[Q_t = 0] = \lambda/\mu(1 - p) > 0$.

The input E_t of the service facility E_t is not homogeneously poissonian. We omit the proof, since it is completely analogous to the previous one.

The output S_t of the service facility is not homogeneously poissonian. Suppose it were, and let ξ be the necessarily strictly positive $\mathscr{F}_t^S$-intensity of S_t. Since $1 - \hat{Z}_{t-}^S(0)$ is another predictable $\mathscr{F}_t^S$-intensity for S_t, it follows by II, T12 that $\mu(1 - \hat{Z}_t^S(0)) = \xi, t \geq 0$, P-a.s. Therefore $\hat{Z}_t^S(0)$ is a constant, necessarily equal to $P[Q_t = 0] = p_0$:

$$\hat{Z}_t^S(0) = p_0, \qquad t \geq 0. \tag{1.29}$$

If we denote by τ_1 the first jump of S_t, then it follows from IV, Equation (2.22) that $\hat{Z}_{\tau_1}^S(0) = (1 - p)\hat{Z}_{\tau_1-}^S(1)/[1 - \hat{Z}_{\tau_1-}^S(0)]$. Hence by (1.29), $Z_{\tau_1-}^S(1) = p_0(1 - p_0)/(1 - p)$. By the representation result for predictable process (III, E4),

$$\hat{Z}_{t-}^S(1)1(t < \tau_1) = f(t)1(t < \tau_1) \quad P\text{-a.s.,} \qquad t \geq 0, \tag{1.30}$$

for some measurable function f. The representation (1.30) being granted, it follows from the fact that τ_1 has a distribution which is equivalent to Lebesgue measure on $[0, \infty)$ that $f(t) = p_0(1 - p_0)/(1 - p)$. Hence, since $\hat{Z}^S_{t-}(1) = \hat{Z}^S_t(1)$ on $t < \tau_1$,

$$\hat{Z}^S_t(1)1(t < \tau_1) = \frac{p_0(1 - p_0)}{1 - p} 1(t < \tau_1) \quad P\text{-a.s.,} \qquad t \geq 0. \qquad (1.31)$$

In particular, $\hat{Z}^S_0(1) = P[Q_0 = 1] = p_0(1 - p_0)/(1 - p)$. But by (1.24), $P[Q_0 = 1] = p = p_0(1 - p_0)$. The last two equalities contradict one another if $0 < p < 1$, as we assumed.

E2 Exercise. Extend the above results to show that the flows F_t, E_t, and S_t are not poissonian, homogeneous or not (here we have only disproved the "homogeneous poissonian" conjecture).

E3 Exercise. Extend the above results to a birth-and-death feedback queue with a recycling probability $r(n)$ [i.e. depending on the number n of customers present in the system; in the case treated above, $r(n) \equiv p$].

Reversibility in M/M/1 Feedback Queues. Consider the $M/M/1$ feedback queue Q_t described in the present section [in particular the traffic intensity $\rho = \lambda/\mu(1 - p)$ is strictly less than one], and define J_t to be the telegraph process associated to the feedback flow F_t: J_t takes two values, 0 and 1, and switches from one value to the other whenever F_t jumps. The process $X_t = (Q_t, J_t)$ is a Markov chain, with an infinitesimal transition matrix given by the following diagram:

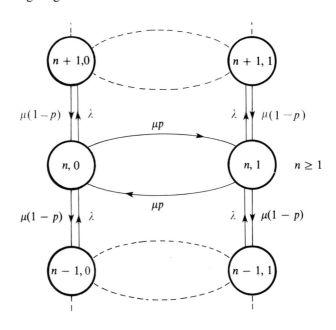

It is easy to check that X_t admits a stationary distribution given by
$$p(n, 0) = p(n, 1) = \tfrac{1}{2}\rho^n(1 - \rho)$$
and that the detailed-balance equations of Appendix A1, §8 are verified. Thus X_t is reversible provided it starts from the equilibrium distribution.

The set of discontinuity points of E_t is the union of the set of discontinuity points of J_t and the set of points of increase of Q_t. The set of discontinuity points of S_t is the union of the set of discontinuity points of J_t and the set of points of decrease of Q_t (or of increase of Q_{-t}). Reich's argument based on these observations shows that the sequence of times of arrival in the service facility and the *reversed* sequence of times of departure from the service facility have the same law.

The two sequences are stationary sequences; therefore their interoccurence times are identically distributed, with the same law for both sequences (see [18]). Moreover the correlation structure of the two sequences is the same. But one *cannot* infer from reversibility that the two sequences have the same law. This would be true, however, if one could prove independently that one of the sequences is a renewal sequence. However, it was shown in [18] that the sequence of departure times from the service facility is *not* renewal.

2. Jackson's Networks

Such networks consist of m stations which are interconnected: some or all of the customers leaving (say) station i are fed into (say) station j or leave the network. (See Figure 4.)

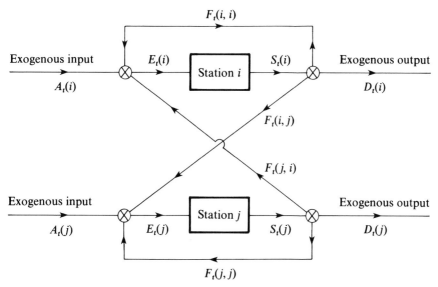

Figure 4 Feedback flows in a network.

The input $E_t(i)$ to the service faciltity i consists of the sum of the exogenous input $A_t(i)$ plus all the feedback flows $F_t(j, i)$ from all stations including station i:

$$E_t(i) = A_t(i) + \sum_{j=1}^{m} F_t(j, i). \tag{2.1}$$

The output $S_t(i)$ from the service facility admits a similar decomposition:

$$S_t(i) = D_t(i) + \sum_{j=1}^{m} F_t(i, j), \tag{2.2}$$

where $D_t(i)$ is the exogenous output from station i. All the processes $A_t(i)$, $D_t(i)$, $F_t(i, j)$, $1 \leq i, j \leq m$, are of course point processes. They have no common jumps and are nonexplosive.

The number of customers in station i at time t is

$$Q_t(i) = Q_0(i) + E_t(i) - S_t(i). \tag{2.3}$$

Let $\bar{\lambda}_i, \mu_i, 1 \leq i \leq m$, be nonnegative real numbers, and let $r_{ij}, 1 \leq i, j \leq m$, also be nonnegative real numbers satisfying in addition

$$\begin{aligned} 0 \leq r_{ij} \leq 1, \quad & 1 \leq i, j \leq m, \\ \sum_{j=1}^{m} r_{ij} \leq 1, \quad & 1 \leq i \leq m. \end{aligned} \tag{2.4}$$

Also define for each i, $1 \leq i \leq m$,

$$r_i = 1 - \sum_{j=1}^{m} r_{ij}. \tag{2.5}$$

Finally let s_i, $1 \leq i \leq m$ be a set of strictly positive integers. Let $\mathscr{H}_t$ be the smallest history common to all the processes of the networks:

$$\mathscr{H}_t = \bigvee_{i,j=1}^{m} (\mathscr{F}_t^Q(i) \vee \mathscr{F}_t^F(i, j)). \tag{2.6}$$

Table 1 describes the dynamics of the network.

Table 1

Flows	$\mathscr{H}_t$-intensities	Terminology	
$A_t(i)$	$\bar{\lambda}_i$	Exogenous input to station i	
$S_t(i)$	$\mu_i \inf(s_i, Q_t(i))$	Output from service facility i	(2.7)
$F_t(i, j)$	$\mu_i \inf(s_i, Q_t(i)) r_{ij}$	Feedback from station i to station j	
$D_t(i)$	$\mu_i \inf(s_i, Q_t(i)) r_i$	Exogenous output from station i	
$E_t(i)$	$\bar{\lambda}_i + \sum_{j=1}^{m} \mu_j \inf(s_j, Q_t(j)) r_{ji}$	Input to service facility i	

2. Jackson's Networks

The parameters of Jackson's networks are to be interpreted as follows:

m = number of stations in the network,

s_i = number of servers at stations i,

$1/\mu_i$ = average service time of each of the s_i independent exponential servers,

$\bar{\lambda}_i$ = input rate at station i, from outside the network,

r_{ij} = routing probability from station i to j,

r_i = probability that a customer leaving station i leaves the network.

A Construction of Jackson's Network

The construction to follow is a generalization of the construction of II, (5.13). It is very useful for the mathematical analysis of Markovian networks of all kinds: jacksonian, processor sharing, etc.

Let $Q_0(i)$, $1 \le i \le m$, be integer-valued nonnegative random variables, and let $A_t(i)$, $\bar{D}_t^k(i)$, $\bar{F}_t^k(i,j)$, $1 \le i, j \le m$, $1 \le k \le s_i$, be Poisson processes with respective intensities $\bar{\lambda}_i$, $\mu_i r_i$, $\mu_i r_{ij}$. Suppose that the above random variables and processes are mutually independent. Then define for all $1 \le i \le m$

$$Q_t(i) = Q_0(i) + A_t(i) + \sum_{j=1}^{m} \sum_{k=1}^{s_j} \int_0^t 1(Q_{s-}(j) \ge k) \, d\bar{F}_s^k(j, i)$$

$$- \sum_{j=1}^{m} \sum_{k=1}^{s_i} \int_0^t 1(Q_{s-}(i) \ge k) \, d\bar{F}_s^k(i, j)$$

$$- \sum_{k=1}^{s_i} \int_0^t 1(Q_{s-}(i) \ge k) \, d\bar{D}_s^k(i). \tag{2.8}$$

E4 Exercise. Define for all $1 \le i \le m$, $1 \le j \le m$

$$F_t(i, j) = \sum_{k=1}^{s_i} \int_0^t 1(Q_{s-}(i) \ge k) \, d\bar{F}_s^k(i, j),$$

$$D_t(i) = \sum_{k=1}^{s_i} \int_0^t 1(Q_{s-}(i) \ge k) \, d\bar{D}_s^k(i). \tag{2.9}$$

Let $Q_t(i)$ and $A_t(i)$ be as above, and let $E_t(i)$ and $S_t(i)$ be defined by (2.1) and (2.2). Let the history $\mathcal{H}_t$ be given by (2.6). Show that the table of $\mathcal{H}_t$-intensities (2.7) is appropriate.

Remark. In (2.8) the i-terms of the two sums $\sum_{j=1}^{m}$ cancel each other: they correspond to the feedback $F_t(i, i)$, which does not influence the trajectories of $Q_t(i)$.

From (2.8) the markovian property of $Q_t = (Q_t(i), \ldots, Q_t(m))$ is clear, since the processes $\bar{A}$, $\bar{D}$, $\bar{F}$ are independent Poisson processes. Before computing the infinitesimal characteristics of Q_t, we introduce a few symbols:

$$S = N_+^m,$$
$$n = (n_1, \ldots, n_m), \qquad (2.10)$$
$$e_i = i\text{th canonical vector of } R^m.$$

By (2.8) and (2.9) the only possible transitions are

$$n \to n + e_i, \qquad n \in S,$$
$$n \to n - e_j, \qquad n \in S, \ n_j > 0,$$
$$n \to n + e_i - e_j, \qquad n \in S, \ n_j > 0.$$

E5 Exercise. Recalling the notation $N_t(n, n')$ for the number of transitions $n \to n'$ in the interval $(0, t]$,

$$N_t(n, n + e_i) = \int_0^t 1(Q_{s-} = n) \, dA_s(i), \qquad n \in S,$$

$$N_t(n, n - e_j) = \int_0^t 1(Q_{s-} = n) \, dD_s(j), \qquad n \in S, \ n_j > 0, \qquad (2.11)$$

$$N_t(n, n + e_i - e_j) = \int_0^t 1(Q_{s-} = n) \, dF_t(j, i), \qquad n \in S, \ n_j > 0, \ i \neq j.$$

Show that the above point processes have the respective $\mathscr{F}_t^Q$-intensities

$$1(Q_t = n)\bar{\lambda}_i,$$
$$1(Q_t = n)\mu_j \inf(Q_s(j), s_j) \, r_j, \qquad (2.12)$$
$$1(Q_t = n)\mu_j \inf(Q_s(j), s_j) \, r_{ji}.$$

These intensities are the *same as if* Q_t had the infinitesimal characteristics

$$q_{n, n+e_i} = \bar{\lambda}_i, \qquad n \in S,$$
$$q_{n, n-e_j} = \mu_j \inf(n_j, s_j) \, r_j, \qquad n \in S, \ n_j > 0, \qquad (2.13)$$
$$q_{n, n+e_j} = \mu_j \inf(n_j, s_j) \, r_{ji}, \qquad n \in S, \ n_j > 0, \ i \neq j.$$

The law of the process Q_t is entirely characterized by the joint law of the counting processes in (2.11). In turn, by the uniqueness theorem (III, T8) the joint law of these counting processes is entirely characterized by their intensities with respect to their smallest common history $\mathscr{F}_t^Q$. Hence, (2.13) indeed gives the infinitesimal characteristics of Q_t.

Let the functional $\mathscr{K}$ be defined by the forward Kolmogorov equation for $P[Q_t = n] = p_t(n)$:

$$\frac{dp_t(n)}{dt} = \mathscr{K}(p_t, n). \qquad (2.14)$$

2. Jackson's Networks

We are going to exhibit a solution, unique in certain circumstances, of the equilibrium equation

$$\mathcal{K}(p, n) = 0 \qquad (2.15)$$

where $p = (p(n), n \in S)$ is a probability distribution.

The Traffic Equations and the Product Theorem

L3 Lemma. *Assume that the Jackson's network above is in equilibrium, i.e. $P[Q_t = n] = p(n), t \geq 0, n \in S$. Then the system of equations in the unknown λ_i, $1 \leq i \leq m$,*

$$\lambda_i = \bar{\lambda}_i + \sum_{j=1}^{m} r_{ji} \lambda_j, \qquad 1 \leq i \leq m, \qquad (2.16)$$

admits a solution such that

$$\frac{\lambda_i}{\mu_i s_i} \leq 1, \qquad 1 \leq i \leq m. \qquad (2.17)$$

Moreover, since i is not a sink (i.e. if either $r_i > 0$ or there exists $j \neq i$ such that $r_{ij} > 0$), then

$$\frac{\lambda_i}{\mu_i s_i} < 1. \qquad (2.18)$$

The equations (2.16) constitute what is called the *flow-conservation equations* or the *traffic equations*.

PROOF. $Q_t(i) = Q_0(i) + E_t(i) - S_t(i)$ and $E[Q_t(i)] = E[Q_0(i)]$; therefore $E[S_t(i)] = E[E_t(i)]$. But $E_t(i) = A_t(i) + \sum_{j=1}^{m} F_t(j, i)$, and, in view of the table of intensities, $E[A_t(i)] = \bar{\lambda}_i t$ and $E[F_t(j, i)] = r_{ji} E[S_t(j)]$; thus $\lambda_i = (1/t) E[S_t(i)]$ solves (2.16). But $E[S_t(i)] = \mu_i \sum_{k=1}^{s_i} P[Q_0(i) \geq k] t$; therefore we have

$$\lambda_i = \mu_i \left(\sum_{k=1}^{s_i} P[Q_0(i) \geq k] \right) \leq \mu_i s_i. \qquad (2.19)$$

We will now show that since i is not a sink, $P[Q_0(i) = 0] > 0$ so that $\sum_{k=1}^{s} P[Q_0(i) \geq k] < s_i$ and therefore (2.18) is verified.

We will prove this in the case where $s_i = 1$, for the sake of notational convenience. We then denote $\bar{F}_t^1(i, j)$ by $\bar{F}_t(i, j)$. We treat the case where for some $j \neq i$, $r_{ij} > 0$ (the case $r_i > 0$ is similar). If $P[Q_0(i) = 0]$ were null, then, necessarily, for some $k \geq 1$, $P[Q_0(i) = k] > 0$. Let t be any strictly positive time, and let τ_n be the ordered sequence of jumps of the processes $A, \bar{D}, \bar{F}$. Define

$$B = \{\Delta F_{\tau_p}(i, j) = 1, 1 \leq p \leq k \text{ and } \tau_k \leq t < \tau_{k+1}\},$$
$$C = \{Q_0(i) = k\}.$$

Because the A, $\bar{D}$, $\bar{F}$ are independent Poisson processes and $\bar{F}_t(i,j)$ has a strictly positive intensity $\mu_i r_{ij}$, we have $P(B) > 0$. Also $P(C) > 0$. Since Q_0 is independent of A, $\bar{D}$, $\bar{F}$, we have $P(B \cap C) = P(B)P(C) > 0$. Now $\{Q_t(i) = 0\} \supset B \cap C$, and therefore $P[Q_0(i) = 0] = P[Q_t(i) = 0] > 0$. $\square$

We will now introduce a few topological definitions:

D4 Definition. The network is said to be *open* if for each i there exists a sequence $k_1, k_2, \ldots, k_q, j$ with

$$r_{ik_1} r_{k_1 k_2} \cdots r_{k_{q-1} k_q} r_{k_q j} r_j > 0. \tag{2.20}$$

It is said to be *exogenously supplied* if for each i there exists a sequence $j, k_q, k_{q-1}, \ldots, k_1$ with

$$\bar{\lambda}_j r_{jk_q} r_{k_q k_{q-1}} \cdots r_{k_2 k_1} r_{k_1 i} > 0. \tag{2.21}$$

L5 Lemma. *If the network is open, the traffic equations have a unique solution.*

PROOF. To the routing matrix $R = \{r_{ij}, 1 \leq i, j \leq m\}$, associate a homogeneous Markov chain $(X_n, n \geq 0)$ with state space $\{0, 1, 2, \ldots, m\}$ and transition matrix

$$P = \begin{pmatrix} 1 & 0 \cdots 0 \\ \hline r_1 & \\ \vdots & R \\ r_m & \end{pmatrix}. \tag{2.22}$$

By (2.20), the set of states $\{1, 2, \ldots, m\}$ is a set of transient states. Therefore, in particular, the series

$$I + R + \cdots + R^n$$

converges to $(I - R)^{-1}$. Now (2.6) reads, in vector form, $\lambda(I - R) = \bar{\lambda}$. Thus it admits a unique solution $\lambda = \bar{\lambda}(I - R)^{-1}$.

L6 Lemma. *If the network is open and exogenously supplied, the Markov chain Q_t is irreducible.*

In particular, all states are transient, or all states are null recurrent, or all states are positive recurrent.

PROOF. It is enough to show that for all $1 \leq i \leq m$ and all $n \in S$, n communicates with $n + e_i$, i.e. $n \leadsto n + e_i$ and $n + e_i \leadsto n$. The proof is in the same vein as the proof of (2.18) in L3, and is left to the reader.

Remark. The proof of $n + e_i \leadsto n$ relies only on the hypothesis that the network is open, where as the proof of $n \leadsto n + e_i$ is based solely on the hypothesis that it is exogenously supplied. Therefore, if we only assume that

2. Jackson's Networks

the network is open, we can still prove the weaker result: there is only one recurrent class. Indeed, for all $n \in S$, $n \rightsquigarrow 0$.

The stations which are not exogenously supplied are not interesting in an open network at equilibrium, since clearly for any such station i, $P[Q_t(i) = 0] = 1$.

We will now proceed to the famed Jackson's product theorem. First we introduce the following notation: let λ, μ be nonnegative numbers, $\mu > 0$, and let s be a strictly positive integer. Assuming that $\lambda/\mu s < 1$, we denote by $p(n, \lambda, \mu, s)$ the stationary probability distribution of an $M/M/s$ queue with parameters λ and μ.

T7 Theorem (Jackson's Product Theorem [13]). *Let Q_t be a Jackson's network with parameters given by (2.7). Assume that it is open and exogenously supplied. There exists a unique stationary distribution if and only if the (unique) solution of the traffic equations (2.16) satisfies the light-traffic condition*

$$\frac{\lambda_i}{\mu_i s_i} < 1, \quad 1 \le i \le m. \tag{2.23}$$

The stationary distribution $(p(n), n \in S)$ is then given by the product

$$p(n) = \prod_{i=1}^{m} p(n_i, \lambda_i, \mu_i, s_i). \tag{2.24}$$

In particular, at each instant $t \ge 0$, the random variables $Q_t(i)$, $1 \le i \le m$, are mutually independent.

PROOF The uniqueness of the equilibrium distribution follows from L6. The necessity of (2.23) was proven in L3.

The sufficiency of (2.23): (2.23) allows us to define $p(n)$ by (2.24) for all $n \in S$. But $(p(n), n \in S)$ is a probability distribution, and it can be checked by computation that it verifies $\mathcal{K}(p, n) = 0$ for all $n \in S$. Hence it is the equilibrium distribution. □

Remark. The formula (2.24) tells us that each station i has an equilibrium distribution which is the same as for a $M/M/s_i$ queue with parameters λ_i and μ_i. Thus, in equilibrium, station i behaves "as if" it were fed by a poissonian stream of intensity λ_i. This is in fact very misleading: the "internal" flows [i.e. all flows besides $A_t(i)$ and $D_t(i)$] are poissonian, at equilibrium, only under special topological circumstances to be identified in §4. For the time being, we will prove a positive result: namely that the output streams $D_t(i)$, $1 \le i \le m$, of a jacksonian network at equilibrium are independent Poisson processes with intensities λ_i, $1 \le i \le m$, respectively. As a matter of fact, a more general result will be proven, which extends T1.

3. Burke's Output Theorem for Networks

The situation is that of Figure 4 and Equations (2.1), (2.2), and (2.3), and $\mathcal{H}_t$ is defined by (2.6). The table of $\mathcal{H}_t$-intensities is now:

Flows	$\mathcal{H}_t$-intensities
$A_t(i)$	$\lambda_i(Q_t)$
$D_t(i)$	$\mu_i(Q_t)r_i(Q_t)$
$F_t(i,j)$	$\mu_i(Q_t)r_{ij}(Q_t)$
$E_t(i)$	$\lambda_i(Q_t) + \sum_{j=1}^{m} r_{ji}(Q_t)\mu_j(Q_t)$
$S_t(i)$	$\mu_i(Q_t)$

where the λ_i's and μ_i's are nonnegative bounded functions of n and $\mu_i(n) = 0$ if $n_i = 0$; the r_i's and r_{ij}'s are also non-negative functions of n satisfying in addition

$$r_i(n) + \sum_{j=1}^{m} r_{ij}(n) = 1. \tag{3.1}$$

The interpretation of the r_i's and r_{ij}'s in terms of routing probabilities is the same as for jacksonian networks.

In the case of the jacksonian network described in §2, we have $\lambda_i(n) = \bar{\lambda}_i$, $\mu_i(n) = \mu_i \inf(n_i, s_i)$, and $r_{ij}(n) = r_{ij}$. But the above setting need not be interpreted in terms of a network of interconnected stations. For instance if $F_{ij}(t) \equiv 0$ for all $1 \leq i, j \leq m$, we find the following situation: one station, m classes of customers, or equivalently, m stations *in parallel*, with interaction of service. Also a mixture of the two above situations fits the general model under consideration: the index i can represent a station–class pair, and the flows $F_t(i, j)$ do not merely embody the transfer from one station to the other, but also the change of class for a given customer.

T8 Theorem (General Output Theorem [19, 6]). *In the situation just described, denote N_+^m by S and let G_i, $1 \leq i \leq h$, be h disjoint subsets of $\{1, 2, \ldots, m\}$. Define*

$$\mathcal{G}_t = \sigma\left(\sum_{j \in G_i} D_s(j), 1 \leq i \leq h, 0 \leq s \leq t\right). \tag{3.2}$$

The following conditions are equivalent:

(a) *The system is in statistical equilibrium and the stationary probability distribution $p = (p(n), n \in S)$, where $p(n) = P[Q_t = n]$, satisfies*

$$p(n)\lambda_{G_i} = \sum_{j \in G_i} \mu_j(n + e_j)r_j(n + e_j)p(n + e_j), \quad 1 \leq i \leq h, \tag{3.3}$$

3. Burke's Output Theorem for Networks

where λ_G is defined for any subset G of $\{1, \ldots, m\}$ by

$$\lambda_G = \sum_{i \in G} \sum_{n \in S} \mu_i(n) r_i(n) 1(n_i > 0) p(n). \tag{3.4}$$

(b) For each $1 \leq i \leq h$ and each time $t \geq 0$, Q_t is independent of $\tilde{\mathscr{F}}_t^D(G_i) = \sigma(\sum_{j \in G_i} D_s(j), 0 \leq s \leq t)$.
(c) For each $1 \leq i \leq h$ and each finite $\mathscr{G}_t$-stopping time τ, Q_τ is independent of $\tilde{\mathscr{F}}_\tau^D(G_i)$.
(d) For each $1 \leq i \leq h$ and each finite $\mathscr{G}_t$-stopping time τ, we have $P[Q_\tau = n] = p(n)$, $n \in S$.

Moreover either of the two conditions (a) or (b) implies that $D_t(G_i) = \sum_{j \in G_i} D_t(j)$, $1 \leq i \leq h$, form independent Poisson processes with intensities λ_{G_i}, $1 \leq i \leq h$, respectively.

PROOF. *Equivalence of* (a), (b), (c) *and* (d): Same as the proof of T1. It is based on the following representation for $\hat{Z}_t(n) = P[Q_t = n | \mathscr{G}_t]$:

$$\hat{Z}_t(n) = \hat{Z}_0(n) + \int_0^t \mathscr{K}(\hat{Z}_s, n)\, ds$$

$$+ \sum_{i=1}^h \int_0^t K_i(\hat{Z}_{s-}, n)(dD_s(G_i) - \lambda_s(G_i)\, ds), \tag{3.5}$$

where

$$K_i(\hat{Z}_t, n) = -\hat{Z}_t(n) + \frac{\sum_{j \in G_i} \mu_j(n + e_j) r_j(n + e_j) \hat{Z}_t(n + e_j)}{\bar{\lambda}_t(G_i)}, \tag{3.6}$$

$$\bar{\lambda}_t(G_i) = \sum_{j \in G_i} \sum_{n \in S} \mu_i(n) r_i(n) 1(n_i > 0) \hat{Z}_t(n),$$

and $dp_t(n)/dt = \mathscr{K}(p_t, n)$ is the forward Kolmogorov equations for the Markov chain Q_t.

Poisson outputs: $\bar{\lambda}_t(G_i)$ is the $\mathscr{G}_t$-intensity of $D_t(G_i)$. Since $\hat{Z}_t(n) = p(n)$, $\bar{\lambda}_t(G_i) = \lambda_{G_i}$, a constant. The conclusion follows from Watanabe's multichannel theorem (II, T6).

Applications

Jackson's Network. Consider the network of §2, assumed at equilibrium. Take, in T8, $h = m$, $G_i = \{i\}$. The condition (3.3) reads, after division by r_i,

$$p(n)\left(\sum_{n \in S} \mu_i \inf(n_i, s_i) p(n)\right) = \mu_i \inf(n_i + 1, s_i) p(n + e_i). \tag{3.7}$$

But $\sum_{n \in S} \mu_i \inf(n_i, s_i) p(n) = (1/t) E[S_t(i)] = \lambda_i$ (see the proof of L3) and therefore (3.7) reduces to

$$p(n)\lambda_i = \mu_i \inf(n_i + 1, s_i) p(n + e_i). \tag{3.8}$$

This equality is granted by (2.24).

Therefore the outputs of a jacksonian network are independent Poisson processes (a result due to Kelly [14]).

Processor-Sharing Station. Suppose that the general network of T8 is not connected, i.e., $r_{ij}(n) \equiv 0$ for all $n \in S$, $1 \leq i, j \leq m$. In other words, the m stations are served in parallel. There is an interaction of service embodied by

$$\mu_i(n) = \mu \frac{n_i}{\sum_{j=1}^m n_j} \quad (=0 \text{ if } n = 0, \text{ by convention}). \tag{3.9}$$

Assume that the inputs are poissonian:

$$\lambda_i(n) = \lambda_i. \tag{3.10}$$

This mathematical model can also be interpreted in terms of a single station with m classes of customers; $Q_t(i)$ is then the number of customers of class i present in the system at time t. The different classes of customers share the processing potential μ in proportion to their ratio of occupation.

It is known (see [1] for instance, or check it directly on the Kolmogorov equations), that under the light-traffic condition

$$\sum_{i=1}^m \frac{\lambda_i}{\mu} < 1, \tag{3.11}$$

the equilibrium distribution exists and is given by

$$p(n) = \frac{p(0)}{(\sum_{j=1}^m n_j)!} \prod_{i=1}^m \rho_i^{n_i}(n_i)!, \tag{3.12}$$

where $\rho_i = \lambda_i/\mu$. We are going to check the condition (3.3) in the case $h = m$, $G_i = \{i\}$. It reads

$$p(n)\lambda_i = p(n + e_i)\mu \frac{n_i + 1}{\sum_{j=1}^m n_j + 1}, \tag{3.13}$$

and it is an immediate consequence of (3.12).

Thus the outputs of a processor-sharing station are independent Poisson processes (provided the inputs are also independent Poisson processes).

Processor-Sharing Network. We have K classes of customers and N stations. Define $m = N \times K$, and let Q_t be a network of the general type described at the beginning of the section, with a different meaning for the word "station," which now stands for a station–class pair.

The vector Q_t is a $N \times K$ vector

$$Q_t = (Q_t(1), \ldots, Q_t(N)), \tag{3.14}$$

where $Q_t(i)$ is the state at station i:

$$Q_t(i) = (Q_t^1(i), \ldots, Q_t^K(i)), \tag{3.15}$$

3. Burke's Output Theorem for Networks

$Q_t^l(i)$ being the number of customers of class l present at station i at time t. The notation

$$n = (n(1), \ldots, n(N)),$$
$$n(i) = (n^1(i), \ldots, n^K(i))$$

need not be explained. We define $e^l(i)$ to be the $(i(K-1)+l)$th canonical basis vector of $R^{N \times K}$.

In the "station–class" context, the routing probabilities are $r_{ij}^{lq}(n)$, where i and j are names of stations, and l and q are names of classes. The number $r_{ij}^{lq}(n)$ is to be interpreted as follows: the state of the network being n, $r_{ij}^{lq}(n)$ is the probability that a customer of class l leaving service facility i will be routed to station j and change his class from l to q. We assume that $r_{ij}^{lq}(n)$ does not depend upon n, and define

$$r_i^l = 1 - \sum_{q=1}^{K} \sum_{j=1}^{N} r_{ij}^{lq}. \tag{3.16}$$

At station–class (i, l), the service potential is given by functions $\mu_i^l(n)$. It is assumed that it is of the processor-sharing type, i.e.

$$\mu_i^l(n) = \mu_i \frac{n_i^l}{\sum_{l=1}^{K} n_j^l}. \tag{3.17}$$

Customers of class l entering the network at station i form a Poisson flow of intensity $\bar{\lambda}_i^l$. The table of intensities contains new notation which should be obvious:

Flow	$\mathcal{H}_t$-intensities
$A_t^l(i)$	$\bar{\lambda}_i^l$
$S_t^l(i)$	$\mu_i^l(Q_t)$
$D_t^l(i)$	$\mu_i^l(Q_t) r_i^l$
$F_t^{lq}(i, j)$	$\mu_i^l(Q_t) r_{ij}^{lq}$

The flow $E_t^l(i)$ is defined by

$$E_t^l(i) = A_t^l(i) + \sum_{j=1}^{N} \sum_{q=1}^{K} F_t^{ql}(j, i). \tag{3.18}$$

The traffic equations for such a network are

$$\lambda_i^l = \bar{\lambda}_i^l + \sum_{j=1}^{N} \sum_{q=1}^{K} r_{ji}^{ql} \lambda_j^q. \tag{3.19}$$

Assume that the light-traffic conditions are verified, i.e.

$$\sum_{l=1}^{K} \frac{\lambda_i^l}{\mu_i} < 1 \quad \text{for all } 1 \leq i \leq N. \tag{3.20}$$

Then (see [1]) there exists a stationary distribution given by:

$$p(n) = \prod_{i=1}^{N} p_i(n(i)), \tag{3.21}$$

where $p_i(n(i))$ is the marginal distribution $P[Q_t(i) = n(i)]$ at station i:

$$p_i(n(i)) = \frac{p_i(0)}{(\sum_{l=1}^{K} n^l(i))!} \prod_{l=1}^{K} (\rho_i^l)^{n^l(i)} (n^l(i))! \tag{3.22}$$

and

$$\rho_i^l = \frac{\lambda_i^l}{\mu_i^l}. \tag{3.23}$$

Here also, it can be shown, by an immediate application of T8, and in view of the above expression for the equilibrium distribution, that the flows $D_t^l(i), 1 \leq i \leq N, 1 \leq l \leq K$, are independent Poisson processes with respective intensities $\lambda_i^l, 1 \leq i \leq N, 1 \leq l \leq K$.

A Counterexample: Poisson Does Not Imply Independence

The Poisson property of the output streams is a consequence of the condition (3.3). A natural question is: does the converse hold, that is to say, if the output streams $D_t(G_i), 1 \leq i \leq m$, of T8 are poissonian, is condition (3.3) necessarily verified? Here is a simple counterexample.

In the model under consideration in T8, take $m = 2$, $\lambda_1(n) = \lambda_2(n) = \lambda$, $\mu_1(n) = \mu 1(n_1 > 0)$, $\mu_2(n) = \mu 1(n_2 > 0)$, $r_{12}(n) = r_{21}(n) = 0$. Thus we have two queues in parallel or two classes of customers with independent poissonian arrivals, a single server with potential μ, and absolute priority of class 1 over class 2.

Assume that $Q_t = (Q_t(1), Q_t(2))$ is in equilibrium. Clearly $Q_t(1)$ is an M/M/1 queue in equilibrium, so that $D_t(1)$ is, by Burke's output theorem T1, poissonian. We want to disprove

$$\lambda p(n_1, n_2) = \mu p(n_1 + 1, n_2), \quad n_1 \geq 0, \ n_2 \geq 0. \tag{3.24}$$

E6 Exercise. Show that if (3.24) is verified, then

$$\lambda p(n_1, n_2) = \mu 1(n_1 = 0) p(n_1, n_2 + 1), \quad n_1 \geq 0, \ n_2 \geq 0, \tag{3.25}$$

and deduce from (3.24) and (3.25) a contradiction with the existence of a stationary distribution $p(n)$.

4. Cascades and Loops in Jackson's Networks

Positive Results: Cascade Configurations

Tandem or Series Networks. Let us consider the simplest network besides the single-station network. It consists of two $M/M/1$ queues in tandem, as in Figure 5. The exogenous input $A_t(1)$ into station 1 is Poisson, intensity λ. There is no other exogenous input, and the output flow from service facility 1 forms the totality of the input flow into service facility 2. The

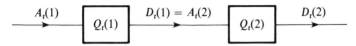

Figure 5   A tandem queue.

servers in stations 1 and 2 are independent exponential servers with service potentials μ_1 and μ_2 respectively. In other words, we are in the jacksonian framework prevailing in §2 with $\bar{\lambda}_1 = \lambda, \bar{\lambda}_2 = 0, r_{12} = 1, r_{21} = 0, s_1 = s_2 = 1$. The solution (λ_1, λ_2) to the flow conservation system is $\lambda_1 = \lambda, \lambda_2 = \lambda$. We assume that

$$\rho_1 = \lambda/\mu_1 < 1 \quad \text{and} \quad \rho_2 = \lambda/\mu_2 < 1, \tag{4.1}$$

so that there exists a stationary distribution given by

$$p(n_1, n_2) = (1 - \rho_1)\rho_1^{n_1}(1 - \rho_2)\rho_2^{n_2}. \tag{4.2}$$

By T8, $D_t(2)$ is poissonian with the intensity λ. Also the interstation flow $D_t(1) \equiv A_t(2)$ is poissonian. The tandem network is the archetype of cascade Jackson's networks, for which all interstation flows are poissonian, as we will soon see.

Before proceeding to the general cascade we will make a small digression to the effect of showing that there exist nonmarkovian queues which have the Poisson-input, Poisson-output property. This is related to a result of Finch {57} saying that if an $M/G/1$ queue has, at equilibrium, a poissonian output, then necessarily $G = M$, i.e., the queue is $M/M/1$. Sometimes, Finch's result is misquoted as "the only Poisson-input, Poisson-output queue is the birth-and-death process with $\lambda_n \equiv \lambda$." We will construct a nonmarkovian queue with the Poisson-input, Poisson-output property. Our example is rather artificial; in fact we will take the Jackson's tandem queue above, at equilibrium, and we will define Q_t to be the sum of $Q_t(1)$ and $Q_t(2)$, i.e. the total number of customers in the system.

Suppose, with a view to contradiction, that Q_t is Markov; since the corresponding input process $A_t = A_t(1)$ is Poisson (intensity λ) and the equilibrium distribution of Q_t is

$$P[Q_t = n] = \begin{cases} (1 - \rho_1)(1 - \rho_2) \dfrac{\rho_2^{n+1} - \rho_1^{n+1}}{\rho_2 - \rho_1} & \text{if } \rho_1 \neq \rho_2, \\ (n + 1)(1 - \rho)^2 \rho^n & \text{if } \rho_1 = \rho_2 = \rho, \end{cases} \quad (4.3)$$

the death parameters μ_n of the supposedly markovian process Q_t are easily computed from

$$\frac{P[Q_t = n]}{P[Q_t = 0]} = \begin{cases} \dfrac{\rho_2^{n+1} - \rho_1^{n+1}}{\rho_2 - \rho_1} = \dfrac{\lambda^n}{\mu_1 \cdots \mu_n} & \text{if } \rho_1 \neq \rho_2, \\ (n+1)\rho^n = \dfrac{\lambda^n}{\mu_1 \cdots \mu_n} & \text{if } \rho_1 = \rho_2 = \rho. \end{cases} \quad (4.4)$$

If Q_t is a birth-and-death process, its departure process has an $\mathscr{F}_t^Q$-intensity $\mu_{Q_t} 1(Q_t > 0)$. But since the departure process D_t of Q_t is $D_t(2)$, it admits the $\mathscr{F}_t^Q(1) \vee \mathscr{F}_t^Q(2)$-intensity $\mu_2 1(Q_t(2) > 0)$, and therefore the $\mathscr{F}_t^Q$-intensity $\mu_2(1 - P[Q_t(2) = 0 | \mathscr{F}_t^Q])$. A contradiction will arise because $\mu_{Q_t} 1(Q_t > 0) \neq \mu_2(1 - P[Q_t(2) = 0 | \mathscr{F}_t^Q])$.

To see this we must compute $P[Q_t(2) = 0 | \mathscr{F}_t^Q]$. More generally, we will compute

$$\hat{Z}_t(n_1, n_2) = P[Q(1) = n_1, Q_t(2) = n_2 | \mathscr{F}_t^Q]. \quad (4.5)$$

Since $\mathscr{F}_t^Q = \sigma(Q_0) \vee \mathscr{F}_t^A(1) \vee \mathscr{F}_t^D(2)$, the innovations theory of Chapter IV applies.

E7 Exercise. Show that

$$\hat{Z}_t(n_1, n_2) = \hat{Z}_0(n_1, n_2) + \int_0^t \mathscr{K}(\hat{Z}_s, n_1, n_2) \, ds$$

$$+ \int_0^t ((\hat{Z}_{s-}(n_1 - 1), n_2) 1(n_1 > 0) - \hat{Z}_{s-}(n_1, n_2))(dA_s - \lambda \, ds)$$

$$+ \int_0^t \left(-\hat{Z}_{s-}(n_1, n_2) + \frac{\hat{Z}_{s-}(n_1, n_2 + 1)}{1 - \hat{Z}_{s-}(0)} \right) (dD_s - \mu_2(1 - \hat{Z}_s(0)) \, ds), \quad (4.6)$$

where $\hat{Z}_t(n_2) = P[Q_t(2) = n_2 | \mathscr{F}_t^Q]$, and $\mathscr{K}(p, n_1, n_2) = 0$ is the forward Kolmogorov equation for $(Q_t(1), Q_t(2))$.

Summing (3.6) over $n_1 > 0$, we see that

$$\hat{Z}_t(n_2) = \hat{Z}_0(n_2) + \int_0^t f_s(n_2) \, ds$$

$$+ \int_0^t \left(-\hat{Z}_{s-}(n_2) + \frac{\hat{Z}_{s-}(n_2 + 1)}{1 - \hat{Z}_{s-}(0)} \right) (dD_s - \mu_2(1 - \hat{Z}_s(0)) \, ds), \quad (4.7)$$

4. Cascades and Loops in Jackson's Networks   145

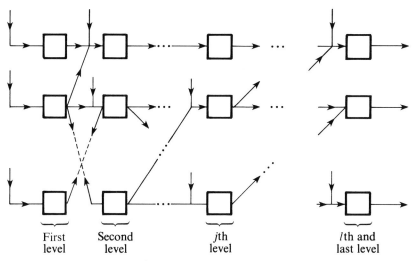

Figure 6 Cascade structure.

where $f_s(n_2) = \hat{Z}_s(n_2 - 1)1(n_2 > 0) + (\hat{Z}_s(0, n_2) - \hat{Z}_s(n_2))\mu_1 + (\hat{Z}_s(n_2 + 1) - Z_s(n_2)1(n_2 > 0))\mu_2$.

The fact to be noted is the absence of an integral with respect to dA_t in (4.7). Thus $\hat{Z}_t(0)$ does not jump with A_t, which it should do if $\mu_{Q_t}1(Q_t > 0) \equiv \mu_2(1 - Z_s(0))$. Hence the announced contradiction.

We now go back to the analysis of the flows in Jackson's network.

Cascade Networks. The positive result concerning the poissonian character of the "internal" flows in a Jackson tandem queue obviously extends to jacksonian networks with a cascade structure.

Rather than formalize the notion of cascade in words, we will use the illustration in Figure 6, where the number of stations in a level can vary with the level.

In a cascade structure all flows go from left to right and there are no flows between two stations in the same level. The tandem queue is a cascade network. The $M/M/1$ feedback queue is not a cascade network. The structure in Figure 7 (call it the "8") is not a cascade structure.

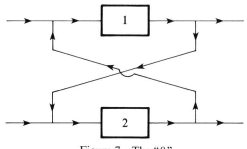

Figure 7 The "8".

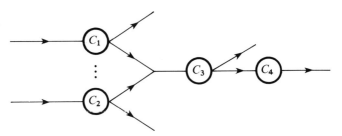

Figure 8   Inter-subnetwork cascade.

Although the above definition of a cascade structure does not satisfy any standard of mathematical precision, we are convinced that the reader—at any rate, any adept at graph theory—can provide himself with a rigorous definition, or, at least can feel in his heart what is meant by a cascade.

Subnetworks and Inter-Subnetwork Flows. We will continue in this mode of informal definitions and describe a result of Kelly [15], using words and pictures. Given any jacksonian network, we can define an equivalence between stations as follows: First we will say that station i is connected to station $j \neq i$ if there exist two ordered strings of stations

$$i, i_1, i_2, \ldots, i_n, j,$$

$$j, j_1, j_2, \ldots, j_l, i$$

such that $r_{ii_1} r_{i_1 i_2} \cdots r_{i_{n-1} i_n} r_{i_n j} r_{j j_1} r_{j_1 j_2} \cdots r_{j_{l-1} j_l} r_{j_l i} > 0$. In other words, i and j belong to a *loop*. This equivalence relation (biconnection) generates a partition of the stations into subnetworks $C_1, C_2, \ldots, C_k$, and these subnetworks form a cascade. (See Figure 8.) If we look in more detail, the arrows in and out any subnetwork in the above Figure 8 can be decomposed, as in Figure 9.

With the help of the result in T8 and of the forward induction method as applied in the case of tandem queues, it can be shown that all the flows

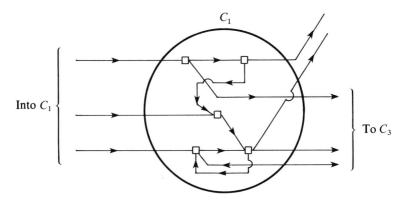

Figure 9   Inside a subnetwork.

4. Cascades and Loops in Jackson's Networks

participating in the arrows of Figure 8 are poissonian, and this is the content of Kelly's decomposition result.

Negative Results: Loop Configurations

As we have seen, cascade structures yield poissonian flows. We will show formally that as soon as there is a loop in a network (i.e. there exist two stations which are biconnected), one will automatically find that not all the streams in the loop are poissonian.

Before producing a formal proof, we will present heuristic argument in the case of the "8" structure of Figure 7, with ultrarapid servers in both queues, say $\mu_1 = \mu_2 = 1000$ (if the unit of time is the second, the average service time is one millisecond), and with very sparse arrivals, say $\lambda_1 = \lambda_2 = 1/2 \times 24 \times 3600$ (on the average one customer per day enters the network); the routing is $r_{12} = r_{21} = \frac{1}{2}$. A typical trajectory looks like this: a customer enters the network, say at about 10:00 A.M., at station 1. Because the service is so fast, he will get out of the service facility very rapidly, about one millisecond after (note also that the average number of customers waiting in station 1 or 2 is very close to 0, so that our lucky customer will be, most likely, served immediately). Now with probability $\frac{1}{2}$ the customer leaves the system, which then enjoys a long period of idleness (of 24 hours in the average, since arrivals are separated by an average time of 24 hours). With probability $\frac{1}{2}$ he goes to service facility 2, from which he exits very very shortly (say 1 millisecond) after, and with probability $\frac{1}{2}$ goes back to station 1, exits a few milliseconds later, and so on. The flow at the exit of service facility 1 looks like Figure 10 (the cross represents the time of arrival

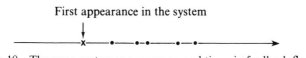

Figure 10 The same customer appears several times in feedback flows.

of the customer, the dots represent the passages of this customer at the exit of service facility 1). There is a cluster of dots after the cross, and the times between consecutive dots or between the cross and the first dot are of the order of a millisecond. Moreover, the number N of dots corresponding to the same customer follows a law which is easily computed (it is of the geometric type). These clusters of dots follow each arrival, so that the flow at the exit of the service facility of station 1 looks like Figure 11. The distance between two clusters is 24 hours in the average. Clearly this is not a Poisson process.

Figure 11 The clustered appearance of the feedback stream.

We now proceed to the statement and proof of the main result.

T9 Theorem [20]. *Let Q_t be the Jackson network of T7, at equilibrium. Let i and j be two stations such that there exist a set of stations $\{k_1, k_2, \ldots, k_p\}$ with*

$$r_{ij}r_{jk_1} \cdots r_{k_{p-1}k_p}r_{k_p i} > 0. \tag{4.8}$$

The feedback flow $F_t(i, j)$ is not poissonian.

PROOF [28]. We do the proof in the case $s_i = 1$, $1 \leq i \leq m$, for notational convenience.

Only the law of the process Q_t matters, so that we are free to use the construction of §2: for all $1 \leq i \leq m$,

$$Q_t(i) = Q_0(i) + A_t(i) + \sum_{j=1}^{m} \int_0^t 1(Q_{s-}(j) > 0) \, d\bar{F}_s(j, i)$$

$$- \sum_{j=1}^{m} \int_0^t 1(Q_{s-}(i) > 0) \, d\bar{F}_s(i, j) - \int_0^t 1(Q_{s-}(i) > 0) \, d\bar{D}_s(i), \tag{4.9}$$

where:

(α) the random variables $Q_0(i)$ and the processes $A_t(j)$, $\bar{D}_t(k)$, $\bar{F}_t(l, q)$ are mutually independent,
(β) $P[Q_0 = n] = p(n) = \prod_{i=1}^{m} \rho_i^{n_i}(1 - \rho_i)$,
(γ) $A_t(j)$, $\bar{D}_t(k)$, $\bar{F}_t(l, q)$ are Poisson processes with intensities $\bar{\lambda}_j$, $\mu_k r_k$, $\mu_l r_{lq}$ respectively.

We write (4.9) symbolically as

$$Q_t = h(t, A, \bar{F}, \bar{D}, Q_0), \tag{4.10}$$

and we define

$$\tilde{Q}_t = h(t, A, \bar{F}, \bar{D}, Q_0 + e_j). \tag{4.11}$$

In other terms, $\tilde{Q}_t$ is constructed exactly in the same way as Q_t, starting with one more customer in station j.

A little reflection shows that for all $t \geq 0$

$$\tilde{Q}_t \geq Q_t \quad \text{(i.e., } \tilde{Q}_t(i) \geq Q_t(i) \text{ for all } 1 \leq i \leq m\text{).} \tag{4.12}$$

In particular

$$P[Q_t(i) > 0, \tilde{Q}_t(i) = 0] = 0. \tag{4.13}$$

It will be shown, as a preliminary, that

$$P[Q_t(i) = 0, \tilde{Q}_t(i) > 0] > 0, \tag{4.14}$$

so that combining (4.13) and (4.14), we obtain

$$P[\tilde{Q}_t(i) > 0] > P[Q_t(i) > 0] \quad \text{whenever } t > 0. \tag{4.15}$$

4. Cascades and Loops in Jackson's Networks

The inequality $\geq$ in (4.15) is clear; it is the strict inequality which is not obvious: it depends very much on the existence of a loop passing through i and j. Indeed, to prove (4.14) we define $A = \{Q_0 = 0\}$ and $B = \{\Delta \bar{F}_{\tau_1}(j, k_1) = 1, \Delta \bar{F}_{\tau_2}(k_1, k_2) = 1, \ldots, \Delta \bar{F}_{\tau_p}(k_{p-1}, k_p) = 1, \Delta \bar{F}_{\tau_{p+1}}(k_p, i) = 1, \tau_{p+1} \leq t < \tau_{p+2}\}$, where τ_n is the sequence of jumps of all processes $A_t(j)$, $\bar{D}_t(k)$, $\bar{F}_t(l, q)$, $1 \leq j, k, l, q \leq m$.

We have

$$P[A] = \prod_{i=1}^{m} (1 - \rho_i) > 0 \tag{4.16}$$

and also

$$P[B] > 0, \tag{4.17}$$

since the processes A, $\bar{D}$, and $\bar{F}$ are independent Poisson processes with strictly positive intensities. Therefore, since A and B are independent,

$$P[A \cap B] = P[A]P[B] > 0. \tag{4.18}$$

Now, the event $A \cap B$ is included in $\{Q_t(i) = 0, \tilde{Q}_t(i) > 0\}$ and thus the inequality (4.14) is proved.

Our next task will be to show that a poissonian assumption for $F_t(i, j)$ results in a contradiction with (4.15).

First, if $F_t(i, j)$ is poissonian, its intensity is necessarily $\lambda_i r_{ij}$, by the conservation law for average flows at equilibrium. But this intensity is also $\mu_i r_{ij} P[Q_t(i) > 0 | \mathcal{F}_t^F(i, j)]$. Therefore, the same arguments as in the proof of the nonpoissonian character of the feedback flow in a $M/M/1$ feedback queue (R2) lead to

$$P[Q_t(i) > 0 | \mathcal{F}_t^F(i, j)] = P[Q_0(i) > 0] \quad P\text{-a.s.}, \quad t \geq 0. \tag{4.19}$$

In particular, for all finite $\mathcal{F}_t^F(i, j)$-stopping times τ,

$$P[Q_\tau(i) > 0 | \mathcal{F}_\tau^F(i, j)] = P[Q_0(i) > 0]. \tag{4.20}$$

We take $\tau = \tau_1 + b$, where τ_1 is now the first jump time of $F_t(i, j)$, and b is a strictly positive real number:

$$P[Q_{\tau_1+b}(i) > 0 | \mathcal{F}_{\tau_1+b}^F(i, j)] = P[Q_0(i) > 0], \tag{4.21}$$

or, since $P[Q_0(i) > 0] = P[Q_b(i) > 0]$ and $\mathcal{F}_{\tau_1+b}^F(i, j) \supset \mathcal{F}_{\tau_1}^F(i, j)$,

$$P[Q_{\tau_1+b}(i) > 0 | \mathcal{F}_{\tau_1}^F(i, j)] = P[Q_b(i) > 0]. \tag{4.22}$$

Therefore, for any $a > 0$,

$$E[1(Q_{\tau_1+b}(i) > 0)1(\tau_1 \leq a)] = P[Q_b(i) > 0]P[\tau_1 \leq a]. \tag{4.23}$$

But

$$E[1(Q_{\tau_1+b}(i) > 0)1(\tau_1 \le a)] = \sum_{n \in S} E[1(Q_{\tau_1+b}(i) > 0)1(Q_{\tau_1} = n)1(\tau_1 \le a)]$$
$$= \sum_{n \in S} E[P[Q_{\tau_1+b}(i) > 0 | Q_{\tau_1} = n, \tau_1 \le a]1(Q_{\tau_1} = n)1(\tau_1 \le a)]$$
$$= \sum_{n \in S} E[P[Q_{\tau_1+b}(i) > 0 | Q_{\tau_1} = n]P[Q_{\tau_1} = n | \mathscr{F}_{\tau_1}^F(i,j)]1(\tau_1 \le a)],$$

(4.24)

where we have used the strong Markov property. We will prove later that

$$P[Q_{\tau_1} = n | \mathscr{F}_{\tau_1}^F(i,j)]1(\tau \le a) = (p(n-e_j) + \Sigma(a))1(\tau \le a), \quad (4.25)$$

where $\Sigma(a) \le Ka$ for some constant K. Hence, the last term in the string of equalities (4.24) is

$$P[\tilde{Q}_b(i) > 0]P(\tau_1 \le a)(1 + \eta(a)), \quad (4.26)$$

where $\eta(a)$ tends to zero as a goes to zero. But if $a > 0$, $P(\tau_1 \le a)$ is strictly positive, since $F_t(i, j)$ is supposedly Poisson with strictly positive intensity. Therefore

$$P[\tilde{Q}_b(i) > 0](1 + \eta(a)) = P[Q_b(i) > 0]. \quad (4.27)$$

Contradiction with (4.15) is obtained by letting a go to zero.

It remains to prove (4.25), and this will be done by means of the filtering expression for $Z_t(n) = P[Q_t = n | \mathscr{F}_t^F(i,j)]$. Indeed, for all $n \in S$, using the assumption that $F_t(i, j)$ is Poisson,

$$\hat{Z}_t(n) = p(n) + \int_0^t \mathscr{K}(n, \hat{Z}_s) \, ds$$
$$+ \int_0^t \left(-\hat{Z}_{s-}(n) + \frac{\hat{Z}_{s-}(n+e_i-e_j)}{\rho_i} \right)(dF_s(i,j) - \lambda_i r_{ij} \, ds) \quad (4.28)$$

where $\dot{p}_t(n) = \mathscr{K}(n, p_t)$ is the forward Kolmogorov equation satisfied by $p_t(n) = P[Q_t = n]$. In particular

$$\hat{Z}_{\tau_1}(n) = \frac{1}{\rho_i} \hat{Z}_{\tau_1-}(n + e_i - e_j)$$
$$= \frac{1}{\rho_i} \left(p(n + e_i - e_j) + \int_0^{\tau_1} \mathscr{K}(n + e_i - e_j, \hat{Z}_s) \, ds \right)$$
$$= p(n - e_j) + \frac{1}{\rho_i} \int_0^{\tau_1} \mathscr{K}(n + e_i - e_j, \hat{Z}_s) \, ds,$$

and (4.25) easily follows from the boundedness of $\mathscr{K}(n, \hat{Z}_s)$.

An Alternative Model. From the discussion preceding T9 it seems that the negative result stating that no stream of a loop can be poissonian is essentially due to a poor modeling of the situation by means of a jacksonian network with just one class of customer. Indeed, in such networks a single customer is allowed to loop, whereas realistic routing strategies will prevent looping. Another way of phrasing this is: in such a network there might be *topological loops* which appear on the graph of the network but do not correspond to *cycling* of a given customer. A possible model which discriminates between topological loops and real loops is the following: one distinguishes K classes of customers, corresponding to different entrance–exit pairs of nodes. In each class, no looping is allowed, that is to say, the network graph of a given class is of the cascade type. Of course, when all the graphs are put together, the new graph obtained *will* exhibit topological loops. (See Figure 12.) However,

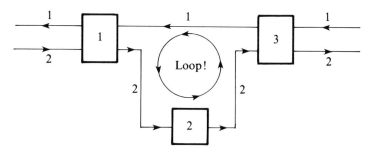

Figure 12   Topological loops but no cycling.

even in the processor-sharing network of §3 with no change of class and no cycling, it appears that topological loops cannot shelter poissonian streams. The reader is referred to [24] for the extension of the above analysis of loops to more complex markovian networks.

5. Independence and Poissonian Flows in Markov Chains

We are now going to generalize the independence result T8, taking the point of view of Melamed [19], who gave conditions for the independence of the state of a Markov chain and of some transition flows inside the chain in terms of local-balance equations, and then obtained specific results concerning markovian queueing systems as special cases.

The notations and terminology of the theorem to follow are those of IV, T7. The proof follows from the same arguments as the proofs of T1 and T8, via filtering, and is based on IV, T7.

T10 Theorem. *Suppose the chain X_t is at equilibrium, with the stationary probability $p(i)$. The following statements are then equivalent:*

(a) *For all $i \in S$ and $1 \leq l \leq h$*

$$p(i)\left(\sum_{\gamma \in C_l} p(\gamma) \sum_{\delta \in D_l(\gamma)} q_{\gamma\delta}\right) = \sum_{\gamma \in C_l} 1(i \in D_l(\gamma))p(\gamma)q_{\gamma i}. \tag{5.1}$$

(b) *For each $t \geq 0$, X_t is independent of $\mathscr{G}_t$.*
(c) *For each finite $\mathscr{G}_t$-stopping time τ, X_τ is independent of $\mathscr{G}_\tau$.*
(d) *For each finite $\mathscr{G}_t$-stopping time τ, and for all $i \in S$,*

$$P[X_\tau = i] = p(i). \tag{5.2}$$

In turn, any of the above four conditions implies that the flows $N_t(l)$ from C_l to D_l are independent Poisson processes.

The quantity $\sum_{\gamma \in C_l} 1(i \in D_l(\gamma))p(\gamma)q_{\gamma i}$ is the average intensity of this part of the flow from C_l to D_l corresponding to a transition *into* i, whereas $\sum_{\gamma \in C_l} \sum_{\delta \in D_l(\gamma)} p(\gamma)q_{\gamma\delta}$ is the average intensity of the *total* flow from C_l to D_l. Theorem T7 then reads: in order for X_t to be independent of, say, $\mathscr{F}_t^N(l)$ at each time t, it is necessary and sufficient that for each state i the proportion of jumps of $N_t(l)$ representing a transition into i equal $p(i)$:

$$p(i) = \frac{\text{flow into } i}{\text{total flow}}.$$

An immediate corollary of T10 is

T11 Corollary. *Suppose that the chain X_t in the statement of T10 is ergodic and at equilibrium. Then for X_t to be independent of the past $\mathscr{F}_t^J$ of $J_t = \sum_{0 < s \leq t} 1(X_s \neq X_{s-})$ at each time $t \geq 0$, it is necessary and sufficient that q_i be independent of $i \in S$.*

PROOF. J_t is the flow from S to S; for such flow Equation (5.1) reads

$$p(i)\left(\sum_{\gamma \in S} p(\gamma)q_\gamma\right) = \sum_{\gamma \in S} p(\gamma)q_{\gamma i}. \tag{5.3}$$

But the forward Kolmogorov equation must be satisfied, since we are at equilibrium:

$$p(i)q_i = \sum_{\gamma \in S} p(\gamma)q_{\gamma i}. \tag{5.4}$$

By the ergodicity assumption, $p(i) > 0$, and therefore by (5.3) and (5.4), $q_i = \sum_{\gamma \in S} p(\gamma)q_\gamma$.

Application: The Output Flows of Kelly's Network are Poissonian

Kelly's network is a generalization of Jackson's network in that it allows for several classes of customers interacting at any station on a first-come, first-served basis. There are N stations and K classes of customers. The exogenous arrivals of customers of class l into station i form a Poisson process of intensity $\bar{\lambda}_i^l$. At each station i service is provided by a single server, and the service time is exponential with the same mean $1/\mu_i$ for all classes of customers. On his way from station i to station j a customer can pass from class l to class p. The routing policy is of the Bernoulli switch type, as in Jackson's network: r_{ij}^{lp} is the probability that a customer of class l exiting from station i will be routed to station j and change from class l to class p; r_{i0}^{lp} is the probability that such a customer leaves the network. Of course,

$$\sum_{j=0}^{N}\sum_{p=1}^{K} r_{ij}^{lp} = 1. \tag{5.5}$$

Consider the traffic equations analogous to (4.9):

$$\lambda_i^l = \bar{\lambda}_i^l + \sum_{j=1}^{N}\sum_{p=1}^{K} \lambda_j^p r_{ji}^{pl}, \tag{5.6}$$

and suppose that there exists a solution in the λ_i^l's satisfying the light-traffic condition

$$\rho_i = \frac{\lambda_i}{\mu_i} < 1 \quad \text{where } \lambda_i = \sum_{l=1}^{K} \lambda_i^l. \tag{5.7}$$

The whole queueing system can be represented as a Markov chain taking its values in the space S a generic element of which is a set of N strings $x = (x_1, \ldots, x_N)$. String x_i, which summarizes the state of station i, has length $k(x_i)$ equal to the number of customers in station i (waiting or being served). If $k(x_i) = 0$, x_i is the empty string $\emptyset$. If $k(x_i) > 0$, the rth entry from the left in x_i is l if the rth customer in line is of class l (recall that the customers respect the order of arrival in the waiting line). Let $k^l(x_i)$ be the number of customers of class l in station i.

T12 Theorem [14]. *There exists an equilibrium distribution of the above Kelly's network; it is given by*

$$p(x_1, \ldots, x_N) = \prod_{i=1}^{N} p_i(x_i),$$

$$p_i(x_i) = \rho_i^{k(x_i)}(1 - \rho_i) \prod_{l=1}^{K} \left(\frac{\lambda_i^l}{\lambda_i}\right)^{k^l(x_i)}. \tag{5.8}$$

A detailed proof is given in [26].

T13 Theorem [16, 26]. *Let $D^l_{i,t}$ be the process counting the departures from the network of customers of class l who visited station i last. At any time $t \geq 0$, the state X_t is independent of the past of the $D^l_{i,t}$'s, and the $D^l_{i,t}$'s form independent Poisson processes.*

PROOF [26]. It is based on T10. For any $x \in S$, let $x_i^{+,l}$ represent the state which is identical to x except that there is one more customer present in the system, of class l, and in the first position at line i (being served). One must verify

$$p(x)\left(\sum_{x \in S} p(x_i^{+,l})q_{x_i^{+},l_x}\right) = p(x_i^{+,l})q_{x_i^{+},l_x}. \quad (5.9)$$

Clearly $q_{x_i^{+},l_x} = \mu_i$, since when the system is in state $x_i^{+,l}$, the line at station i is not empty and the customer of class l who is in front of the line is served by an exponential server of mean service time $1/\mu_i$. The verification of (5.8) is then immediate from $p(x_i^{+,l}) = p(x)\lambda_i^l$ [by (5.7)].

References

[1] Baskett, F., Chandy, K. M., Muntz, R. R., and Palacios, F. G. (1975) Open, closed and mixed networks of queues with different classes of customers, J.A.C.M. **22**, pp. 248–260.
[2] Beutler, F., Melamed, B., and Zeigler, B. (1977) Equilibrium properties of arbitrarily interconnected queueing networks, *Multivariate Analysis IV*, Krishnaïah, ed., North-Holland, pp. 351–370.
[3] Brémaud, P. (1975) Estimation de l'état d'une file d'attente et du temps de panne d'une machine par la méthode des semi-martingales, Adv. Appl. Probab. **7**, pp. 845–863.
[4] Brémaud, P. (1978) On the output theorem of queuing theory, via filtering, J. Appl. Probab. **15**, pp. 397–405.
[5] Brémaud, P. (1978) Streams of a $M/M/1$ feedback queue, Z. für W. **45**, pp. 21–33.
[6] Brémaud, P. (1978) Local balance and innovations gain. Report.
[7] Brémaud, P. (1980) A counterexample to a converse of the Burke–Reich theorem. Report.
[8] Burke, P. J. (1956). The output of a queueing system, Op. Res. **4**, pp. 699–704.
[9] Burke, P. J. (1972) Output processes and tandem queues, *Proc. Symp. on Computer Communications Networks and Teletraffic*, Wiley, New York, pp. 419–428.
[10] Burke, P. J. (1976) Proof of a conjecture on the interarrival time distribution in a $M/M/1$ queue with feedback IEEE Transactions **COM-24**, pp. 175–176.
[11] Daley, D. (1976) Queueing output processes, Adv. Appl. Probab. **8**, pp. 395–415.
[12] Hadidi, N. (1972) On the output process of a state dependent queue, Skandinavisk Aktuarietidskrift, pp. 182–186.
[13] Jackson, J. (1957) Networks of waiting lines, Op. Res. **15**, pp. 254–265.
[14] Kelly, F. (1975) Networks of queues with customers of different types, J. Appl. Probab. **12**, pp. 542–554.
[15] Kelly, F. (1976) Networks of queues, Adv. Appl. Probab. **8**, pp. 416–432.
[16] Kelly, F. (1979) *Reversibility and Stochastic Networks*, Wiley, London.

[17] Kolmogorov, A. (1935) Zur Theorie der Markoffschen Ketten, Mathematische Annalen **112**, pp. 155–160.
[18] Labetoulle, J., Pujolle, G., and Soula, C. (1979) Distributions of the flows in a general Jackson network, Rapport de Recherche No. 341, Iria/Laboria.
[19] Melamed, B. (1979) On Poisson Traffic processes in discrete state markovian systems with applications to queueing theory, Adv. Appl. Probab. **11**, pp. 218–239.
[20] Melamed, B. (1979) Characterization of Poisson traffic streams in Jackson queueing networks, Adv. Appl. Probab. **11**, pp. 422–438.
[21] Muntz, R. R. (1972) Poisson departure processes and queueing networks, Rep. RC 4145, IBM Res. Lab., Yorktown Heights, N.Y.
[22] Reich, E. (1957) Waiting times when queues are in tandem, Ann. Math. Stat. 28, pp. 768–773.
[23] Reich, E. (1965) Departure processes, *in Proc. Symp. on Congestion Theory*, W. L. Smith and W. Wilkinson, ed., Univ. of North Carolina Press, Chapel Hill, pp. 439–457.
[24] Sismaïl, K. (1981) Représentation intégrale des réseaux markoviens de files d'attente, Thèse de 3ème cycle, Université de Paris IX.
[25] Varaiya, P., Walrand, J. (1978) When is a flow in a Jacksonian network Poisson? Electronics Res. Lab. Memo ERL M78/59, Dept. of EECS, Univ. of Calif., Berkeley.
[26] Varaiya, P. and Walrand, J. (1978) The outputs of Jacksonian networks are Poissonian, Electronics Res. Lab. Memo. ERL M78/60, Dept. of EECS, Univ. of Calif., Berkeley.
[27] Varaiya, P. and Walrand, J. (1980) Interconnections of Markov chains and quasi-reversible queueing networks, Stochastic Processes and their Applications **10**, pp. 209–219.
[28] Varaiya, P. and Walrand, J. (1979) Flows in queueing networks: a martingale approach, Math Operat. Res., to appear.
[29] Varaiya, P. and Walrand, J. (1980) Sojourn times and the overtaking condition in Jacksonian networks, Adv. Appl. Probab. **12**, pp. 1000–1018.
[30] Whittle, P. (1968) Equilibrium distributions for an open migration process, J. Appl. Probab. **5**, pp. 567–571.

Solutions to Exercises, Chapter V

E1. By A1, §8:

$$p(n) = \begin{cases} \rho^n \dfrac{1-\rho}{1-\rho^{K+1}} & \text{if } 0 \leq n \leq K, \\ 0 & \text{otherwise.} \end{cases} \qquad (1)$$

If Q_t were independent of $\mathscr{F}_t^S$, then $\hat{Z}_t(n) = P[Q_t = n | \mathscr{F}_t^S] = p(n)$ for all $n \geq 0$. But the equation giving $\hat{Z}_t(n)$ has an innovations gain given by

$$K_t(n) = -\hat{Z}_{t-}(n) + \frac{\hat{Z}_{t-}(n+1)}{1 - \hat{Z}_{t-}(0)}, \qquad n \geq 0, \qquad (2)$$

and for $\hat{Z}_t(n)$ to be a constant for all $n \geq 0$, it is necessary that $K_t(n) \equiv 0$ for all $n \geq 0$. Thus independence implies

$$-p(n) + \frac{p(n+1)}{1 - p(0)} = 0, \qquad n \geq 0. \qquad (3)$$

In particular $-p(K) + p(K + 1)/[1 - p(0)] = 0$. Therefore, in view of (1), $p(K) = 0$; and this is a contradiction with (1), since $\rho > 0$.

Let $K_t(n)$ be the innovations gain for $\hat{Z}_t(n) = P[Q_t = n | \mathscr{F}_t^D]$. The usual computations lead to

$$K_t(n) = -\hat{Z}_{t-}(n) + \frac{\mu \hat{Z}_{t-}(n+1)1(n \le K-1) + \lambda \hat{Z}_{t-}(K)1(n=K)}{\mu(1 - \hat{Z}_{t-}(0)) + \lambda \hat{Z}_{t-}(K)}. \qquad (4)$$

It remains to verify that if we set $\hat{Z}_t(n) = p(n)$ in (4), then $K_t(n) \equiv 0$. But $\mu(1 - p(0)) + \lambda p(K) = \lambda$, and therefore, for $0 \le n \le K - 1$,

$$-p(n) + \frac{\mu p(n+1)}{\mu(1 - p(0)) + \lambda p(K)} = -p(n) + \frac{\mu}{\lambda} p(n+1)$$

$$= -p(n) + \frac{p(n+1)}{\rho} = 0, \qquad (5)$$

and for $n = K$,

$$-p(K) + \frac{\lambda p(K)}{\mu(1 - p(0)) + \lambda p(K)} = -p(K) \frac{\lambda p(K)}{\lambda} = 0. \qquad (6)$$

E2. If the flow S_t (for instance) were poissonian with intensity $\lambda(t)$, then for all $t \ge 0$, $E[S_t] = E[\int_0^t \lambda(s) \, ds] = \int_0^t \lambda(s) \, ds$. But $E[S_t] = E[\int_0^t \mu 1(Q_s > 0) \, ds] = \mu \int_0^t (1 - p(0)) \, ds = t\mu(1 - p(0))$. Hence $\lambda(t)$ must be a constant. Therefore S_t would necessarily be homogeneously poissonian, and this has already been disproven.

E3. F_t is not homogeneously poissonian: the same line of argument as in the case $r(n) \equiv p$ can be used. Indeed, $K_t^F(0) = -\hat{Z}_{t-}(0)$, and this is really all we need to carry on the proof.

S_t is not homogeneously poissonian: same proof as for $r(n) \equiv p$, with p replaced by $r(1)$ in the expression for $\hat{Z}_{t-}(0)$.

E4. Define $\mathscr{F}_t = \sigma(Q_0, \bar{F}_s^{k_i}(i,j), \bar{D}_s^{k_i}(i), A_s(i); s \in [0,t], 1 \le i, j \le m, 1 \le k_i \le s_i)$. Clearly $\mathscr{F}_t \supset \mathscr{H}_t$ for all $t \ge 0$. Also by the independence assumption relative to $Q_0, A, \bar{D}, \bar{F}$, the $\mathscr{F}_t$-intensities of the processes $A, \bar{D}, \bar{F}$ are the same as their internal intensities (i.e. the intensities with respect to their internal history). If the $\mathscr{F}_t$-intensities of A, D, F are adapted to $\mathscr{H}_t$, they are also the $\mathscr{H}_t$-intensities. This is the case, as we shall see. Indeed, for $F_t(i,j)$, for instance,

$$E\left[\int_0^\infty C_t \, dF_t(i,j)\right] = E\left[\sum_{k=1}^{s_i} \int_0^\infty C_t 1(Q_{t-}(i) \ge k) \, d\bar{F}_t^k(i,j)\right]$$

$$= E\left[\sum_{k=1}^{s_i} \int_0^\infty C_t 1(Q_{t-}(i) \ge k) \mu_i r_{ij} \, dt\right] \qquad (1)$$

for all nonnegative $\mathscr{F}_t$-predictable (and a fortiori $\mathscr{H}_t$-predictable) processes C_t. [The last equality of (1) follows from the assumption that the internal intensity—and therefore, as we mentioned before, the $\mathscr{F}_t$-intensity—of $\bar{F}_t^k(i,j)$ is $\mu_i r_{ij}$.] Hence the $\mathscr{F}_t$-intensity of $F_t(i,j)$ is $\sum_{k=1}^{s_i} \mu_i r_{ij} 1(Q_{t-}(i) \ge k) = \mu_i r_{ij} \inf(s_i, Q_t(i))$.

Solutions to Exercises, Chapter V   157

E5. Take $N_t(n, n + e_i)$ for instance. Since $\bar{\lambda}_i$ is—as we saw in E4—the $\mathcal{H}_t$-intensity of $A_t(i)$, we have for all $\mathcal{H}_t$-predictable nonnegative processes C_t

$$E\left[\int_0^\infty C_s \, dN_t(n, n + e_i)\right] = E\left[\int_0^\infty C_t 1(Q_{t-} = n) \, dA_t(i)\right]$$

$$= E\left[\int_0^\infty C_t 1(Q_{t-} = n)\bar{\lambda}_i \, dt\right],$$

and this shows (definition of intensity) that $\bar{\lambda}_i 1(Q_t = n)$ is an $\mathcal{H}_t$-intensity of $N_t(n, n + e_i)$.

E6. The forward stationary Kolmogorov equation for the system $(Q_t(1), Q_t(2))$ is

$$p(n_1, n_2)(2\lambda + \mu(1(n_1 > 0) + 1(n_1 = 0)1(n_2 > 0)))$$
$$= p(n_1 + 1, n_2)\mu + p(n_1, n_2 + 1)\mu$$
$$+ p(n_1 - 1, n_2)1(n_1 > 0)\lambda + p(n_1, n_2 - 1)1(n_2 > 0)\lambda. \quad (1)$$

From (1) and the assumption (3.24), it follows immediately that (3.25) holds. This is in fact more general: for any markovian queueing system $(Q_t(1), Q_t(2))$ in equilibrium, local balance for station (or class) 1 implies local balance for station (or class) 2.

From (3.24) we obtain

$$p(n_1, n_2) = \left(\frac{\lambda}{\mu}\right)^{n_1} p(0, n_2). \quad (2)$$

From (3.25) we obtain

$$p(0, n_2) = 0, \quad (3)$$

and therefore by combining (2) and (3), $p(n_1, n_2) = 0$ for all n_1, n_2. And this is a contradiction, because p is a probability distribution.

E7. This is a simple exercise. A remark, however: the gain with respect to A_t [i.e. $\hat{Z}_{t-}(n_1 - 1, n_2)1(n_1 > 0) - \hat{Z}_{t-}(n_1, n_2)$] has the same form as if we had filtered with respect to A_t only. The same for the gain with respect to D_t. This is a general situation when you filter with respect to a k-variate point process $(N_t(1), \ldots, N_t(k))$ where the $N_t(i)$'s have no common jumps.

CHAPTER VI

Likelihood Ratios

1. Radon–Nikodym Derivatives and Tests of Hypotheses
2. Changes of Intensities "à la Girsanov"
3. Filtering by the Method of the Probability of Reference
4. Applications
5. The Capacity of a Point-Process Channel
6. Detection Formula
References
Solutions to Exercises, Chapter VI

1. Radon–Nikodym Derivatives and Tests of Hypotheses

Elements of Decision Theory

A point process is apt to carry information about another process, and one might be interested in recovering part or all of the information content from the observation of the sequence of points. As an example, consider a laser source (see Figure 1) modulated by an electric current which provides the energy of activation and contains the information. Call the modulation signal S_t; the emission of photons is proportional to some nonnegative functions $f(S_t)$ of this signal satisfying $f(0) = 0$. At the receiving end of this communication system there is a photoelectronic conversion device: the output is a stream of electrons which would be roughly proportional to the mean flux of incoming photons, at least if there were no conversion noise.

The flux of electrons is observable; the associated counting process N_t has statistics which can be described in terms of its intensity: N_t is an $\mathscr{F}_t$-conditional process with the intensity $\lambda_t = \mu + B_t f(S_t)$. Here μ is the intensity corresponding to the conversion noise: if $S_t \equiv 0$, some electrons

1. Radon–Nikodym Derivatives and Tests of Hypotheses

are emitted from the cathode, and form a Poisson process with the intensity μ. B_t is a multiplicative noise, completely independent of the signal S_t, and due to absorption and turbulence in the atmosphere between the laser source and the receiver. Of course, since we are describing a doubly stochastic Poisson model, λ_t must be $\mathscr{F}_0$-measurable (II, D1), that is to say $\mathscr{F}_0$ contains $\mathscr{F}^B_\infty$ and $\mathscr{F}^S_\infty$. The most economical choice of $\mathscr{F}_t$ is

$$\mathscr{F}_t = \mathscr{F}^B_\infty \vee \mathscr{F}^S_\infty \vee \mathscr{F}^N_t.$$

The above model may not be very accurate from a physical point of view, and the reader is advised to consult the relevant literature on the subject of laser transmission; however, it provides a useful guideline for our discussion of decision theory, and in particular of detection theory. The interested reader will find in Snyder {140} a few examples relative to the detection and estimation problems arising in optical communications.

Detection problems arise when there is a choice between two possibilities (say, the occurence or the nonoccurence of a given event) based on incomplete

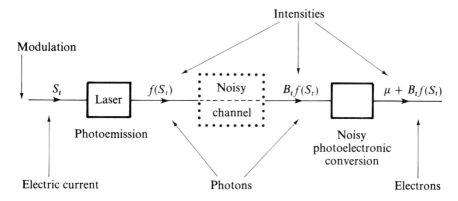

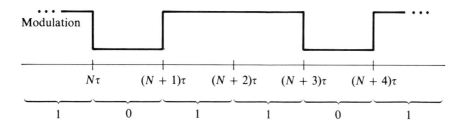

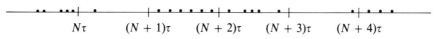

Figure 1 Schematic picture of a laser communication device.

observation. For instance, suppose that in a normal situation, the modulating signal S_t takes a fixed value μ_1. The corresponding probability which governs the statistics of N_t is P_1, and N_t has the $(P_1, \mathscr{F}_t)$-intensity $\mu + a_1 B_t$, where $a_1 = f(\mu_1)$.[1]

Now suppose the laser ceases to operate (breaks down). The probability governing the statistics of N_t in such a situation is P_2, and N_t has the $(P_2, \mathscr{F}_t)$-intensity μ; it is a Poisson process. The problem that arises is that of detecting the faulty state of the laser on the basis of the observation of a stretch of trajectory $(N_s, s \in [t_1, t_2])$. Of course, the conclusion that will be drawn ("laser functions" or "laser does not function") will be followed by some action if needed (repair). If the decision to repair is taken when in fact the laser is still operating, a cost will be incurred, and the strategy of decision ("repair" or "do not repair") should take such costs into consideration.

Let us now examine the two above situations (P_1 or $\lambda_t = \mu + \mu_1 B_t$, and P_2 or $\lambda_t = \mu$) in a different light (see Figure 1): P_1 corresponds to the transmission of the bit of information 0, whereas P_2 corresponds to the bit of information 1. Also $t_1 = N\tau$ and $t_2 = (N+1)\tau$. The interpretation is in terms of data transmission: every τ seconds a bit of information is transmitted, either 0 or 1, and the observation of $(N_s, s \in [N\tau, (N+1)\tau)$ corresponds to the Nth transmission. On the basis of such information the receiver is to decide whether 0 or 1 has been transmitted, and to each strategy of decision, there corresponds a cost, say, the probability of error, to be minimized.

The difference between the above two interpretations of P_1 and P_2 is fundamental, although we have not yet stressed it. Indeed, in the detection situation, there is no *a priori* probability available: one does not know beforehand the probability of breakdown of the laser source (this breakdown might be due to a meteorite falling on the satellite which is equipped with such a source). On the other hand, in the data-transmission setting, the *a priori* probabilities of 0 and 1 are usually known: they characterize the message source, which is, in general, well identified from the statistical point of view. Statisticians would say that data transmission falls into the bayesian setting, whereas detection is not a bayesian problem.

We wish to present very briefly the two corresponding theories in the case where the costs are measured in terms of probability of error. Our main motivation here is to exhibit the fundamental role of Radon–Nikodym derivatives or likelihood ratios.[2] We will then turn to the theory of likelihood ratios in point-process systems, especially the theorem of separation of detection and filtering (§3 and §4).

---

[1] In this chapter it will become necessary to mention the probability measures with respect to which intensities are defined, since we will be concerned mainly with *changing* such probability measures.

[2] For more details on estimation theory and the role of likelihood ratios in statistics, consult a standard textbook, for instance Ferguson {56}. However, all the statistical background that we need in the rest of this chapter is contained in the present section.

Hypothesis Testing: the Bayesian Approach

Comparison of the Likelihood Ratio with a Threshold. Let H be a set consisting of two elements h_1 and h_2, and define $\mathcal{H} = \mathcal{P}(H)$ to be the trivial σ-field on H. Define on $(H, \mathcal{H})$ a probability measure μ by

$$\mu(\{h_1\}) = \mu_1, \quad \mu(\{h_2\}) = \mu_2, \tag{1.1}$$

where μ_1 and μ_2 are strictly positive and $\mu_1 + \mu_2 = 1$. H is called the *hypotheses space*. We will have to tell which of the two hypotheses h_1 or h_2 is in force, our judgment being based on *observations* ω lying in some set Ω equipped with a σ-field of observable events $\mathcal{F}$. Hypotheses h_1 and h_2 influence the statistics governing the observations. More precisely, there are two probabilities on $(\Omega, \mathcal{F})$: P_1 corresponds to h_1, and P_2 corresponds to h_2. The probabilities μ, P_1, P_2 determine an overall probability on $(\Omega', \mathcal{F}')$ $= (H \times \Omega, \mathcal{H} \otimes \mathcal{F})$ as follows:

$$P'(\{h_i\}, d\omega) = \mu_i P_i(d\omega), \quad i = 1, 2. \tag{1.2}$$

A strategy of discrimination between h_1 and h_2 on the basis of the observation ω consists of a set $A_1 \in \mathcal{F}$, the interpretation of which is: if $\omega \in A_1$, guess that hypothesis h_1 is in force, and if $\omega \in A_2 = A_1^c$, guess that hypothesis h_2 is in force.

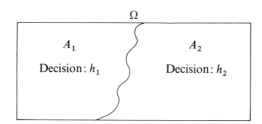

Figure 2   Decision regions in the observation space.

When a strategy A_1 is adopted two kinds of errors may occur: the *error of first kind* is guessing h_1 when actually $h = h_2$; the other type of error, taking h_2 when $h = h_1$, is the *error of second kind*. The probability of error of first kind is

$$P_2(A_1) = 1 - P_2(A_2) \tag{1.3}$$

(such an error occurs when $h = h_2$ and $\omega \in A_1$). Similarly, the probability of error of second kind is

$$P_1(A_2) = 1 - P_1(A_1). \tag{1.4}$$

The above two types of error are exhaustive and exclusive; therefore the probability of error P_E resulting from strategy A_1 is

$$P_E(A_1) = \mu_1 + \mu_2 P_2(A_1) - \mu_1 P_1(A_1). \tag{1.5}$$

Let us suppose that
$$P_2 \ll P_1 \quad \text{on } (\Omega, \mathscr{F}), \tag{1.6}$$
and let L be the corresponding Radon–Nikodym derivative dP_2/dP_1. Then
$$P_2(A_1) = \int_{A_1} L(\omega) P_1(d\omega), \tag{1.7}$$
and
$$P_E(A_1) = \mu_1 + \int_{A_1} (\mu_2 L(\omega) - \mu_1) P_1(d\omega). \tag{1.8}$$
From this formula it is clear that A_1^* defined by
$$A_1^* = \{\omega \,|\, L(\omega) \le \mu_1/\mu_2\} \tag{1.9}$$
minimizes $P_E(A_1)$. The optimal strategy consists therefore of comparing the *likelihood ratio* $L = dP_2/dP_1$ with a *threshold* $\sigma = \mu_1/\mu_2$, and deciding for hypothesis h_2 (h_1) when L is greater than (less than) σ.

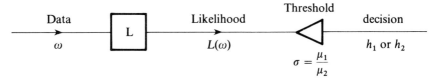

Figure 3   A minimum-probability-of-error bayesian detector.

Restricting the Observations. Note that we have defined a strategy to be a set A_1 lying in $\mathscr{F}$, and not just any set. The physical meaning of this restriction is that the events of $\mathscr{F}$ consist of the observable events, and therefore if we want our decision to be based on observable events, A_1 must be in $\mathscr{F}$. We verify that, since L is $\mathscr{F}$-measurable, A_1^* is indeed in $\mathscr{F}$.

Let now $\mathscr{G}$ be a sub-σ-field of $\mathscr{F}$, and suppose that the observable events are those in $\mathscr{G}$. A strategy A_1 is now a $\mathscr{G}$-measurable event. We still have for $P_E(A_1)$ the expression (1.8), but now, since $A_1 \in \mathscr{G}$,
$$\int_{A_1} L \, dP_1 = \int_{A_1} E_1[L|\mathscr{G}] \, dP_1, \tag{1.10}$$
and therefore
$$P_E(A_1) = \mu_1 + \int_{A_1} (\mu_2 E_1[L|\mathscr{G}](\omega) - \mu_1) P_1(d\omega), \tag{1.11}$$
and the optimal strategy is
$$\tilde{A}_1^* = \left\{ \omega \,\Big|\, E_1[L|\mathscr{G}](\omega) \le \frac{\mu_1}{\mu_2} \right\}. \tag{1.12}$$

1. Radon–Nikodym Derivatives and Tests of Hypotheses

Note that nothing essentially new has been said relative to (1.9), since $E[L|\mathcal{G}]$ is the Radon–Nikodym derivative of P_2 with respect to P_1, both probabilities being restricted to $(\Omega, \mathcal{G})$, the new observation space. Denote A_1^* in (1.9) by $A_1^*(\mathcal{F})$, and $\tilde{A}_1^*$ in (1.12) by $A_1^*(\mathcal{G})$. It is expected that

$$P_E(A_1^*(\mathcal{F})) \leq P_E(A_1^*(\mathcal{G})) \tag{1.13}$$

since there is less information in $\mathcal{G}$ than in $\mathcal{F}$. This can be proven

$$P_E(A_1^*(\mathcal{F})) = \mu_1 + \int_{A_1^*(\mathcal{F})} (\mu_2 L - \mu_1) \, dP_1$$

$$= \mu_1 + \int_\Omega (\mu_2 L - \mu_1) 1\left(L \leq \frac{\mu_1}{\mu_2}\right) dP_1,$$

$$P_E(A_1^*(\mathcal{G})) = \mu_1 + \int_{A_1^*(\mathcal{G})} (\mu_2 E_1[L|\mathcal{G}] - \mu_1) \, dP_1$$

$$= \mu_1 + \int_\Omega (\mu_2 E_1[L|\mathcal{G}] - \mu_1) 1\left(E_1[L|\mathcal{G}] \leq \frac{\mu_1}{\mu_2}\right) dP_1.$$

The inequality (1.13) then follows by Jensen's inequality applied to the convex function

$$\phi(x) = (\mu_2 x - \mu_1) 1\left(x \leq \frac{\mu_1}{\mu_2}\right). \tag{1.14}$$

E1 Exercise. Denote $P_E(A_1^*(\mathcal{F})) = P_E(\mu_1)$, and show that $P_E(\mu_1)$ reaches its maximum for some value of μ_1 in the open interval $(0, 1)$. Also show that at this value of μ_1, P_E is equal to the error of second kind, and $1 - P_E$ to the error of first kind.

Hypothesis Testing: the Neyman–Pearson Approach

It may happen that no *a priori* probability is available for the hypotheses to be tested. In this situation, Neyman and Pearson thought of dissymmetrizing the roles of h_1 and h_2, by minimizing the error of second kind while retaining the error of the first kind below a certain *level* α. This point of view is that of radar systems engineers, for whom h_1 corresponds to a clear sky and h_2 corresponds to the presence of a target. There is no *a priori* probability available for such events, at least in situations that are not identified from the statistical point of view. In such a context, the error probability of the first kind is called the *probability of false alarm*, and that of the second kind is the *probability of nondetection*. False alarms must not be too numerous, otherwise the control-station computers might reach saturation; the level α of the test is set as a function of the capacity of the station. Here is the basic result of nonbayesian hypothesis testing:

T1 Theorem (Neyman–Pearson Lemma). *Suppose that for some $\alpha \in (0, 1)$ there exists a number σ such that*

$$P_1[\{\omega | L(\omega) \geq \sigma\}] = \alpha \tag{1.15}$$

where L is the Radon–Nikodym derivative of P_2 with respect to P_1 both probabilities on $(\Omega, \mathscr{F})$. Then the test corresponding to

$$\complement A_1^* = A_2^* = \{L \geq \sigma\} \tag{1.16}$$

is optimal for $\mathscr{F}$-observations in the sense that for any $A_2 = \complement A_1 \in \mathscr{F}$ such that $P_1(A_2) \leq \alpha$, we have $P_2(A_2^*) \geq P_2(A_2)$.

Remark. The Neyman–Pearson lemma states that the strategy of choosing hypothesis h_2 when $L(\omega) \geq \sigma$ is optimal in the sense that it minimizes the probability of error of the second kind among all strategies corresponding to an error of the first kind which does not exceed the level α.

PROOF Write:

$$P_2(A_2^*) - P_2(A_2) = P_2(A_2^* \cap A_1) - P_2(A_1^* \cap A_2). \tag{1.17}$$

By the definition of L and of A_2,

$$P_2(A_2^* \cap A_1) = \int_{A_2^* \cap A_1} dP_2 = \int_{A_2^* \cap A_1} L\, dP_1$$

$$\geq \sigma \int_{A_2^* \cap A_1} dP_1 = \sigma P_1(A_2^* \cap A_1). \tag{1.18}$$

Similarly

$$P_2(A_1^* \cap A_2) < \sigma P_1(A_1^* \cap A_2), \tag{1.19}$$

and therefore

$$P_2(A_2^*) - P_2(A_2) \geq \sigma(P_1(A_2^*) - P_1(A_2)) \geq 0, \tag{1.20}$$

since $P_1(A_2^*) = P_1[L \geq \sigma] = \alpha$ and $P_1(A_2) \leq \alpha$. □

Remark. If the distribution function of the random variable L is continuous, then for any $\alpha \in (0, 1)$ there does exist a threshold σ satisfying (1.15). This is not true in general if the continuity property of the distribution function of L is not satisfied. However, in this book, either the continuity assumption is verified, or the thresholds are not points of discontinuities of L. Otherwise, a coin must be tossed for those particular experiments ω where $L(\omega)$ is exactly equal to the threshold, in order to decide which hypothesis to adopt; for more details, see Ferguson {56}.

E2 Exercise. Let X be a random variable taking the values $0, 1, 2, \ldots$ and defined on some $(\Omega, \mathscr{F})$. Let P_1 and P_2 be two probability measures on $(\Omega, \sigma(X))$ such that for all $k \geq 0$, if $P_1(X = k) = 0$, then $P_2(X = k) = 0$. Show that $P_2 \ll P_1$ and compute $L = dP_2/dP_1$.

E3 Exercise. Let N_t be a point process on $(\Omega, \mathscr{F})$, and for $i = 1, 2$, let P_i be a probability measure on $(\Omega, \mathscr{F})$ such that N_t is Poisson with the intensity $\lambda_i > 0$. Show that for all $t \geq 0$, $P_{2,t} \ll P_{1,t}$, where $P_{i,t}$ is the restriction of P_i to $\sigma(N_t)$, and that

$$\frac{dP_{2,t}}{dP_{1,t}} = \left(\frac{\lambda_2}{\lambda_1}\right)^{N_t} \exp\{(\lambda_2 - \lambda_1)t\}. \tag{1.21}$$

E4 Exercise. Same situation as in E2. Suppose that hypothesis P_i occurs with probability μ_i where $\mu_1 + \mu_2 = 1$. Compute the probability of error resulting from a minimal-error bayesian test based on the observation of N_T.

2. Changes of Intensities "à la Girsanov"

In this section the relation between a certain type of absolutely continuous change of probability measures and the change of intensity that it induces will be examined. It turns out that such changes of probability concerning point processes with an intensity are quite general, as will be seen in §4, where a kind of converse to T3 will be proven. The result to follow extends E3: an absolutely continuous change of probability measure $P \to \tilde{P}$ is described by its Radon–Nikodym derivative in terms of the Radon–Nikodym derivative relative to the change of measure $dN_t\, dP \to dN_t\, d\tilde{P}$, that is to say, the change of intensity $\lambda_t \to \tilde{\lambda}_t$.

We first describe a fundamental martingale, a particular form of which has already been encountered in III, §4. It is a special case of Doléans-Dade's exponential martingale [7].

The Exponential Supermartingale

T2 Theorem [7]. *Let $(N_t(1), \ldots, N_t(k))$ be a k-variate point process adapted to some history $\mathscr{F}_t$, and let $\lambda_t(i)$, $1 \leq i \leq k$, be the predictable $(P, \mathscr{F}_t)$-intensities of $N_t(i)$, $1 \leq i \leq k$, respectively. Let $\mu_t(i)$, $1 \leq i \leq k$, be $\mathscr{F}_t$-predictable processes, nonnegative, and such that for all $t \geq 0$ and all $1 \leq i \leq k$*

$$\int_0^t \mu_s(i)\lambda_s(i)\, ds < \infty \quad P\text{-a.s.} \tag{2.1}$$

Define the process L_t by

$$L_t = \prod_{i=1}^{k} L_t(i), \tag{2.2}$$

where

$$L_t(i) = \begin{cases} \exp\left\{\int_0^t (1 - \mu_s(i))\lambda_s(i)\,ds\right\} & \text{if } t < T_1(i), \\ \left(\prod_{n \geq 1} \mu_{T_n}(i)1(T_n(i) \leq t)\right) \exp\left\{\int_0^t (1 - \mu_s(i))\lambda_s(i)\,ds\right\} & \text{if } t \geq T_1(i). \end{cases} \tag{2.3}$$

The notation of (2.3) will be henceforward simplified to

$$L_t(i) = \left(\prod_{n \geq 1} \mu_{T_n}(i)1(T_n(i) \leq t)\right) \exp\left\{\int_0^t (1 - \mu_s(i))\lambda_s(i)\,ds\right\}.$$

Then L_t is a $(P, \mathscr{F}_t)$-nonnegative local martingale and a $(P, \mathscr{F}_t)$-supermartingale.

In particular, for all $1 \leq i \leq k$, $L_t(i)$ is a $(P, \mathscr{F}_t)$-local martingale and a $(P, \mathscr{F}_t)$-supermartingale.

PROOF By application of Appendix A4, T4,

$$L_t = 1 + \sum_{i=1}^{k} \int_0^t L_{s-}(\mu_s(i) - 1)\,dM_s(i), \tag{2.4}$$

where $M_t(i) = N_t(i) - \int_0^t \lambda_s(i)\,ds$. Define

$$S_n = \begin{cases} \inf\left\{t \mid L_{t-} \geq n \text{ or } \sum_{i=1}^{k} \int_0^t (\mu_s(i) + 1)\lambda_s(i)\,ds \geq n\right\} & \text{if } \{\cdots\} \neq \emptyset, \\ +\infty & \text{otherwise,} \end{cases}$$

and apply II, T8 to obtain that $L_{t \wedge S_n}$ is a $(P, \mathscr{F}_t)$-martingale. Now, (2.1) and $\int_0^t \lambda_s(i)\,ds < \infty$, P-a.s., $t \geq 0$, $1 \leq i \leq k$, and the fact that L_{t-} is a finite left-continuous process imply that $S_n \uparrow \infty$, P-a.s.; therefore L_t is a $(P, \mathscr{F}_t)$-local martingale. It is obviously nonnegative and therefore a $(P, \mathscr{F}_t)$-supermartingale by I, E8.

T3 Theorem (Direct Radon–Nikodym-derivative theorem [3, 4]). *Same notation as in T2. Suppose moreover that*

$$E[L_1] = 1. \tag{2.5}$$

Define the probability measure $\tilde{P}$ by

$$\frac{d\tilde{P}}{dP} = L_1. \tag{2.6}$$

2. Changes of Intensities "à la Girsanov"

Then, for each $1 \leq i \leq k$, $N_t(i)$ has the $(\tilde{P}, \mathscr{F}_t)$-intensity $\tilde{\lambda}_t(i) = \mu_t(i)\lambda_t(i)$ over $[0, 1]$.

PROOF. First observe that by (2.5) and I, E5, L_t is a $(P, \mathscr{F}_t)$-martingale over $[0, 1]$, and it is moreover of bounded variation. By definition of $\tilde{P}$,

$$\tilde{E}\left[\int_0^1 C_s \, dN_s(i)\right] = E\left[L_1 \int_0^1 C_s \, dN_s(i)\right] \tag{2.7}$$

and

$$\tilde{E}\left[\int_0^1 C_s \tilde{\lambda}_s(i) \, ds\right] = E\left[L_1 \int_0^1 C_s \tilde{\lambda}_s(i) \, ds\right]. \tag{2.8}$$

From a result of Dellacherie stated in Appendix A2, T19,

$$E\left[L_1 \int_0^1 C_s \, dN_s(i)\right] = E\left[\int_0^1 L_s C_s \, dN_s(i)\right] \tag{2.9}$$

and

$$E\left[L_1 \int_0^1 C_s \tilde{\lambda}_s(i) \, ds\right] = E\left[\int_0^1 L_{s-} C_s \tilde{\lambda}_s(i) \, ds\right]. \tag{2.10}$$

Therefore, in order to prove that $\tilde{\lambda}_t(i)$ is the $(\tilde{P}, \mathscr{F}_t)$-intensity of $N_t(i)$ over $[0, 1]$ it suffices to show that

$$E\left[\int_0^1 L_s C_s \, dN_s(i)\right] = E\left[\int_0^1 L_{s-} C_s \tilde{\lambda}_s(i) \, ds\right] \tag{2.11}$$

for all nonnegative $\mathscr{F}_t$-predictable processes C_t. Now, this follows from the observation that $E[\int_0^1 L_s C_s \, dN_s(i)] = E[\int_0^1 L_{s-} C_s \mu_s(i) \, dN_s(i)]$ and the definition of $\lambda_t(i)$ as the $(P, \mathscr{F}_t)$-intensity of $N_t(i)$. □

E5 Exercise. Consider the situation described in T3 with $k = 1$, $\mathscr{F}_t = \mathscr{F}_t^N$. Let $g^{(n+1)}$ and $\tilde{g}^{(n+1)}$ be the conditional densities of S_{n+1} given $\mathscr{F}_{T_n}^N$ relative to P and $\tilde{P}$ respectively, i.e.

$$P[S_{n+1} \leq x | \mathscr{F}_{T_n}^N] = \int_0^x g^{(n+1)}(s) \, ds = G^{(n+1)}([0, x]),$$

$$\tilde{P}[S_{n+1} \leq x | \mathscr{F}_{T_n}^N] = \int_0^x \tilde{g}^{(n+1)}(s) \, ds = \tilde{G}^{(n+1)}([0, x]). \tag{2.12}$$

Show that L_t has the form

$$L_t = \left(\prod_{n \geq 0} \frac{\tilde{g}^{(n+1)}(S_{n+1})}{g^{(n+1)}(S_{n+1})} 1(T_{n+1} \leq t)\right) \frac{1 - \int_0^{t-T_{N_t}} \tilde{g}^{(N_t+1)}(s) \, ds}{1 - \int_0^{t-T_{N_t}} g^{(N_t+1)}(s) \, ds}, \tag{2.13}$$

where the product $\prod_{n \geq 0}$ is taken to be 1 if $T_1 > t$, according to our previous convention.

Remark. If (and only if) $\mu_{T_n(i)}(i) > 0$ on $\{T_n(i) < \infty\}$, P-a.s., for all $n \geq 0$ and all $1 \leq i \leq k$, then $L_t > 0$, P-a.s., $t \geq 0$. In particular $L_1 > 0$, P-a.s., and P is absolutely continuous with respect to $\tilde{P}$, with

$$\frac{dP}{d\tilde{P}} = \frac{1}{L_1}. \tag{2.14}$$

The next result is a sufficient condition for (2.5) to hold. We will see a more general sufficient condition in VIII, T11.

T4 Theorem [3, 4]. *Same notation as in T2. Suppose moreover that for $1 \leq i \leq k$, $\lambda_t(i) = 1$ and $\mu_t(i)$ is bounded. Then $E[L_1] = 1$.*

PROOF. We treat the case $k = 1$ for notational convenience. Also we write μ_t, $\tilde{\lambda}_t$, N_t instead of $\mu_t(1)$, $\tilde{\lambda}_t(1)$, $N_t(1)$. Clearly

$$L_t \leq K^{N_t} e^{Kt}, \tag{2.15}$$

where K is the upper bound of μ_t. Therefore:

$$E\left[\int_0^t |L_{s-}(\mu_s - 1)| \, ds\right] \leq E[(K+1)K^{N_t}e^K] < \infty, \tag{2.16}$$

where the last inequality follows from the fact that N_t is a Poisson random variable with parameter λ by Watanabe's theorem (II, T5). From (2.16) and (II, T8), we see that L_t is a martingale. □

T4 can be viewed as an *existence theorem*, when combined with T3. Indeed, given the existence of a probability measure P for which N_t is $\mathscr{F}_t$-Poisson, we can construct, for any $\mathscr{F}_t$-predictable nonnegative *bounded* process μ_t, a probability $\tilde{P}$ for which N_t has the $(\tilde{P}, \mathscr{F}_t)$-intensity μ_t (take $d\tilde{P}/dP = L_1$).

Case of Queueing Processes

A simple queue Q_t can be viewed as a bivariate point process (A_t, D_t), where A_t and D_t are the input and output processes respectively. Suppose that under probability P, Q_t is $M/M/1$ with $\mathscr{F}_t$-parameters λ and μ. Let now λ_t and μ_t be two *bounded* nonnegative $\mathscr{F}_t$-predictable processes. Then, by T4 and T3,

$$L_t = \exp\left\{\int_0^t \log \frac{\lambda_s}{\lambda} dA_s - \int_0^t (\lambda_s - \lambda) \, ds\right\}$$

$$\times \exp\left\{\int_0^t \log \frac{\mu_s}{\mu} dD_s - \int_0^t (\mu_s - \mu) 1(Q_s > 0) \, ds\right\} \tag{2.17}$$

is a $(P, \mathscr{F}_t)$-martingale, and if we define $\tilde{P}$ by $d\tilde{P}/dP = L_1$, Q_t is a queue with the $(\tilde{P}, \mathscr{F}_t)$-parameters λ_t and μ_t over $[0, 1]$.

2. Changes of Intensities "à la Girsanov"

E6 Exercise. What is the $\tilde{P}$-distribution of Q_0 (in comparison with its P-distribution)?

E7 Exercise. Let A_t and D_t be two $(P, \mathscr{F}_t)$-Poisson processes with the intensities λ and μ respectively. Show that L_1 defined by

$$L_1 = 1(A_s \geq D_s, s \in [0, 1]) \exp\left\{\mu \int_0^1 1(A_s = D_s)\, ds\right\} \quad (2.18)$$

describes a change of probability $P \to \tilde{P}$ by $d\tilde{P}/dP = L_1$. Moreover, under $\tilde{P}$, $Q_t = A_t - D_t$ is an M/M/1 queue with the $\mathscr{F}_t$-parameters λ and μ.

On the Necessity of Choosing a Predictable Version of the Intensity

E8 Exercise. Consider the situation described in T4 with the following particularity: $k = 1$ [univariate case, notation $N_t = N_t(1)$, $\lambda_t = \lambda_t(1)$, etc.]. One therefore starts with a probability measure P such that N_t is a $(P, \mathscr{F}_t)$-standard Poisson process. Suppose that one wants to obtain in the way described in T2 and T3 a probability $\tilde{P}$ such that N_t has the $(\tilde{P}, \mathscr{F}_t)$-intensity

$$\tilde{\lambda}_t = a1(t \leq T_1) + 1(t > T_1). \quad (2.19)$$

What is the correct choice of μ_t:

$$\mu_t = a1(t \leq T_1) + 1(t > T_1) \quad (2.20)$$

or

$$\mu_t = a1(t < T_1) + 1(t \geq T_1)? \quad (2.21)$$

(Note that in both cases $\int_0^t \tilde{\lambda}_s\, ds = \int_0^t \mu_s \lambda_s\, ds$ since the two choices of μ_t differ only at an isolated point T_1.)

Insensitivity of the Operational Characteristics of a Queuing System

Let Q_t be a simple queue defined on $(\Omega, \mathscr{F})$ and adapted to a history $\mathscr{F}_t$. Suppose that under probability P, Q_t is M/M/1 with parameters λ and μ. Let now $\lambda_t^{(n)}$ and $\mu_t^{(n)}$ be two sequences of nonnegative $\mathscr{F}_t$-predictable processes that converge boundedly (t, ω)-pointwise to the bounded $\mathscr{F}_t$-predictable processes λ_t and μ_t respectively. Define for each $n \geq 1$ the probability $\tilde{P}^{(n)}$ by

$$\frac{d\tilde{P}^{(n)}}{dP} = L_1^{(n)} \quad (2.22)$$

and

$$L_1^{(n)} = \exp\left\{\int_0^t \log \frac{\lambda_s^{(n)}}{\lambda}\, dA_s - \int_0^t (\lambda_s^{(n)} - \lambda)\, ds\right\}$$
$$\times \exp\left\{\int_0^t \log \frac{\mu_s^{(n)}}{\mu}\, dD_s - \int_0^t (\mu_s^{(n)} - \mu)1(Q_s > 0)\, ds\right\}. \quad (2.23)$$

Also define $\tilde{P}$ by means of the Radon–Nikodym derivative

$$\frac{d\tilde{P}}{dP} = L_1, \qquad (2.24)$$

where L_t is given by (2.10). In view of the proof of T4, the family $(L_1^{(n)}, n \geq 1)$ is, with respect to P, uniformly integrable. Also $L_1^{(n)} \to L_1$, P-a.s., and therefore, under P, $L_1^{(n)} \to L_1$ in $L^1(P)$ (Appendix A1, T26). Therefore, for any nonnegative random variable X

$$\lim_{n \uparrow \infty} E^{(n)}[X] = \tilde{E}[X], \qquad (2.25)$$

where $E^{(n)}$ and $\tilde{E}$ denote integration with respect to $\tilde{P}^{(n)}$ and $\tilde{P}$ respectively.

Suppose now that for each n, the expected value $E^{(n)}[X]$ depends on n only as a continuous function f of a parameter $\mu^{(n)}$:

$$E^{(n)}[X] = f(\mu^{(n)}), \qquad (2.26)$$

and suppose also that $\lim_{n \uparrow \infty} \mu^{(n)} = \mu$. Then

$$\tilde{E}[X] = f(\mu). \qquad (2.27)$$

Such a situation has been exploited by Kelly {87} and Barbour {1} to prove the "insensitivity" of the stationary distribution of certain queueing systems. More precisely, they showed that the distributions were a given function of the mean service times. This is the case for instance in jacksonian networks (generalized to nonexponential service distributions). In Kelly {87} the $\tilde{P}^{(n)}$'s correspond to service times with a mixed erlangian (coxian) distribution. The procedure works because any distribution can be approximated by a sequence of mixed erlangian distributions. The details can be found in {87} or {1}.

3. Filtering by the Method of the Probability of Reference

Bayes' Formula

There exists an alternative to the innovations method of IV, §2. The presentation will be made in terms of a univariate point-process observation, for reasons of clarity in the notation. The extension to the multivariate case is not difficult, and left to the reader.

In the method of the reference probability, probability P actually governing the statistics of the observation N_t is obtained by an absolutely continuous

3. Filtering by the Method of the Probability of Reference

change of measure $Q \to P$. The *reference* probability Q is such that N_t is a $(Q, \mathscr{F}_t)$-Poisson process of intensity 1, where

$$\mathscr{F}_t = \mathscr{F}_t^N \vee \mathscr{F}_\infty^X \tag{3.1}$$

and $\mathscr{F}_t^X$ is the history of the state process X_t taking its values in $(U, \mathscr{U})$.

For each $t \geq 0$, the restrictions P_t and Q_t of P and Q respectively to $(\Omega, \mathscr{F}_t)$ are such that

$$P_t \ll Q_t, \tag{3.2}$$

and the corresponding Radon–Nikodym derivative is given by

$$L_t = \frac{dP_t}{dQ_t} = \left(\prod_{0 < s \leq t} \lambda_s \, \Delta N_s \right) \exp\left\{ \int_0^t (1 - \lambda_s) \, ds \right\}, \tag{3.3}$$

where λ_t is nonnegative bounded measurable process, $\mathscr{F}_t$-predictable (see T3, T4). In particular, if $\lambda_t = \lambda(X_t)$, the doubly stochastic Poisson process N_t is driven by X_t (see Chapter II, after D7, where this terminology was introduced). Let us now examine a few consequences of the above assumptions:

(i) By T3 and T4, L_t is a $(Q, \mathscr{F}_t)$-martingale, and

$$N_t - \int_0^t \lambda_s \, ds \text{ is a } (P, \mathscr{F}_t)\text{-martingale.} \tag{3.4}$$

(ii) The restrictions of Q and P to $(\Omega, \mathscr{F}_\infty^X)$ are the same. This is because Q_0 and P_0, the restrictions of Q and P respectively to $(\Omega, \mathscr{F}_0) = (\Omega, \mathscr{F}_\infty^X)$, are such that, by (3.3), $dP_0/dQ_0 = 1$.

(iii) $\mathscr{F}_\infty^N$ and $\mathscr{F}_\infty^X$ are Q-independent. Indeed, the hypothesis that N_t is a $(Q, \mathscr{F}_t)$-Poisson process implies, by Watanabe's theorem (II, T6), that for all $0 \leq s \leq t$, $N_t - N_s$ is Q-independent of $\mathscr{F}_s$. In particular, letting $s = 0$ in the previous statement, we see that N_t is Q-independent of $\mathscr{F}_0 = \mathscr{F}_\infty^X$ for all $t \geq 0$, that is to say, $\mathscr{F}_\infty^N$ and $\mathscr{F}_\infty^X$ are Q-independent.

The goal is to find, for any real-valued bounded process Z_t adapted to $\mathscr{F}_t$, an expression for the estimate $E_P[Z_t | \mathscr{F}_t^N]$.

The following lemma replaces estimation with respect to P by estimation with respect to the reference probability Q. It gives a sophisticated avatar of Bayes' formula:

L5 Lemma. *Let P and Q be two probabilities, and $\mathscr{F}_t$ be some history on $(\Omega, \mathscr{F})$. Suppose that P_t and Q_t, the restrictions of P and Q respectively to $(\Omega, \mathscr{F}_t)$, are such that $P_t \ll Q_t$, and define $L_t = dP_t/dQ_t$. Let Z_t be a real-valued bounded process adapted to $\mathscr{F}_t$. Then, for all histories $\mathscr{G}_t$ such that $\mathscr{G}_t \subset \mathscr{F}_t$, $t \geq 0$,*

$$E_Q[L_t | \mathscr{G}_t] E_P[Z_t | \mathscr{G}_t] = E_Q[Z_t L_t | \mathscr{G}_t] \quad Q\text{-a.s.}, \tag{3.5}$$

or equivalently,

$$E_P[Z_t | \mathscr{G}_t] = \frac{E_Q[Z_t L_t | \mathscr{G}_t]}{E_Q[L_t | \mathscr{G}_t]}, \quad P\text{-a.s.} \tag{3.5'}$$

PROOF. By definition of conditional expectation, for any $A \in \mathcal{G}_t$,

$$\int_A Z_t \, dP = \int_A E_P[Z_t | \mathcal{G}_t] \, dP. \qquad (3.6)$$

Also, by definition of L_t, and since A and Z_t are $\mathcal{F}_t$-measurable,

$$\int_A Z_t \, dP = \int_A Z_t L_t \, dQ. \qquad (3.7)$$

Similarly

$$\int_A E_P[Z_t | \mathcal{G}_t] \, dP = \int_A E_P[Z_t | \mathcal{G}_t] L_t \, dQ. \qquad (3.8)$$

The right-hand sides of (3.7) and (3.8) can be rewritten as $\int_A E_Q[Z_t L_t | \mathcal{G}_t] \, dQ$ and $\int_A E_P[Z_t | \mathcal{G}_t] E_Q[L_t | \mathcal{G}_t] \, dQ$ respectively (by the definition of conditional expectation). Therefore (3.6) becomes

$$\int_A E_Q[Z_t L_t | \mathcal{G}_t] \, dQ = \int_A E_P[Z_t | \mathcal{G}_t] E_Q[L_t | \mathcal{G}_t] \, dQ. \qquad (3.9)$$

The set A being arbitrary in $\mathcal{G}_t$, and the integrands in the above equality being $\mathcal{G}_t$-measurable, we thus obtain (3.5). The equality (3.5') follows from (3.5) by remarking that the set $C_t = \{\omega | E_Q[L_t | \mathcal{G}_t](\omega) = 0\}$ has P-measure 0 (indeed, $P(C_t) = \int_{C_t} L_t \, dQ$, and C_t is in $\mathcal{G}_t$; therefore

$$P(C_t) = \int_{C_t} E_Q[L_t | \mathcal{G}_t] \, dQ = 0).$$

Applying Lemma L5 to our situation, where $\mathcal{G}_t = \mathcal{F}_t^N$, we obtain

$$E_P[Z_t | \mathcal{F}_t^N] = \frac{E_Q[Z_t L_t | \mathcal{F}_t^N]}{E_Q[L_t | \mathcal{F}_t^N]} \quad P\text{-a.s.} \qquad (3.10)$$

The estimation problem under P is therefore replaced by an estimation problem under Q: namely, one has to find expressions for quantities such as $E_Q[Z_t L_t | \mathcal{F}_t^N]$.

Remark. Clearly (3.10) holds when Z_t is replaced by a bounded real-valued random variable Z, $\mathcal{F}_\infty^X$-measurable, because $Z_t = Z$ is adapted to $\mathcal{F}_t = \mathcal{F}_\infty^X \vee \mathcal{F}_t^N$.

E9 Exercise. Suppose that the change $Q \to P$ described at the beginning of this section is given by $\mathcal{F}_t = \sigma(\Lambda) \vee \mathcal{F}_t^N$ and

$$\frac{dP_t}{dQ_t} = \Lambda^{N_t} \exp\{(1 - \Lambda)t\} \qquad (3.11)$$

3. Filtering by the Method of the Probability of Reference

where Λ is a nonnegative random variable of Q-distribution $F(d\lambda)$. In other words, by T3 and T4, N_t is an $\mathscr{F}_t$-doubly stochastic Poisson process with the intensity Λ. Show that

$$E_P[\Lambda|\mathscr{F}_t^N] = \frac{\int_0^\infty \lambda^{N_t+1} e^{-\lambda t} F(d\lambda)}{\int_0^\infty \lambda^{N_t} e^{-\lambda t} F(d\lambda)}. \tag{3.12}$$

(Note that $E_P[\Lambda|\mathscr{F}_t^N]$ is the $(P, \mathscr{F}_t^N)$- intensity of N_t and has the form $g(t, N_t)$.)

An Abstract Filtering Formula

A definition is needed in order to state the main result of this section:

L6 Lemma. *Let H_t be a measurable process defined on some probability space $(\Omega, \mathscr{F}, P)$ and such that*

$$E_P\left[\int_0^t |H_s|\, ds\right] < \infty, \qquad t \geq 0. \tag{3.13}$$

Let $\mathscr{F}_t$ be a history on $(\Omega, \mathscr{F})$. There exists an $\mathscr{F}_t$-predictable process, denoted $\tilde{E}_P[H_t|\mathscr{F}_t]$, such that for all nonbounded $\mathscr{F}_t$-predictable processes C_t

$$E_P\left[\int_0^t C_s H_s\, ds\right] = E_P\left[\int_0^t C_s \tilde{E}_P[H_s|\mathscr{F}_s]\, ds\right]. \tag{3.14}$$

Moreover, this process is defined up to an $\mathscr{F}_t$-predictable set of $dP \times dt$-measure 0.

The proof is similar to that of II, T13 and is left to the reader.

We can now state the central result of this section. It is an abstract formula but a useful one. Applications will be harvested in §4.

T7 Theorem (General Projection Formula [3, 5]). *Under the conditions prevailing in this section,*

$$E_Q[L_t Z|\mathscr{F}_t^N] = E_Q[Z] + \int_0^t E_Q[L_{s-}(\lambda_s - 1)Z|\mathscr{F}_s^N]\, d(N_s - s)$$

$$P\text{-a.s.}, \qquad t \geq 0, \quad (3.15)$$

for any bounded random variable Z that is $\mathscr{F}_\infty^X$-measurable.

PROOF. By T2, L_t can be written

$$L_t = 1 + \int_0^t L_{s-}(\lambda_s - 1)\, d(N_s - s),$$

and therefore

$$L_t Z = Z + \int_0^t L_{s-}(\lambda_s - 1) Z\, d(N_s - s). \tag{3.16}$$

In order to prove that the right-hand side of (3.15), denoted R_t, is actually $E_Q[L_t Z | \mathscr{F}_t^N]$, it is enough to show that

$$E_Q[L_t Z M_t] = E_Q[R_t M_t], \quad t \geq 0, \quad (3.17)$$

for every bounded $(Q, \mathscr{F}_t^N)$-martingale M_t. Indeed, for any $t_0 \geq 0$ and any $A \in \mathscr{F}_{t_0}^N$, let $M_t = E_Q[1_A | \mathscr{F}_t^N]$ and $t = t_0$ in (3.17). Hence

$$E_Q[L_{t_0} Z M_{t_0}] = E_Q[R_{t_0} M_{t_0}], \quad (3.18)$$

that is to say, since $M_{t_0} = 1_A$,

$$E_Q[L_{t_0} Z 1_A] = E_Q[R_{t_0} 1_A]. \quad (3.19)$$

A being arbitrary in $\mathscr{F}_{t_0}^N$, this implies $R_{t_0} = R_Q[L_{t_0} Z | \mathscr{F}_{t_0}^N]$.

By III, T12, a bounded $(Q, \mathscr{F}_t^N)$-martingale M_t has the form

$$M_t = M_0 + \int_0^t H_s \, d(N_s - s), \quad (3.20)$$

where H_t is an $\mathscr{F}_t^N$-predictable process such that $E_Q[\int_0^t H_s^2 \, ds] < \infty$, $t \geq 0$. Now in view of (3.20) and of the definition of R_t, the equality (3.17) is implied by

$$E_Q\left[\int_0^t L_{s-}(\lambda_s - 1) Z \, d(N_s - s) \int_0^t H_s \, d(N_s - s)\right]$$

$$= E_Q\left(\int_0^t E_Q[L_{s-}(\lambda_s - 1) Z | \mathscr{F}_s^N] \, d(N_s - s)\right) \int_0^t H_s \, d(N_s - s). \quad (3.21)$$

By the isometry of III, T13, the above equality is equivalent to

$$E_Q\left[\int_0^t L_{s-}(\lambda_s - 1) Z H_s \, ds\right] = E_Q\left[\int_0^t E_Q[L_{s-}(\lambda_s - 1) Z | \mathscr{F}_s^N] H_s \, ds\right], \quad (3.22)$$

which follows directly from the definition of $E_Q[\cdots | \mathscr{F}_t^N]$ in Lemma L5. □

4. Applications

Separation of Detection and Filtering

R8 Result (Detection Formula [12, 3]). *Under the conditions prevailing in this section,*

$$E_Q[L_t | \mathscr{F}_t^N] = \left(\prod_{0 < s \leq t} \hat{\lambda}_s \Delta N_s\right) \exp\left\{\int_0^t (1 - \hat{\lambda}_s) \, ds\right\}, \quad (4.1)$$

where $\hat{\lambda}_t$ is an $\mathscr{F}_t^N$-predictable version of the $(P, \mathscr{F}_t^N)$-intensity of N_t.

4. Applications

PROOF. Apply Theorem T7 to $Z = 1$ to obtain

$$E_Q[L_t|\mathcal{F}_t^N] = 1 + \int_0^t E_Q[L_{s-}(\lambda_s - 1)|\mathcal{F}_s^N] \, d(N_s - s). \tag{4.2}$$

In view of (4.2) it suffices, in order to prove Result R8, to show that

$$E_Q[L_{t-}(\lambda_t - 1)|\mathcal{F}_t^N] = \hat{L}_{t-}(\hat{\lambda}_s - 1), \tag{4.3}$$

where $\hat{L}_t = E_Q[L_t|\mathcal{F}_t^N]$; indeed, if (4.3) is true, (4.2) then reads

$$\hat{L}_t = 1 + \int_0^t \hat{L}_{s-}(\hat{\lambda}_s - 1) \, d(N_s - s), \tag{4.4}$$

that is to say,

$$\hat{L}_t = \left(\prod_{s \le t} \hat{\lambda}_s \Delta N_s\right) \exp\left\{\int_0^t (1 - \hat{\lambda}_s) \, ds\right\}, \tag{4.5}$$

which is the announced result. We must now prove the equality (4.3).
Integration by parts yields

$$\int_0^t C_s L_{s-}(\lambda_s - 1) \, ds = L_t \int_0^t C_s(\lambda_s - 1) \, ds$$
$$- \int_0^t \left(\int_0^s C_u(\lambda_u - 1) \, du\right) dL_s. \tag{4.6}$$

Now

$$\int_0^t \left(\int_0^s C_u(\lambda_u - 1) \, du\right) dL_s = \int_0^t \left(\int_0^s C_u(\lambda_u - 1) \, du\right) L_{s-}(\lambda_s - 1) \, d(N_s - s) \tag{4.7}$$

is a $(Q, \mathcal{F}_t^N)$-martingale, and therefore

$$E_Q\left[\int_0^t L_{s-} C_s(\lambda_s - 1) \, ds\right] = E_Q\left[L_t \int_0^t C_s(\lambda_s - 1) \, ds\right], \tag{4.8}$$

or, by definition of L_t,

$$E_Q\left[\int_0^t L_{s-} C_s(\lambda_s - 1) \, ds\right] = E_P\left[\int_0^t C_s(\lambda_s - 1) \, ds\right]. \tag{4.9}$$

By the definition of $\hat{\lambda}_t$ as the $(P, \mathcal{F}_t^N)$-intensity of N_t,

$$E_P\left[\int_0^t C_s \hat{\lambda}_s \, ds\right] = E_P\left[\int_0^t C_s \, dN_s\right], \tag{4.10}$$

and by the definition of λ_t as $(P, \mathcal{F}_t)$-intensity,

$$E_P\left[\int_0^t C_s \lambda_s \, ds\right] = E_P\left[\int_0^t C_s \, dN_s\right]. \tag{4.11}$$

The above two equalities yield

$$E_P\left[\int_0^t C_s(\lambda_s - 1)\, ds\right] = E_P\left[\int_0^t C_s(\hat{\lambda}_s - 1)\, ds\right]. \tag{4.12}$$

By definition of L_t,

$$E_P\left[\int_0^t C_s(\hat{\lambda}_s - 1)\, ds\right] = E_Q\left[L_t \int_0^t C_s(\hat{\lambda}_s - 1)\, ds\right]. \tag{4.13}$$

By the same arguments as those leading to (4.8).

$$E_Q\left[L_t \int_0^t C_s(\hat{\lambda}_s - 1)\, ds\right] = E_Q\left[\int_0^t L_s C_s(\hat{\lambda}_s - 1)\, ds\right]. \tag{4.14}$$

By Fubini,

$$E_Q\left[\int_0^t L_s C_s(\hat{\lambda}_s - 1)\, ds\right] = E_Q\left[\int_0^t \hat{L}_s C_s(\hat{\lambda}_s - 1)\, ds\right]. \tag{4.15}$$

Now, since the discontinuity points of $\hat{L}_t$ have Q-a.s. Lebesgue measure 0,

$$E_Q\left[\int_0^t L_s C_s(\hat{\lambda}_s - 1)\, ds\right] = E_Q\left[\int_0^t L_s C_s(\hat{\lambda}_s - 1)\, ds\right]. \tag{4.16}$$

From the above string of equalities

$$E_Q\left[\int_0^t L_{s-} C_s(\lambda_s - 1)\, ds\right] = E_Q\left[\int_0^t \hat{L}_{s-} C_s(\hat{\lambda}_s - 1)\, ds\right]; \tag{4.17}$$

therefore, by the definition of $E_Q[\cdots]$, (4.3) is proven. $\square$

Result R8 has a special meaning for communications engineers. Indeed, if one wants to discriminate the two hypotheses

(Q) N_t has the $\mathscr{F}_t$ intensity λ_t,
(P) N_t has the $\mathscr{F}_t$ intensity 1

on the basis of observations that are events of $\mathscr{F}_t$, then all standard statistical tests require that the likelihood ratio L_t be compared with some threshold (see §1). If the discrimination is to be made in view of observations lying in $\mathscr{F}_t^N$ only, then the relevant likelihood ratio to be compared with a threshold is $\hat{L}_t$.

Let us use the name *detector* for the operator which associates to a nonnegative process ϕ_t the process $D(\phi)_t$ defined by

$$D(\phi)_t = \left(\prod_{s\leq t} \phi_s\, \Delta N_s\right) \exp\left\{\int_0^t (1 - \phi_s)\, ds\right\}. \tag{4.18}$$

Then T8 reads

$$L_t = D(\lambda)_t, \qquad \hat{L}_t = D(\hat{\lambda})_t. \tag{4.19}$$

4. Applications

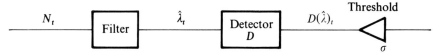

Figure 4 Separation of filtering and detection.

This means that $\hat{L}_t$ is constructed in two steps: first one filters λ_t to obtain $\hat{\lambda}_t$, and then one feeds $\hat{\lambda}_t$ into the detector D. Hence the terminology "separation of detection and filtering" used in the engineering literature on communications. (See Figure 4.)

We will see in §5 an alternative proof and a generalization of R8 based on the martingale representations of Chapter III.

Filtering for Doubly Stochastic Poisson Processes Driven by a Feller Process

In Theorem T7, Z is just any $\mathscr{F}_\infty^X$-measurable bounded variable. We suppose now that $Z = f(X_t)$, where f is a bounded measurable function from U to R, and we assume that

X_t is a U-valued Feller[3] process with semigroup $(P_t, t \geq 0)$. (4.20)

Then theorem T7 takes the following form:

R9 Result. *The conditions are those prevailing in Theorem T5, with the additional assumptions*

(i) X_t *is a U-valued Feller process with transition semigroup* $(P_t, t \geq 0)$,
(ii) $Z = f(X_t)$, *where* $f: U \to R$ *is bounded and continuous. We have*

$$E_Q[L_t f(X_t)|\mathscr{F}_t^N] = E_Q[f(X_t)]$$

$$+ \int_0^t E_Q[L_{s-}(\lambda_s - 1)P_{t-s}f(X_s)|\mathscr{F}_s^N]\, d(N_s - s) \quad P\text{-a.s.}, \quad t \geq 0. \quad (4.21)$$

PROOF. It remains to prove that

$$E_Q[L_{s-}(\lambda_s - 1)f(X_t)|\mathscr{F}_s^N] = E_Q[L_{s-}(\lambda_s - 1)P_{t-s}f(X_{s-})|\mathscr{F}_s^N],$$

$$0 \leq s \leq t, \quad (4.22)$$

or equivalently,

$$E_Q\left[\int_0^t C_s L_{s-}(\lambda_s - 1)f(X_t)\, ds\right] = E_Q\left[\int_0^t C_s L_{s-}(\lambda_s - 1)P_{t-s}f(X_s)\, ds\right] \quad (4.23)$$

---

[3] By definition of a Feller process, if f is continuous, then $P_t f$ is continuous. This assumption is always met in practical cases and is here to guarantee that processes such as $P_t f(X_t)$ will be progressive, being right-continuous.

for all bounded $\mathscr{F}_t^N$-predictable processes C_t. Now

$$P_{t-s}f(X_s) = E_Q[f(X_t)|\sigma(X_s)] = E_Q[f(X_t)|\mathscr{F}_s^N \vee \mathscr{F}_s^X] \quad (4.24)$$

(where we have used the Q-independence of N_t and X_t and the markovian property); therefore

$$E_Q\left[\int_0^t V_s f(X_t)\, ds\right] = E_Q\left[\int_0^t V_s P_{t-s} f(X_s)\, ds\right] \quad (4.25)$$

for all measurable process V_t adapted to $\mathscr{F}_t^X \vee \mathscr{F}_t^N$, such that $E[\int_0^t V_s\, ds] < \infty$. By specializing V_t, the latter equality yields (4.23). □

R9 leads to explicit formulas for the estimates of a Markov chain with point-process observations. In order to give them in a case general enough for practical purposes, we must consider k-variate observations; that is to say, N_t is a k-variate point process with the $(P, \mathscr{F}_t)$-intensity $(\lambda_t(1), \ldots, \lambda_t(k))$. Equations (3.15) and (4.21) are still valid if we read $E_Q[L_{s-}(\lambda_s - 1)Z|\mathscr{F}_s^N] \times d(N_s - s)$ and $E_Q[L_{s-}(\lambda_s - 1)P_{t-s}f(X_s)|\mathscr{F}_s^N]\, d(N_s - s)$ as scalar products; for instance (4.21) becomes

$$E_Q[L_t f(X_t)|\mathscr{F}_t^N] = E_Q[f(X_t)]$$
$$+ \sum_{j=1}^k \int_0^t E_Q[L_{s-}(\lambda_s(j) - 1)P_{t-s}f(X_t)|\mathscr{F}_s^N](dN_s(j) - ds). \quad (4.26)$$

Here

$$L_t = 1 + \sum_{j=1}^k \int_0^t L_{s-}(\lambda_s(j) - 1)(dN_s(j) - ds) \quad (4.27)$$

summarizes the change of probability $Q \to P$, where Q is such that N_t is a k-variate standard Poisson process.

Result R9 will be specialized to the case where X_t is a homogeneous Markov chain with finite space $(1, 2, \ldots, m)$ and with infinitesimal characteristics q_{ij} such that

$$q_{ii} = -\sum_{j \neq i} q_{ij}. \quad (4.28)$$

Let $\mathscr{L}$ be the associated infinitesimal operator defined by

$$(\mathscr{L}f)(t, i) = \sum_{j \neq i} q_{ij} f(t, j). \quad (4.29)$$

Also suppose that

$$\lambda_s(j, \omega) = \lambda(j, s, \omega, X_s(\omega)) \quad (4.30)$$

where the mappings $t, \omega \to \lambda(j, t, \omega, i)$ are $\mathscr{F}_t^N$-predictable. We use the notation

$$P_{ij}(t) = P[X_{t+s} = j | X_s = i], \quad P(t, i) = P[X_t = i]. \quad (4.31)$$

Then:

R10 Result [15]. *Under the above assumptions the processes $V(t, i)$ defined by*

$$V(t, i) = E_Q[L_t 1(X_t = i) | \mathscr{F}_t^N], \quad 1 \le i \le m, \quad (4.32)$$

have right-continuous left-hand-limited versions which satisfy

$V(t, i) = P(t, i)$
$$+ \sum_{j=1}^{k} \sum_{l=1}^{m} \int_0^t (V(s-, l)(\lambda(j, s, l) - 1)P_{li}(t - s)) \, d(N_s(j) - s) \quad (4.33)$$

and

$$V(t, i) = P(0, i) + \int_0^t \mathscr{L}V(s, i) \, ds$$
$$+ \sum_{j=1}^{k} \int_0^t V(s-, i)(\lambda(j, s, i) - 1) \, d(N_s(j) - s). \quad (4.34)$$

Remark. $V(t, i)$ is the *pseudoconditional density* of X_t given $\mathscr{F}_t^N$; by L3, $P[X_t = i | \mathscr{F}_t^N] = V(t, i)/\sum_{i=1}^{m} V(t, i)$.

PROOF. That $V(t, i)$ admits right-continuous left-hand-limited versions follows from (4.26) applied to $f(x) = 1(x = i)$. Equation (4.33) is just a transcription of (4.26). We will prove (4.34) in the case where $k = 1$ for simplicity. In this case (4.33) reads, if we let $N_t(1) = N_t$, $\lambda(1, s, l) = \lambda(s, l)$,

$$V(t, i) = P(t, i) + \sum_{l=1}^{m} \int_0^t (V(s-, l)(\lambda(s, l) - 1)P_{li}(t - s)) \, d(N_s - s). \quad (4.35)$$

At a jump of N_t

$$\Delta V(t, i) = V(t, i) - V(t-, i)$$
$$= \sum_{l=1}^{m} V(t-, l)(\lambda(s, l) - 1)P_{li}(0)$$
$$= V(t-, i)(\lambda(s, i) - 1), \quad (4.36)$$

since $P_{li}(0) = \delta_{li}$. At a time t strictly between two jumps of N_t, we have, using the properties of the infinitesimal characteristics q_{ij},

$$\frac{dV(t, i)}{dt} = \frac{dP(t, i)}{dt} + \sum_{l=1}^{m} V(l, t)(\lambda(t, l) - 1)P_{li}(0)$$

$$+ \sum_{l=1}^{m} \int_0^t V(s-, l)(\lambda(s, l) - 1)\frac{dP_{li}(t - s)}{dt} ds$$

$$= \sum_{j=1}^{m} q_{ij} P(t, j) + V(t, i)(\lambda(t, i) - 1)$$

$$+ \sum_{j=1}^{m} q_{ij} \sum_{l=1}^{m} \int_0^t V(s-, l)(\lambda(s, l) - 1)P_{lj}(t - s) ds$$

$$= \mathscr{L}V(t, i) + V(t, i)(\lambda(t, i) - 1). \tag{4.37}$$

The announced equation follows by combining (4.36) and (4.37). $\square$

E10 Exercise. Apply Result R10 to the situation where the observation is univariate N_t, with the intensity $\lambda(X_t)$, and where X_t is a telegraph signal, i.e., X_t takes the values 0 and 1, is Markov, and has the infinitesimal parameters q_{01} and q_{10}. Compare with IV, E11.

5. The Capacity of a Point-Process Channel

In order to transmit a random time-varying signal θ_t, a point-process channel is used. In mathematical terms: on a probability space $(\Omega, \mathscr{F}, P)$ two processes θ_t and N_t are given, the latter being a point process. With respect to the history

$$\mathscr{G}_t = \mathscr{F}_\infty^\theta \vee \mathscr{F}_t^N \tag{5.1}$$

N_t has an intensity λ_t of the form

$$\lambda_t = \lambda + \mu_t, \tag{5.2}$$

where for some positive constant c

$$0 \leq \mu_t \leq c, \tag{5.3}$$

and where

$$\mu_t \text{ is adapted to } \mathscr{F}_t^\theta \vee \mathscr{F}_t^N. \tag{5.4}$$

The communications-theoretical interpretation of this transmission scheme is the following: θ_t is the *signal*, μ_t is a *coding* of the signal, and N_t is the *noisy observation* of θ_t.

There are two kinds of noises. One is *background noise*: when $\mu_t \equiv 0$, N_t is Poisson with intensity λ. The other noise is essential to point-process transmission in that the intensity of λ_t of a point process N_t can never be exactly recovered from the observation of a finite stretch of trajectory $(N_s, s \in [0, t])$. The condition (5.3) is a power *constraint* on the coding modulation. As to (5.4), it says two things: coding does not anticipate on the message [at each time t, μ_t depends on the process θ_t only through $(\theta_s, s \in [0, t])$], and nonanticipative feedback is allowed [i.e., at time t, μ_t may depend on $(N_s, s \in [0, t])$].

Since μ will vary, P will be denoted by P_μ. P_0 then represents a probability measure under which N_t is a Poisson process with intensity λ.

For fixed μ and fixed t, let us denote by $P_{\mu,t}$, $P_{\mu,t}^\theta$, $P_{\mu,t}^N$ the restrictions of P_μ to $\mathscr{G}_t$, $\mathscr{F}_t^\theta$, $\mathscr{F}_t^N$ respectively. From the Girsanov-type theorems T3 and R8, we see that $P_{\mu,t} \ll P_{0,t}$ and $P_{\mu,t}^N \ll P_{\mu,t}^N$, and that the corresponding Radon–Nikodym derivatives are given by

$$\frac{dP_{\mu,t}}{dP_{0,t}} = \exp\left\{\int_0^t \log\left(1 + \frac{\mu_s}{\lambda}\right) dN_s - \int_0^t \mu_s \, ds\right\}, \tag{5.5}$$

$$\frac{dP_{\mu,t}^N}{dP_{0,t}^N} = \exp\left\{\int_0^t \log\left(1 + \frac{\hat{\mu}_s}{\lambda}\right) dN_s - \int_0^t \hat{\mu}_s \, ds\right\}, \tag{5.6}$$

where $\hat{\mu}_t$ is the predictable P_μ-projection of μ_t on $\mathscr{F}_t^N$. Also, as was already discussed in §3,

$$P_{\mu,t}^\theta \equiv P_{0,t}^\theta, \tag{5.7}$$

and θ_t and N_t are P_0-independent processes.

By definition, the P_μ-mutual information between θ_t and N_t on the interval $[0, t]$, denoted $I_\mu^t(\theta, N)$, is given symbolically by

$$I_\mu^t(\theta, N) = E_\mu\left[\log \frac{dP_{\mu,t}^{\theta,N}}{dP_{\mu,t}^\theta \, dP_{\mu,t}^N}\right] \tag{5.8}$$

with the following interpretation:

$$\frac{dP_{\mu,t}^{\theta,N}}{dP_{\mu,t}^\theta \, dP_{\mu,t}^P} = \frac{dP_{\mu,t}}{dP_{0,t}} \bigg/ \left(\frac{dP_{\mu,t}^\theta}{dP_{0,t}^\theta} \frac{dP_{\mu,t}^N}{dP_{0,t}^\theta}\right). \tag{5.9}$$

E11 Exercise. Check that the expression for $dP_{\mu,t}^{\theta,N}/dP_{\mu,t}^\theta \, dP_{\mu,t}^N$ does not depend upon P_0 in the following sense: if P_0' is any propability measure on $(\Omega, \mathscr{F})$ such that $P_{0,t}'$ is equivalent to $P_{0,t}$, and θ_t and N_t are P_0'-independent, then

$$\frac{dP_{\mu,t}}{P_{0,t}} \bigg/ \left(\frac{dP_{\mu,t}^\theta}{dP_{0,t}^\theta} \frac{dP_{\mu,t}^N}{dP_{0,t}^N}\right) = \frac{dP_{\mu,t}}{dP_{0,t}'} \bigg/ \left(\frac{dP_{\mu,t}^\theta}{dP_{0,t}'^\theta} \frac{dP_{\mu,t}^N}{dP_{0,t}'^N}\right).$$

In view of (5.7), the mutual information between θ_t and N_t takes the simple form

$$I_\mu^t(\theta, N) = E_\mu\left[\log \frac{dP_{\mu,t}}{dP_{0,t}} - \log \frac{dP_{\mu,t}^N}{dP_{0,t}^N}\right], \tag{5.10}$$

and therefore, using the expressions (5.5) and (5.6) and the integration theorem (II, T8),

$$I_\mu^t(\theta, N) = E_\mu\left[\int_0^t (\varphi(\mu_s) - \varphi(\hat{\mu}_s))\, ds\right], \tag{5.11}$$

where

$$\varphi(x) = (\lambda + x)\log(\lambda + x) - \lambda \log \lambda. \tag{5.12}$$

The capacity C of the above point-process channel with the constraints (5.2), (5.3), and (5.4) is defined by

$$C = \frac{1}{t} \sup_\theta \sup_\mu I_\mu^t(\theta, N), \tag{5.13}$$

where the suprema involved are taken in the following sense: $\sup_\theta$ is over all the signals θ_t with no restriction whatsoever on the state space, the form of the trajectories, or the statistics [the constraints (5.2), (5.3), and (5.4) bear on the coding modulation μ_t and not on the signal θ_t itself]; the $\sup_\mu$ is over all coding modulations μ_t satisfying (5.4).

We will not recall the physical interpretation of the capacity of a channel given by Shannon {134}. In the case of continuous sources, such as those considered here, the interpretation is that of the rate-distortion theory. In imprecise words, the capacity is the maximum rate of transmission of the signal for which there exists an admissible coding μ_t ($0 \le \mu_t \le c$) such that the distortion between θ_t and some suitable decoding $\hat{\theta}_t$ can be made as small as wished.

We now turn to the computation of capacity, following Kabanov [10] very closely. It is given by

$$C = \lambda\left[\frac{1}{e}\left(1 + \frac{e}{\lambda}\right)^{1 + \lambda/c} - \left(1 + \frac{\lambda}{c}\right)\log\left(1 + \frac{c}{\lambda}\right)\right]. \tag{5.14}$$

To prove this, an upper estimate of C is obtained via Jensen's inequality:

$$I_\mu^t(\theta, N) \le \int_0^t (E_\mu[\varphi(\mu_s)] - \varphi(E_\mu[\mu_s]))\, ds, \tag{5.15}$$

where we have used the equality $E_\mu[\mu_t] = E_\mu[\hat{\mu}_t]$.

Let $\mathscr{A}$ be the set of probability measures Q on $([0, c], B([0, c]))$, and denote $\int_0^c f(x)Q(dx)$ by $Q(f)$; also, let i be the identity function. The quantity $E_\mu[\varphi(\mu_s)] - \varphi(E_\mu[\mu_s])$ is of the form $Q(\varphi) - \varphi(Q(i))$, and therefore

$$C \le \sup_{Q \in \mathscr{A}} (Q(\varphi) - \varphi(Q(i))). \tag{5.16}$$

Denote the right-hand side of (5.16) by U; then $U = \max_{0 \le k \le c} U(k)$, where

$$U(k) = \max_{Q \in \mathscr{A},\, Q(i) = k} (Q(\varphi) - \varphi(k)). \tag{5.17}$$

5. The Capacity of a Point-Process Channel

It therefore remains to compute the quantity $\max_{Q \in \mathscr{A}, Q(i)=k} Q(\varphi)$, which is equal to

$$\max_{Q \in \mathscr{A}, Q(i)=k} (Q(\varphi - bi)) + bk, \quad (5.18)$$

where

$$b = \frac{\varphi(c)}{c}. \quad (5.19)$$

In fact the equality holds for any b, but the particular choice (5.19) is convenient. Indeed, the function

$$x \to \varphi(x) - bx = \varphi(x) - \frac{\varphi(c)}{c} x$$

then has the shape shown in Figure 5.

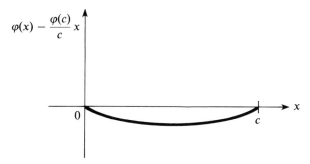

Figure 5

It is then clear that for fixed k, the element Q of $\mathscr{A}$ which maximizes $U(k)$ subject to the constraint $Q(i) = k$ is given by

$$Q(\{0\}) = 1 - \frac{k}{c}, \qquad Q(\{c\}) = \frac{k}{c}. \quad (5.20)$$

Therefore

$$U = \max_{0 \le k \le c} \left(k \frac{\varphi(c)}{c} - \varphi(k) \right). \quad (5.21)$$

Hence

$$U = \lambda \left[\frac{1}{e} \left(1 + \frac{c}{\lambda} \right)^{1+\lambda/c} - \left(1 + \frac{\lambda}{c} \right) \log \left(1 + \frac{c}{\lambda} \right) \right], \quad (5.22)$$

and the value of k for which the maximum of $U(k)$ is attained is

$$k_m = \lambda \left(\frac{1}{e} \left(1 + \frac{c}{\lambda} \right)^{1+\lambda/c} - 1 \right). \quad (5.23)$$

So far, we have found that $C \leq U$, and it remains to show that $C = U$. For this we will find a sequence $\theta_t^{(n)}$ of signal processes and a coding μ_t such that

$$\lim_{n\uparrow\infty} \frac{1}{t} I_\mu^t(\theta^{(n)}, N) = U. \tag{5.24}$$

The coding will be chosen linear, i.e.

$$\mu_t = c\theta_t. \tag{5.25}$$

Each $\theta_t^{(n)}$ will be a telegraph signal, that is to say, a right-continuous Markov chain taking two values 0 or 1 (such process was already considered in E10) with the infinitesimal parameters

$$q_{01} = n, \qquad q_{10} = n(1-p)/p, \tag{5.26}$$

where

$$p = \frac{k_m}{c}. \tag{5.27}$$

Also, for all n, $\theta_0^{(n)}$ is assumed to have the distribution

$$P[\theta_0^{(n)} = 1] = p, \tag{5.28}$$

so that the processes $\theta_t^{(n)}$ are in statistical equilibrium. Omitting the superscript n, the process $\theta_t^{(n)}$ admits the representation

$$\theta_t = \theta_0 + n \int_0^t 1(\theta_s = 0)\, ds$$
$$- \frac{n(1-p)}{p} \int_0^t 1(\theta_s = 1)\, ds + m_t, \tag{5.29}$$

where m_t is a martingale. Equivalently, since θ_t takes only the values 0 or 1,

$$\theta_t = \theta_0 + \frac{n}{p} \int_0^t (p - \theta_s)\, ds + m_t. \tag{5.30}$$

Let $\hat{\theta}_t$ be a right-continuous version of $E[\theta_t | \mathscr{F}_t^N]$; by (IV, E11),

$$\hat{\theta}_t = p + \frac{n}{p} \int_0^t (p - \hat{\theta}_s)\, ds$$
$$+ \int_0^t \frac{c\hat{\theta}_{s-}(1 - \hat{\theta}_{s-})}{1 + c\hat{\theta}_{s-}} (dN_s - (1 + c\hat{\theta}_s)\, ds). \tag{5.31}$$

Let us now compute $E_\mu[\varphi(c\theta_t)] - E_\mu[\varphi(c\hat{\theta}_t)]$. Clearly

$$E_\mu[\varphi(c\theta_t)] = \frac{k_m}{c} \varphi(c), \tag{5.32}$$

since $\varphi(c\theta_t) = \theta_t\varphi(c) + (1 - \theta_t)\varphi(0) = \theta_t\varphi(c)$ and $E_\mu[\theta_t] = p = k_m/c$. Therefore it remains to show that

$$\lim_{n\uparrow\infty} E_\mu[\varphi(c\hat{\theta}_t^{(n)})] = \varphi(k_m) \tag{5.33}$$

(where we have reintroduced upper indices). Since the derivative of φ is bounded on $[0, c]$, say by a constant M, we have the bound

$$|\varphi(c\hat{\theta}_t) - \varphi(k_m)| \le M|c\hat{\theta}_t - k_m| \le Mc|\hat{\theta}_t - p|, \tag{5.34}$$

and further, using the Schwarz inequality,

$$E_\mu[|\varphi(c\hat{\theta}_t) - \varphi(k_m)|] \le McE_\mu[|\hat{\theta}_t - p|^2]^{1/2}. \tag{5.35}$$

Therefore we need only prove that

$$\lim_{n\uparrow\infty} E_\mu[\widehat{\theta_t^{(n)}}^2] = p^2. \tag{5.36}$$

To show this we start from (5.31) and apply Stieltjes calculus to obtain

$$\hat{\theta}_t^2 = p^2 + \frac{2n}{p}\int_0^t \hat{\theta}_s(p - \hat{\theta}_s)\,ds$$

$$+ \int_0^t \frac{c^2[\hat{\theta}_s(1 - \hat{\theta}_s)]^2}{1 + c\hat{\theta}_s}\,ds + \tilde{m}_t, \tag{5.37}$$

where $\tilde{m}_t$ is a martingale. Hence

$$E_\mu[\hat{\theta}_t^2] = p^2 + \frac{2n}{p}E_\mu\left[\int_0^t \hat{\theta}_s(p - \hat{\theta}_s)\,ds\right]$$

$$+ E_\mu\left[\int_0^t \frac{c^2[\hat{\theta}_s(1 - \hat{\theta}_s)]^2}{1 + c\hat{\theta}_s}\,ds\right]. \tag{5.38}$$

Denote $E_\mu[\hat{\theta}_s^2]$ by Z_s; (5.38) becomes

$$Z_t = p^2 + \frac{2n}{p}\int_0^t (p^2 - Z_s^2)\,ds + \int_0^t g_s\,ds, \tag{5.39}$$

where

$$g_s = E_\mu\left[\frac{c^2[\hat{\theta}_s(1 - \hat{\theta}_s)]^2}{1 + c\hat{\theta}_s}\right]. \tag{5.40}$$

Observe that g_t is bounded. The solution to (5.39) is

$$Z_t^{(n)} = p^2 e^{-2(n/p)t} + e^{-2(n/p)t}\int_0^t (g_s^{(n)} + 2np)e^{2(n/p)s}\,ds \tag{5.41}$$

(where the superscripts have been reintroduced). Hence, in view of the boundedness of $g_t^{(n)}$, $\lim_{n\uparrow\infty} Z_t^{(n)} = p^2$.

Now the proof is terminated, but much more has been proven than the equality (5.14). Indeed, the coding μ_t given by (5.25) does not involve feedback, in the sense that

$$\mu_t \text{ is adapted to } \mathscr{F}_t^\theta \tag{5.42}$$

(μ_t does not depend of N_t directly). Hence:

T11 Theorem [10]. *Feedback does not improve the capacity of the point-process channel* (5.2), (5.3) *This capacity is*

$$C = \lambda \left[\frac{1}{e}\left(1 + \frac{c}{\lambda}\right)^{1+\lambda/c} - \left(1 + \frac{\lambda}{c}\right)\log\left(1 + \frac{c}{\lambda}\right) \right]. \tag{5.43}$$

An analogous result has been proven for the white gaussian channel by Kadota, Zakai, and Ziv {80}. The capacity of the above channel subjected to an additional constraint on the energy:

$$\int_0^t \mu_s \, ds \leq p_0 t \tag{5.44}$$

has been computed in [6]. Define

$$h(x) = \frac{x}{c}\varphi(c) - \varphi(x). \tag{5.45}$$

Then, from (5.11),

$$I_\mu^t(\theta, N) \leq \int_0^t h(E_\mu[\mu_s]) \, ds. \tag{5.46}$$

Two cases arise. First suppose that $p_0 \geq k_m$; then since $h(x)$ is maximum for $x = k_m$, it is clear that the maximum of $\int_0^t h(E_\mu[\mu_s]) \, ds$ is attained when $E\mu[\mu_s] = k_m$, and the energy constraint (5.44) is respected. But the coding μ_t described by (5.25), (5.28) does satisfy $E_\mu[\mu_s] = k_m$, and therefore, we find that the additional energy constraint is not operative in case $p_0 \geq k_m$.

In case $p_0 < k_m$, however, it is the energy constraint which is predominant. Indeed, the maximum $E_\mu[\mu_s]$ that (5.44) allows is $E_\mu[\mu_t] = p_0$, and $h(x)$ is maximized on $[0, p_0]$ at $x = p_0$; therefore $I_\mu^t(\theta, N) \leq th(p_0)$. Now a sequence of signals $\theta_t^{(n)}$ and a coding μ_t do achieve this extremal value: it is in fact the same coding and the same sequence as described in (5.25)–(5.28) with $p = k_m/c$ replaced by $p = p_0/c$.

In summary [6]:

(α) If $p_0 \geq k_m$, the peak constraint (5.3) predominates over the energy constraint (5.44) and C is given by (5.43).
(β) If $p_0 < k_m$, the energy constraint (5.44) predominates and the capacity is

$$C = \frac{p_0}{c}\varphi(c) - \varphi(p_0)$$

[φ being given by (5.12)].

6. Detection Formula

In §2, we have described an absolutely continuous change of probability measure $P \to \tilde{P}$ in terms of an exponential martingale. In this section we will see that such changes, as described in T3, are the most general ones in some sense. More precisely, in the univariate case, let N_t be a point process defined on $(\Omega, \mathcal{F})$, and let P and $\tilde{P}$ be two probability measures on $(\Omega, \mathcal{F})$ such that for all $t \geq 0$,

$$\tilde{P}_t \ll P_t, \tag{6.1}$$

where P_t and $\tilde{P}_t$ are the restrictions of P and $\tilde{P}$ respectively to $\mathcal{F}_t^N$. Suppose moreover that N_t admits the $(P, \mathcal{F}_t^N)$-intensity λ_t. It will be shown that:

(α) N_t admits a $(\tilde{P}, \mathcal{F}_t^N)$-intensity $\hat{\lambda}_t$ of the form $\hat{\lambda}_t = \mu_t \lambda_t$, where μ_t is $\mathcal{F}_t^N$-predictable,

(β) $d\tilde{P}_t / dP_t = (\prod_{n \geq 1} \mu_{T_n} 1(T_n \leq t)) \exp\{\int_0^t (1 - \mu_s) \lambda_s \, ds\}$.

Such a result can be considered as the converse of the direct Radon–Nikodym theorem T3. Note however that the history $\mathcal{F}_t$ in T3 is more general than the internal history $\mathcal{F}_t^N$ to which we shall be restricted in this section.

Here are a generalization and the proof of the above result.

T12 Theorem (Converse Radon–Nikodym Theorem [2, 3, 4, 11]). *Let $(N_t(1), \ldots, N_t(k))$ be a k-variate point process defined on $(\Omega, \mathcal{F})$ together with two probability measures P and $\tilde{P}$ such that, for all $t \geq 0$,*

$$\tilde{P}_t \ll P_t, \tag{6.2}$$

where P_t and $\tilde{P}_t$ are the restrictions to $\mathcal{F}_t^N$ of P and $\tilde{P}$ respectively. Suppose that N_t admits the $(P, \mathcal{F}_t^N)$-intensity $(\lambda_t(1), \ldots, \lambda_t(k))$. Then there eixts a $\mathcal{F}_t^N$-predictable process $(\mu_t(1) \ldots, \mu_t(k))$ with nonnegative components such that, for each $1 \leq i \leq k$ and all $t \geq 0$, $(\mu_t(1)\lambda_t(1), \ldots, \mu_t(k)\lambda_t(k))$ is the $(\tilde{P}, \mathcal{F}_t^N)$-intensity of N_t. Moreover

$$\frac{d\tilde{P}_t}{dP_t} = L_t = \prod_{i=1}^{k} L_t(i), \tag{6.3}$$

where, for all $1 \leq i \leq k$,

$$L_t(i) = \left(\prod_{n \geq 1} \mu_{T_n(i)}(i) 1(T_n(i) \leq t) \right) \exp\left\{ \int_0^t (1 - \mu_s(i))\lambda_s(i) \, ds \right\} \tag{6.4}$$

[as usual, the product in (6.4) is taken to be 1 if $T_1(i) > t$].

PROOF. We carry out the proof in the univariate case with the symbols $N_t, \lambda_t, \mu_t, T_n$ replacing $N_t(1), \lambda_t(1), \mu_t(1), T_n(1)$ respectively.

As was seen in I, E4, $L_t = d\tilde{P}_t/dP_t$ is a $(P, \mathscr{F}_t^N)$-martingale. Hence, by III, T17, it admits a representation

$$L_t = 1 + \int_0^t K_s(dN_s - \lambda_s \, ds), \tag{6.5}$$

where K_t is a $\mathscr{F}_t^N$-predictable process such that for all $t \geq 0$

$$\int_0^t |K_s| \lambda_s \, ds < \infty \quad P\text{-a.s.} \tag{6.6}$$

Define now μ_t by

$$\mu_t = 1(L_{t-} > 0)\left(\frac{K_t}{L_{t-}} - 1\right). \tag{6.7}$$

Clearly μ_t is an $\mathscr{F}_t^N$-predictable process and

$$K_t = (\mu_t - 1)L_{t-}. \tag{6.8}$$

Therefore

$$L_t = 1 + \int_0^t L_{s-}(\mu_s - 1)(dN_s - \lambda_s \, ds), \tag{6.9}$$

which is equivalent to (6.4) written for the univariate case (see proof of T2). Now by T3, N_t has the $(\tilde{P}, \mathscr{F}_t^N)$-intensity $\hat{\tilde{\lambda}}_t = \mu_t \lambda_t$, and this terminates the proof. □

Theorem T12 can also be viewed as a generalization of Result R8 on the separation of detection and filtering. Indeed let P and $\tilde{P}$ be two probability measures on $(\Omega, \mathscr{F})$, and let N_t be a (say, univariate) point process with a history $\mathscr{F}_t$; suppose that $\tilde{P} \ll P$ on $\mathscr{F}_t$ for all $t \geq 0$, and that N_t admits a $(P, \mathscr{F}_t)$-intensity λ_t. Then:

(i) $\tilde{P} \ll P$ on $\mathscr{F}_t^N$ for all $t \geq 0$.
(ii) N_t admits a $(P, \mathscr{F}_t^N)$-intensity $\hat{\lambda}_t$ (II, T14) and a $(\tilde{P}, \mathscr{F}_t^N)$-intensity $\hat{\tilde{\lambda}}_t$.
(iii) $\hat{\tilde{\lambda}}_t = \hat{\mu}_t \hat{\lambda}_t$ for some $\mathscr{F}_t^N$-predictable process μ_t.
(iv) $d\tilde{P}_t/dP_t = (\prod_{n \geq 1} \hat{\tilde{\lambda}}_{T_n}/\hat{\lambda}_{T_n} 1(T_n \leq t)) \exp\{\int_0^t (\hat{\lambda}_s - \hat{\tilde{\lambda}}_s) \, ds\}$.

If it is known that N_t admits a $(\tilde{P}, \mathscr{F}_t)$-intensity $\tilde{\lambda}_t$, then $\hat{\tilde{\lambda}}_t$ is the projection in the sense of II, T14 of $\tilde{\lambda}_t$ on $\mathscr{F}_t^N$, just as $\hat{\lambda}_t$ is the projection of λ_t on $\mathscr{F}_t^N$. Hence the likelihood ratio $d\tilde{P}_t/dP_t$ has the form

$$\frac{d\tilde{P}_t}{dP_t} = D(\hat{\lambda}, \hat{\tilde{\lambda}})_t, \tag{6.10}$$

where

$$D(\phi, \tilde{\phi})_t = \left(\prod_{n \geq 1} \frac{\tilde{\phi}_{T_n}}{\phi_{T_n}} 1(T_n \leq t)\right) \exp\left\{\int_0^t (\phi_s - \tilde{\phi}_s) \, ds\right\}, \tag{6.11}$$

and this is what is meant by separation of detection and filtering (see discussion after R8): one first filters λ_t and $\tilde{\lambda}_t$ to obtain $\hat{\lambda}_t$ and $\hat{\tilde{\lambda}}_t$ respectively, and then one enters those estimates into the detector D. It is very important to note that $\hat{\lambda}_t$ is the "P-filter" of λ_t, whereas $\hat{\tilde{\lambda}}_t$ is the "$\tilde{P}$-filter" of $\tilde{\lambda}_t$:

$$\hat{\lambda}_t = E[\lambda_t | \mathscr{F}_t^N], \quad P\text{-predictable version,}$$
$$\hat{\tilde{\lambda}}_t = \tilde{E}[\tilde{\lambda}_t | \mathscr{F}_t^N], \quad \tilde{P}\text{-predictable version} \tag{6.12}$$

(the meaning of a phrase such as "P-predictable version" is given by II, T14).

E12 Exercise. In the case where $\lambda_t = \lambda$, devise a proof of R8 based on the innovations theory of filtering developed in Chapter IV.

References

[1] Beutler, F. J. and Dolivo, F. B. (1976) Recursive integral equations for the detection of counting processes, J. Appl. Math. and Opt. **3**, pp. 65–72.
[2] Boel, R., Varaiya, P., and Wong, E. (1975) Martingales on jump processes. Part I: Representation results. Part II: Applications, SIAM J. Control **13**, 5, pp. 999–1061.
[3] Brémaud, P. (1972) A martingale approach to point processes, Memo ERL M 345, Ph.D. Thesis, Dept. of EECS, Univ. of Calif., Berkeley.
[4] Brémaud, P. (1974) The martingale theory of point processes over the real half line, *Control Theory, Numerical Methods, and Computers System Modelling*, Lect. Notes in Econ. and Math. Syst., 107, Springer, pp. 519–542.
[5] Brémaud, P. (1976) Prediction, filtrage et détection pour une observation mixte: méthode de la probabilité de référence, Thèse Doctorat Math., Université de Paris VI.
[6] Davis, M. H. A. (1979) Capacity and cutoff rates for Poisson type channels, preprint.
[7] Doléans-Dade, C. (1970) Quelques applications de la formule de changement de variables pour les semi-martingales, Z. für W. **16**, pp. 181–194.
[8] Duncan, T. E. (1967) Probability densities for diffusion processes with application to non-linear filtering theory and detection theory, Ph.D. Thesis, Stanford Univ.
[9] Girsanov, I. (1960) On transforming a certain class of stochastic processes by absolutely continuous changes of probability measure, Theory of Probability and its Applications, **5**, pp. 285–301.
[10] Kabanov, Y. (1978) The capacity of a channel of the Poisson type, Theory of Probability and Its Applications **23**, 1, pp. 143–147.
[11] Kailath, T. and Segall, A. (1975) Radon–Nikodym derivatives with respect to measures induced by discontinuous independent increment processes, Ann. Probab. **3**, pp. 449–464.
[12] Rubin, I. (1972) Regular point processes and their detection, IEEE Trans. **IT-18**, pp. 547–557.
[13] Rudemo, M. (1972) Doubly stochastic Poisson processes and process control, Adv. Appl. Probab. **4**, 2, pp. 318–338.
[14] Snyder, D. L. (1972) Filtering and detection for doubly stochastic Poisson processes, IEEE Trans. **IT-18**, pp. 97–102.
[15] Yashin, A. L. (1970) Filtering of jump processes, Avtomat. i Telemeh. **5**, pp. 52–58.

Solutions to Exercises, Chapter VI

E1. We use the following notation: $\mu_1 = y$, $\mu_2 = 1 - y$; of course, $0 \le y \le 1$. Also, let the function $f(y, x)$ be defined on $[0, 1) \times [0, \infty)$ by

$$f(y, x) = y + ((1 - y)x - y)1\left(x \le \frac{y}{1 - y}\right). \tag{1}$$

Therefore

$$P_E(y) = \int_\Omega f(y, L) \, dP_1. \tag{2}$$

Formally

$$\frac{dP_E(y)}{dy} = \int_\Omega \frac{\partial f}{\partial y}(y, L) \, dP_1 \tag{3}$$

$\left[x \text{ being fixed, } \dfrac{\partial f}{\partial y}(x, y) \text{ is defined everywhere except at an isolated point given by } x = y/(1 - y)\right]$. But

$$\frac{\partial f}{\partial y}(y, x) = 1\left(x > \frac{y}{1 - y}\right) - x1\left(x \le \frac{y}{1 - y}\right), \tag{4}$$

so that

$$\frac{dP_E(y)}{dy} = \int_\Omega 1\left(L > \frac{y}{1 - y}\right) dP_1 - \int_\Omega L1\left(L \le \frac{y}{1 - y}\right) dP_1. \tag{5}$$

Now $dP_2 = L \, dP_1$, and therefore

$$\frac{dP_E}{dy} = P_1\left(L > \frac{y}{1 - y}\right) - P_2\left(L \le \frac{y}{1 - y}\right). \tag{6}$$

The function $y \to y/(1 - y)$ is strictly increasing from 0 to $+\infty$; therefore if we exclude the trivial case $P_2 \equiv P_1$, dP_E/dy takes the value 0 once and only once in the open interval $(0, 1)$. The mapping $y \to P_E(y)$ is concave and continuous, and $P_E(0) = 0 = P_E(1)$. The value $y_{\max}$ at which $P_E(y)$ attains its maximum is such that

$$P_1\left(L > \frac{\mu_1}{\mu_2}\right) = P_2\left(L \le \frac{\mu_1}{\mu_2}\right), \tag{7}$$

that is to say,

$$P_1(A_2^*) = P_2(A_1^*). \tag{8}$$

Hence

$$P_E = \mu_1 P_1(A_2^*) + \mu_2 P_2(A_1^*) = P_1(A_2^*) = P_2(A_1^*). \tag{9}$$

E2. One must show that for all $A \in \sigma(X)$,

$$P_1(A) = 0 \quad \Rightarrow \quad P_2(A) = 0. \tag{1}$$

Let $\mathscr{G}$ be the family consisting of $\varnothing$ and of all sets $\{X = k\}$, $k \ge 0$, and let $\mathscr{S}$ be the class of sets $A \in \sigma(X)$ such that either (1) is true or $P_1(A) > 0$. By the

Solutions to Exercises, Chapter VI

sequential continuity of probability measures, $\mathscr{S}$ is a d-system. It contains the π-system $\mathscr{G}$ which generates $\sigma(X)$; hence $\mathscr{S} = \sigma(X)$ by the monotone class theorem (Appendix A1, T1).

Define

$$L = \sum_{k=0}^{\infty} 1(X=k) \frac{P_2(X=k)}{P_1(X=k)}. \tag{2}$$

First observe that L is $\sigma(X)$-measurable. It remains to show that for all $A \in \sigma(X)$

$$P_2(A) = \int_A L \, dP_1. \tag{3}$$

Let $\mathscr{T}$ be the collection of sets $A \in \sigma(X)$ such that (3) is true; $\mathscr{T}$ is a d-system which contains the π-system $\mathscr{G}$; hence by A1, T1 again, $\mathscr{T} = \sigma(\mathscr{G}) = \sigma(X)$.

E3. Follows immediately from E2, since

$$P_i(N_t = k) = \exp(-\lambda_i t) \frac{(\lambda_i t)^k}{k!}, \quad k \geq 0. \tag{1}$$

E4. The bayesian test which is optimal with respect to the probability error criterion is:

if $\dfrac{dP_{2,T}}{dP_{1,T}} \geq \dfrac{\mu_1}{\mu_2}$, choose hypothesis P_2;

if $\dfrac{dP_{2,T}}{dP_{1,T}} < \dfrac{\mu_1}{\mu_2}$, choose hypothesis P_1,

or, taking logarithms,

if $\log \dfrac{dP_{2,T}}{dP_{1,T}} = N_T \log \dfrac{\lambda_2}{\lambda_1} + (\lambda_2 - \lambda_1)T \geq \log \dfrac{\mu_1}{\mu_2}$,

choose P_2;

if $\log \dfrac{dP_{2,T}}{dP_{1,T}} = N_T \log \dfrac{\lambda_2}{\lambda_1} + (\lambda_2 - \lambda_1)T < \log \dfrac{\mu_1}{\mu_2}$,

choose P_1.

The probability of error of first kind (choose P_2 when P_1 is in force) is

$$P_1\left(N_T \geq \frac{\log(\mu_1/\mu_2) - (\lambda_2 - \lambda_1)T}{\log(\lambda_2/\lambda_1)}\right) = P_{E,1}. \tag{1}$$

The probability of error of second kind is

$$P_2\left(N_T < \frac{\log(\mu_1/\mu_2) - (\lambda_2 - \lambda_1)T}{\log(\lambda_2/\lambda_1)}\right) = P_{E,2}. \tag{2}$$

Also

$$P_E = \mu_1 P_{E,1} + \mu_2 P_{E,2}. \tag{3}$$

Note: The remark after the proof of T1 tells us that we can apply the simplified theory described in §1 only if $(\log(\mu_1/\mu_2) - (\lambda_2 - \lambda_1)T)/\log(\lambda_2/\lambda_1)$ is not an integer.

E5. Follows directly from the expression for λ_t in terms of the $g^{(n)}$'s given in Chapter III.

E6. Let P_t and $\tilde{P}_t$ be the restrictions of P and $\tilde{P}$ respectively to $\mathscr{F}_t \supset \mathscr{F}_t^Q$. Then

$$\frac{d\tilde{P}_0}{dP_0} = 1. \tag{1}$$

Therefore for all $A \in \mathscr{F}_0^Q = \sigma(Q_0)$, $\tilde{P}(A) = P(A)$; Q_0 has the same distribution under $\tilde{P}$ as under P.

E7. Clearly if $A_t < D_t$ for some $t \in [0, 1]$, then $L_1 = 0$ and therefore, for all $t \in [0, 1]$,

$$\tilde{P}(A_t < D_t) = \int 1(A_t < D_t) L_1 \, dP = 0. \tag{1}$$

Hence

$$Q_t = A_t - D_t \leq 0 \quad \tilde{P}\text{-a.s.}, \quad t \in [0, 1], \tag{2}$$

that is to say, Q_t is a queueing process under $\tilde{P}$. Also L_1 can be written:

$$L_1 = \begin{cases} 1 & \text{if } 1 < T_1^D, \\ \left(\prod_{n \geq 1} 1(Q_{T_n^D} > 0) 1(T_n^D < 1)\right) \exp\left\{\int_0^1 \mu(1 - 1(Q_s > 0)) \, ds\right\} & \text{otherwise,} \end{cases}$$

where T_n^D is the sequence of jumps of D_t. By T3, it follows that A_t has the $(\tilde{P}, \mathscr{F}_t)$-intensity λ [the same as its $(P, \mathscr{F}_t)$-intensity], and D_t has the $(\tilde{P}, \mathscr{F}_t)$-intensity $\mu 1(Q_t > 0)$. Therefore Q_t is, under $\tilde{P}$, M/M/1, with the $(\tilde{P}, \mathscr{F}_t)$-parameters λ and μ.

E8. The correct choice is

$$\mu_t = a 1(t \leq T_1) + 1(t > T_1) \tag{1}$$

because such μ_t is $\mathscr{F}_t$-predictable (being adapted to $\mathscr{F}_t$ and left-continuous). The other choice (the right-continuous version) is wrong; indeed, the corresponding L_t would be

$$L_t = \exp\left\{\int_0^{t \wedge T_1} (1 - a) \, ds\right\}, \tag{2}$$

whereas with the correct choice (1) it would be

$$L_t = \begin{cases} \exp \int_0^t (1 - a) \, ds & \text{if } t < T_1, \\ a \exp \int_0^{t \wedge T_1} (1 - a) \, ds & \text{if } t \geq T_1. \end{cases} \tag{3}$$

Clearly the expression (2) does not define L_t as a martingale [L_t given by (2) either decreases almost surely or increases almost surely according to whether $a > 1$ or $a < 1$].

E9. By Bucy's lemma L5,

$$E_P[\Lambda | \mathscr{F}_t^N] = \frac{E_Q[\Lambda^{N_t+1} \exp\{(1 - \Lambda)t\} | \mathscr{F}_t^N]}{E_Q[\Lambda^{N_t} \exp\{(1 - \Lambda)t\} | \mathscr{F}_t^N]}. \tag{1}$$

Solutions to Exercises, Chapter VI

The exercise will be solved if we can prove that for any nonnegative random variable of the form $f(\Lambda, N_t)$ where f is measurable,

$$E_Q[f(\Lambda, N_t)|\mathscr{F}_t^N] = \int_0^\infty f(\lambda, N_t)F(d\lambda). \tag{2}$$

This is true because under Q, Λ and N_t are independent; indeed, (2) is true whenever

$$f(\Lambda, N_t) = f_1(\Lambda)f_2(N_t), \tag{3}$$

since then, by the Q-independence of Λ and N_t,

$$E_Q[f_1(\Lambda)f_2(N_t)|\mathscr{F}_t^N] = f_2(N_t)E_Q[f_1(\Lambda)|\mathscr{F}_t^N]$$
$$= f_2(N_t)E_Q[f_1(\Lambda)]. \tag{4}$$

The passage to the general case (when f is not necessarily in product form) follows by monotonicity arguments.

E10. Letting $a_0 = q_{01}$ and $a_1 = q_{10}$,

$$V(t, 0) = P(X_0 = 0) + \int_0^t (a_0 V(s, 0) + (1 - a_0)V(s, 1))\, ds$$
$$+ \int_0^t V(s-, 0)(\lambda(0) - 1)(dN_s - ds),$$
$$V(t, 1) = P(X_0 = 1) + \int_0^t (a_1 V(s, 1) + (1 - a_1)V(s, 0))\, ds \tag{1}$$
$$+ \int_0^t V_s(s-, 1)(\lambda(1) - 1)(dN_s - ds).$$

Between jumps of N_t,

$$\frac{dV}{dt}(t, 0) = V(t, 0)(a_0 + 1 - \lambda(0)) + V(t, 1)(1 - a_0),$$
$$\frac{dV}{dt}(t, 1) = V(t, 0)(1 - a_1) + V(t, 1)(a_1 + 1 - \lambda(1)). \tag{2}$$

At a jump T_n of N_t,

$$V(T_n, 0) = \lambda(0)V(T_n-, 0),$$
$$V(T_n, 1) = \lambda(1)V(T_n-, 1). \tag{3}$$

E11. Under P_0 or P_0', θ_t and N_t are *independent* processes, and therefore

$$\frac{dP_{0,t}'}{dP_{0,t}} = \frac{dP_{0,t}'^\theta}{dP_{0,t}^\theta} \frac{dP_{0,t}'^N}{dP_{0,t}^N} \quad P_\mu\text{-a.s.} \tag{1}$$

Indeed, for any $A \in \mathscr{F}_t^\theta$, $B \in \mathscr{F}_t^N$

$$P_0'(A \cap B) = \int_{A \cap B} \frac{dP_{0,t}'}{dP_{0,t}} dP_0. \tag{2}$$

By the P_0'-independence of A and B,

$$P_0'(A \cap B) = P_0'(A)P_0'(B) = \int_A \frac{dP_{0,t}'^\theta}{dP_{0,t}^\theta} dP_0 \int_B \frac{dP_{0,t}'^N}{dP_{0,t}^N} dP_0. \tag{3}$$

By the P_0-independence of $1_A \, dP_{0,t}'^\theta/dP_{0,t}^\theta$ and $1_B \, dP_{0,t}'^N/dP_{0,t}^N$, the rightmost side of (3) is $\int_{A\cap B} (dP_{0,t}'^\theta/dP_{0,t}^\theta)(dP_{0,t}'^N/dP_{0,t}^N) \, dP_0$, so that, by comparison of (2) and (3),

$$\int_{A\cap B} \frac{dP_{0,t}'}{dP_{0,t}} dP_0 = \int_{A\cap B} \frac{dP_{0,t}'^\theta}{dP_{0,t}^\theta} \frac{dP_{0,t}'^N}{dP_{0,t}^N} dP_0. \tag{4}$$

By a monotone class argument, this equality is true when $A \cap B$ is replaced by any $C \in \mathscr{F}_t^\theta \vee \mathscr{F}_t^N$; hence the equality (1), which is P_0-a.s. and therefore P_μ-a.s., since $P_\mu \ll P_0$.

Now

$$\frac{dP_{\mu,t}^{\theta,N}}{dP_{\mu,t}^\theta \, dP_{\mu,t}^N} = \frac{\dfrac{dP_{\mu,t}}{dP_{0,t}}}{\dfrac{dP_{\mu,t}^\theta \, dP_{\mu,t}^N}{dP_{0,t}^\theta \, dP_{0,t}^N}} = \frac{\dfrac{dP_{\mu,t}}{dP_{0,t}'} \dfrac{dP_{0,t}'}{dP_{0,t}}}{\dfrac{dP_{\mu,t}^\theta \, dP_{\mu,t}^N}{dP_{0,t}'^\theta \, dP_{0,t}'^N} \dfrac{dP_{0,t}'^\theta \, dP_{0,t}'^N}{dP_{0,t}^\theta \, dP_{0,t}^N}}, \tag{5}$$

and the announced result then follows from (1).

E12. L_t is a $(P, \mathscr{F}_t)$-martingale; therefore $\hat{L}_t = E[L_t | \mathscr{F}_t^N]$ is a $(P, \mathscr{F}_t^N)$-martingale. By the representation theorem (III, T17)

$$\hat{L}_t = 1 + \int_0^t K_s(dN_s - \lambda \, ds), \tag{1}$$

where K_t is a predictable process such that for all $t \geq 0$

$$\int_0^t |K_s| \lambda \, ds < \infty \quad P\text{-a.s.} \tag{2}$$

Under $\tilde{P}$, N_t has the $\mathscr{F}_t^N$-intensity $\hat{\tilde{\lambda}}_t$; hence for all bounded $\mathscr{F}_t^N$-predictable processes C_t

$$\tilde{E}\left[\int_0^t C_s \, dN_s\right] = \tilde{E}\left[\int_0^t C_s \hat{\tilde{\lambda}}_s \, ds\right], \tag{3}$$

or equivalently,

$$E\left[\hat{L}_t \int_0^t C_s \, dN_s\right] = E\left[\hat{L}_t \int_0^t C_s \hat{\tilde{\lambda}}_s \, ds\right]. \tag{4}$$

Now, using Stieltjes calculus, the representation (1), and the definition of stochastic intensity, we obtain after a few manipulations

$$E\left[\int_0^t C_s K_s \lambda \, ds\right] = E\left[\int_0^t C_s \hat{L}_{s-}(\hat{\tilde{\lambda}}_s - \lambda) \, ds\right]. \tag{5}$$

This equality being true for all bounded $\mathscr{F}_t^N$-predictable processes C_t we obtain

$$K_t \lambda = \hat{L}_{t-}(\hat{\hat{\lambda}}_t - \lambda), \tag{6}$$

$P(d\omega)\, dN_t(\omega)$- or $P(d\omega)\lambda\, dt$-almost everywhere. Hence

$$\hat{L}_t = 1 + \int_0^t \hat{L}_{s-}\left(\frac{\hat{\hat{\lambda}}_s}{\lambda} - 1\right)(dN_s - \lambda\, ds). \tag{7}$$

CHAPTER VII
Optimal Control

1. Modeling Intensity Controls
2. Dynamic Programming for Intensity Controls: Complete-Observation Case
3. Input Regulation. A Case Study in Impulsive Control
4. Attraction Controls
5. Existence via Likelihood Ratio
References
Solutions to Exercises, Chapter VII

1. Modeling Intensity Controls

A point process can be controlled in several ways, each way corresponding to a qualitatively different solution. One can distinguish three types of control: intensity control, optimal stopping, and impulsive control. In this chapter we will consider mainly intensity-control problems in which the generation of points, or rather the power to generate points, is at the disposition of a controller who can modulate the intensity but cannot directly add or erase a point. A large class of problems belong to this category, for instance the dynamical control of the service potential in a queue or the dynamical assignment of priorities without switching costs. The elementary dynamic-programming theory for such optimal-control problems will be developed using the martingale approach. It will lead to sufficient conditions of the *Hamilton–Jacobi* type. As is well known, such equations are usually difficult to solve numerically, due to the large dimension of the system. We will not be concerned with computational problems, our goal being to exhibit the martingale structure which is essential to all problems of dynamical control.

Impulsive controls can add or erase points of a given point process, and in general lead to dynamic-programming conditions different from the Hamilton–Jacobi equations. Bensoussan and Lions [2] have developed the

1. Modeling Intensity Controls

theory of quasivariational inequations in relation to impulsive controls for systems driven by stochastic differential equations of the Ito type. Although we will not develop the similar theory for point-process systems, we will nevertheless consider impulsive-control problems, but of a special kind: the decision can be taken *only at a point* (for instance: the decision of erasing a point). This particular class can be reduced to intensity controls, as is shown in §3 and §4, with the help of examples representative of a large category of problems.

The Terminology

Control, Cost, and Dynamics. Let $N_t = (N_t(1), \ldots, N_t(k))$ be a k-variate point process defined on a measurable space $(\Omega, \mathscr{F})$, and let $\mathscr{U}$ be a set called the *set of admissible controls*. To each *control* $u \in \mathscr{U}$, we associate a probability P_u on $(\Omega, \mathscr{F})$ such that N_t admits the $(P_u, \mathscr{F}_t)$-intensity $\lambda_t(u) = (\lambda_t(1, u), \ldots, \lambda_t(k, u))$, where $\mathscr{F}_t$ is some history of N_t. Also, to each $u \in \mathscr{U}$ corresponds a nonnegative number $J(u)$:

$$J(u) = E_u\left[\int_0^T C_s(u)\,ds + \phi_T(u)\right] < \infty, \tag{1.1}$$

where T is a positive time, $C_t(u)$ is a nonnegative $\mathscr{F}_t$-progressive process, and $\phi_T(u)$ is a nonnegative $\mathscr{F}_T$-measurable random variable. The functional J is called a *criterion*, $J(u)$ is the *cost* associated to u, $C_t(u)$ represents the *cost per unit of time*, and $\phi_T(u)$ is the *final cost*. The time T is referred to as the *terminal time*, and the set of probabilities $(P_u, u \in \mathscr{U})$ is called the *control dynamics*.

Any $u^* \in \mathscr{U}$ such that

$$J(u^*) = \inf_{u \in \mathscr{U}} J(u) \quad \left[\text{respectively} \sup_{u \in \mathscr{U}} J(u)\right] \tag{1.2}$$

is called an *optimal solution*, or *optimal control*, relative to the minimization problem $\inf J(u)$ [respectively, the maximization problem $\sup J(u)$], and

$$J^* = \inf_{u \in \mathscr{U}} J(u) \quad \left[\text{respectively,} \sup_{u \in \mathscr{U}} J(u)\right] \tag{1.3}$$

is the *value* of the control problem. Needless to say, it may happen that no optimal control exists, or, in other words, that no control of $\mathscr{U}$ attains the value J^*. The question of existence will be addressed in §5.

Information Patterns. Suppose, as will always be the case in the sequel, that $\mathscr{U}$ consists of R^l-valued processes defined on $(\Omega, \mathscr{F})$:

$$u_t = (u_t(1), \ldots, u_t(l)), \quad t \in [0, T], \tag{1.4}$$

where for each i, $u_t(i)$ is a measurable process adapted to $\mathscr{F}_t^N$, the internal history of N_t. Moreover we suppose that for each t, u_t is constrained to take its values in a given set U_t:

$$u_t \in U_t \subset R^l. \tag{1.5}$$

In this framework, we call $\mathscr{F}_t$ the *global* history, $\mathscr{F}_t^N$ the *observed* history, or *information pattern*, and N_t the *observation*. The family $(U_t, t \in [0, T])$ represents the *control constraints*. In the case where the observed history is strictly smaller than the global history, the control problem is said to be with *partial information pattern*, whereas if $\mathscr{F}_t \equiv \mathscr{F}_t^N$, it is called a problem with *complete information pattern*.

Local Controls. A control problem with partial information pattern can be formally converted into an equivalent problem with complete information pattern by rewriting the cost $J(u)$ as

$$J(u) = E_u\left[\int_0^T \hat{C}_s(u) \, ds + \hat{\phi}_T(u)\right], \tag{1.6}$$

where $\hat{\phi}_T(u) = E_u[\phi_T(u) | \mathscr{F}_T^N]$ and $\hat{C}_t(u)$ is, for instance, the $(P_u, \mathscr{F}_t^N)$-predictable projection $\hat{C}_t(u)$ of $C_t(u)$.[1] However, this transcription is purely formal and does not produce any simplification in concrete situations. Indeed, in most cases of interest, the cost per unit of time $C_t(u)$ has the form

$$\begin{aligned} C_t(\omega, u) &= C_t(\omega, u_t(\omega)), \\ \phi_T(\omega, u) &= \Psi_T(\omega), \end{aligned} \tag{1.7}$$

and only those complete-information control problems with a cost structure satisfying (1.7) will be treated. We then say that the controls *locally* influence the cost.

Partial-information control problems that arise in applications have a cost structure of the type (1.7); however, the corresponding $\hat{C}_t(u)$ and $\hat{\phi}_T(u)$ will *not* in general be of the form (1.7); in other words, (1.7) does not imply

$$\begin{aligned} \hat{C}_t(\omega, u) &= \hat{C}_t(\omega, u_t(\omega)), \\ \hat{\phi}_T(\omega, u) &= \hat{\Psi}_T(\omega). \end{aligned} \tag{1.8}$$

For instance $E_u[\psi_T]$ will most likely depend upon u. The partial-information case is in general more difficult, and its solution requires the tools of filtering theory developed in Chapter IV. We will not treat it in this book.

In what follows, we suppose that the cost structure (1.7) is granted and that for each admissible control u

$$\lambda_t(\omega, u) = \lambda_t(\omega, u_t(\omega)), \tag{1.9}$$

where the mappings $(t, \omega, u) \to C(t, \omega, u)$ and $(t, \omega, u) \to \lambda(t, \omega, u)$ are $\text{prog}(\mathscr{F}_t) \otimes \mathscr{B}^l$-measurable. The value of the intensity and the value of the

---

[1] That is, $E_u[\int_0^t H_s C_s(u) \, ds] = E_u[\int_0^t H_s \hat{C}_s(u) \, ds]$ for all bounded $\mathscr{F}_t^N$-predictable processes H_t.

1. Modeling Intensity Controls

cost per unit of time at t depend upon u only through u_t. In other words, the control u acts *locally* on the dynamics.

In the *complete-observation case*, we will suppose moreover that

$$\begin{align}
u_t(\omega) &= u(t, \omega, N_t(\omega)), \\
\lambda_t(\omega, u) &= \lambda_t(t, \omega, N_t(\omega), u), \\
C_t(\omega, u) &= C(t, \omega, N_t(\omega), u), \\
\Psi_T(\omega) &= \Psi(T, \omega, N_T(\omega)),
\end{align} \quad (1.10)$$

where for each $n \geq 0$ the mappings $(t, \omega) \to u(t, \omega, n)$ and $(t, \omega) \to \psi(t, \omega, n)$ are $\mathscr{P}(\mathscr{F}_t^N)$-measurable, and the mappings $(t, \omega, u) \to \lambda(t, \omega, n, u)$ and $(t, \omega, u) \to C(t, \omega, n, u)$ are $\mathscr{P}(\mathscr{F}_t^N) \otimes \mathscr{B}^l$-measurable.

Remarks

(α) *Notion of state.* Before we proceed to examples illustrating the above terminology, we want to comment about the notion of *state* in control theory. In practice the global history is defined in terms of a process X_t by $\mathscr{F}_t = \mathscr{F}_t^X$, and such a process is called *the* state of the "system," although there might be arbitrarily many processes X_t that satisfy $\mathscr{F}_t = \mathscr{F}_t^X$. In the complete-information case, where $\mathscr{F}_t = \mathscr{F}_t^N$, then an obvious choice for X_t is N_t, but it might not always be the most appealing choice (see Example 2 to follow).

(β) *Other cost structures.* Cost structures of the type

$$J(u) = E_u \left[\int_0^T C_s(u) \, ds + \sum_{0 < T_n \leq T} k_{T_n}(u) + \phi_T(u) \right], \quad (1.1')$$

where for each $u \in \mathscr{U}$, $k_t(u)$ is a $\mathscr{F}_t^N$-measurable nonnegative process, can be accommodated by (1.1) if $k_t(u)$ is of the form

$$k_t(\omega, u) = k(t, \omega, N_t(\omega), u),$$

where for each $u \in \mathscr{U}$ and each $n \geq 0$, $(t, \omega) \to k(t, \omega, n, u)$ is $\mathscr{F}_t^N$-predictable. Indeed, in the univariate case for instance, by the integration theorem II, T8,

$$\begin{align}
E_u \left[\sum_{0 < T_n \leq T} k_{T_n}(u) \right] &= E_u \left[\sum_{0 < T_n \leq T} k(T_n, N_{T_n-} + 1, u) \right] \\
&= E_u \left[\int_0^T k(s, N_s + 1, u) \lambda_s(u) \, ds \right],
\end{align}$$

so that consideration of cost structures of type (1.1') does not increase generality.

Queueing Examples

EXAMPLE 1 (Controlling the Service, Complete Information). Let (A_t, D_t) be a one-channel queue with state $Q_t = D_t - A_t$, and let $\mathscr{U}$ be the set of measurable processes u_t adapted to $\mathscr{F}_t^Q$ and taking values in $U = \{0, 1, \ldots, N\}$. Suppose that, for each $u \in \mathscr{U}$, the probability P_u is such that the $(P_u, \mathscr{F}_t^Q)$-parameters of Q_t are λ and μu_t for some nonnegative real numbers λ and μ. In other words, the input process is poissonian, of intensity λ, and at time t there are u_t servers, each with a service potential μ. We are in the general situation of §1.1, with

$$N_t = (A_t, D_t), \qquad N_t(1) = A_t, \qquad N_t(2) = D_t,$$
$$\lambda_t(1, u) \equiv \lambda, \qquad \lambda_t(2, u) = \mu u_t 1(Q_t > 0), \qquad (1.11)$$
$$U_t \equiv U = \{0, 1, \ldots, N\}, \qquad u_t \text{ is } \mathscr{F}_t^Q\text{-measurable.}$$

N is the maximum of servers available at a given time, and u_t is to be chosen optimally relative to some criterion, for instance

$$J(u) = E_u\left[\int_0^T (K_1 u_t + K_2 Q_t) \, dt + K_3 Q_T\right]. \qquad (1.12)$$

The coefficient K_1 is interpreted as the cost per unit of time of an active server, K_2 is the waiting cost per unit of time and per customer, and K_3 is the loss incurred when a customer has not been served at terminal time T. The process Q_t can be thought of as the state of the system.

EXAMPLE 2 (Controlling the Service, Partial Information). The setting is the same as above, except that now u_t must be adapted to $\mathscr{F}_t^D$. In other words the observation process is the departure process D_t. Thus, we are in the general situation, where the observation N_t is a univariate point process, namely $N_t = D_t$, and the global history $\mathscr{F}_t \equiv \mathscr{F}_t^Q$ actually differs from the observed history $\mathscr{F}_t^D$. In the notation of §1.1,

$$N_t = N_t(1) = D_t, \qquad \mathscr{F}_t = \mathscr{F}_t^Q,$$
$$\lambda_t(1, u) = \mu u_t 1(Q_t > 0), \qquad (1.13)$$
$$U_t \equiv U = \{0, \ldots, N\}, \qquad u_t \text{ is } \mathscr{F}_t^D\text{-measurable.}$$

Here again a possible choice of the state will be Q_t.

The difference with a complete-information control problem is that the intensity $\lambda_t(1, u) = \mu u_t 1(Q_t > 0)$ is the $(P_u, \mathscr{F}_t^Q)$-intensity; the cost per unit of time, $C_t(u) = K_1 u_t + K_2 Q_t$, is $\mathscr{F}_t^Q$-measurable; and the final cost $K_3 Q_T$ is $\mathscr{F}_T^Q$-measurable, where $\mathscr{F}_t^Q$ is the *global* history, which is *greater* than the observed history $\mathscr{F}_t^D$. As has already been mentioned, the above partial-information control problem can be converted into a complete-information problem by projection on the observed history. If we let

$$\hat{Z}_t^u(n) = P_u[Q_t = n | \mathscr{F}_t^D], \qquad (1.14)$$

1. Modeling Intensity Controls

then the $(P_u, \mathscr{F}_t^D)$-intensity of $N_t(1) = D_t$ is

$$\hat{\lambda}_t(1, u) = E_u[\mu u_t 1(Q_t > 0) | \mathscr{F}_t^D] = \mu u_t(1 - \hat{Z}_t^u(0)), \quad (1.15)$$

and the cost $J(u)$ can be rewritten as

$$J(u) = E_u\left[\int_0^T \left(K_1 u_t + K_2 \sum_{i=1}^\infty i\hat{Z}_t^u(i)\right) dt + K_3 \sum_{i=1}^\infty i\hat{Z}_T^u(i)\right]. \quad (1.16)$$

Note the dependence of the final cost $\hat{\psi}$ upon u:

$$\hat{\Psi}_T(u) = K_3 \sum_{i=1}^\infty i\hat{Z}_T^u(i),$$

a dependence which is materialized by the filtering equations of IV, E5:

$$\hat{Z}_t^u(n) = \hat{Z}_0^u(n) + \int_0^t \mathscr{K}(\hat{Z}_s^u, n)\, ds$$

$$+ \int_0^t \left(-\hat{Z}_{s-}^u(n) + \frac{\hat{Z}_{s-}^u(n+1)}{1 - \hat{Z}_{s-}^u(0)}\right)(dD_s - \mu u_s(1 - \hat{Z}_s^u(0))\, ds). \quad (1.17)$$

It is not difficult to show that if a trajectory $(D_t(\omega), u_t(\omega), t \geq 0)$ is given, then (1.17) considered as an ordinary differential equation with jump disturbances, parametrized by $\omega \in \Omega$, admits a unique solution $Z_t^u = (Z_t^u(1), Z_t^u(2), \ldots)$. Also, if we define $\mathscr{F}_t^Z = \sigma(Z_s^u(i), s \in [0, t], i \geq 0)$, then $\mathscr{F}_t^Z \equiv \mathscr{F}_t^D$. Therefore Z_t^u can be thought of as the state process. This choice of Z_t^u as state process might be interesting, since the cost $J(u)$ is directly expressible in terms of Z_t^u, as well as the $\mathscr{F}_t^D$-intensity of the departure process [Equation (1.15)].

EXAMPLE 3 (Dynamic Assignment of Priority Without Switching Costs). Suppose that two types of customers enter a service system where only one server is available. One wishes to know how this server should be assigned to each type of customer (the server switches from one class of customers to the other) on the basis of his observation of traffic and congestion, and in order to minimize some operating cost.

The above situation can be modelled as follows: let $Q_t(1)$ and $Q_t(2)$ be the two queues defined on some measurable space. For each $i = 1, 2$, the process $Q_t(i) = Q_0(i) + A_t(i) - D_t(i)$ represents the number of customers of class i waiting for service or being served. The observation N_t is the 4-variate point process $(A_t(1), D_t(1), A_t(2), D_t(2))$, and the class $\mathscr{U}$ of admissible controls consists of the measurable processes u_t adapted to $\mathscr{F}_t^N$ and $\{0, 1\}$-valued. The dynamics $(P_u, u \in \mathscr{U})$ are such that for each $u \in \mathscr{U}$, $Q_t(1)$ has the $(P_u, \mathscr{F}_t^N)$-parameters λ_1 and $\mu_1 u_t 1(Q_t(1) > 0)$, and $Q_t(2)$ has the $(P_u, \mathscr{F}_t^N)$-parameters λ_2 and $\mu_2(1 - u_t)1(Q_t(2) > 0)$. Here λ_i is interpreted as the arrival rate of customers of class i, μ_i is the "potential" of the server when assigned to a customer of class i (μ_1 and μ_2 might differ because the services required by the two types of customers are not the same). Finally $u_t = 1$ (0) means that the server is assigned to the first (the second) category of customers. As an example, we will take for criterion J

$$J(u) = E_u \left[\int_0^T \{K_1 u_t + K_2(1 - u_t) + L_1 Q_t(1) + L_2 Q_t(2)\} \, dt \right. \\ \left. + M_1 Q_T(1) + M_1 Q_T(2) \right],$$

letting the reader give a physical meaning to the coefficients K_i, L_i, and M_i. We will see later some other ways of modeling priority assignments problems which are more realistic, taking into account the switching cost (§4).

2. Dynamic Programming for Intensity Controls: Complete-Observation Case

Stochastic Hamilton–Jacobi Sufficient Conditions

We recall the problem: $N_t = (N_t(1), \ldots, N_t(k))$ is a k-variate point process on $(\Omega, \mathcal{F})$; $\mathcal{U}$ is the set of R^l-valued measurable processes u_t of the form

$$u_t(\omega) = u(t, \omega, N_t(\omega)) \qquad (2.1)$$

where for each $n \in N_+^k$, the mapping $(t, \omega) \to u(t, \omega, n)$ is $\mathcal{F}_t^N$-predictable and

$$u_t(\omega) \in U_t, \qquad t \geq 0, \qquad \omega \in \Omega, \qquad (2.2)$$

for some family $(U_t, t \in [0, T])$ of borelian sets of R^l. The dynamics are $(P_u, u \in \mathcal{U})$, and for each $u \in \mathcal{U}$, N_t admits a $(P_u, \mathcal{F}_t^N)$-intensity $\lambda_t(u) = (\lambda_t(1, u), \ldots, \lambda_t(k, u))$ of the form

$$\lambda_t(\omega, u) = \lambda(t, \omega, N_t(\omega), u_t(\omega)), \qquad (2.3)$$

where for each $n \in N_+^k$, the mapping $(t, \omega, v) \to \lambda(t, \omega, n, v)$ is $\mathcal{P}(\mathcal{F}_t^N) \otimes \mathcal{B}^l$-measurable. The cost function is

$$J(u) = E_u \left[\int_0^T C_t(u) \, dt + \phi_T \right], \qquad (2.4)$$

in which, for each $u \in \mathcal{U}$,

$$C_t(\omega, u) = C(t, \omega, N_t(\omega), u_t(\omega)), \\ \phi_T(\omega) = \Psi(T, \omega, N_T(\omega)), \qquad (2.5)$$

where for each $n \in N_+^k$ the mapping $(t, \omega) \to \Psi(t, \omega, n)$ is $\mathcal{F}_t^N$-predictable and nonnegative, and the mapping $(t, \omega, v) \to C(t, \omega, n, v)$ is $\mathcal{P}(\mathcal{F}_t^N) \otimes \mathcal{B}^l$-measurable and nonnegative.

The problem is that of minimizing $J(u)$. The following is a *sufficient condition* for optimality.

T1 Theorem (Hamilton–Jacobi Sufficient Conditions [3]). *Suppose there exists, for each $n \in N_+^k$, a differentiable bounded $\mathscr{F}_t^N$-progressive mapping $(t, \omega) \to V(t, \omega, n)$ such that for all $\omega \in \Omega$ and all $n \in N_+^k$*

$$\frac{\partial V}{\partial t}(t, \omega, n) + \inf_{v \in U_t} \left\{ \sum_{i=1}^k \lambda(i, t, \omega, n, v)[V(t, \omega, n + e_i) - V(t, \omega, n)] + C(t, \omega, n, v) \right\} = 0, \quad (2.6)$$

$$V(T, \omega, n) = \Psi(T, \omega, n),$$

and suppose there exists for each $n \in N_+^k$ an $\mathscr{F}_t^N$-predictable mapping $(t, \omega) \to u^*(t, \omega, n)$ such that for each $n \in N_+^k$, $\omega \in \Omega$, $t \in [0, T]$

$$u^*(t, \omega, n) = \arg\min_{v \in U_t} \left\{ \sum_{i=1}^k \lambda(i, t, \omega, n, v)[V(t, \omega, n + e_i) - V(t, \omega, n)] + C(t, \omega, n, v) \right\} \quad (2.7)$$

[in other words, $u^*(t, \omega, n)$ achieves the minimum of the quantity inside the braces]. Then u_t^* defined by

$$u_t^*(\omega) = u^*(t, \omega, N_t(\omega)) \quad (2.8)$$

is an optimal control.

The equations (2.6) are called the stochastic Hamilton–Jacobi equations for intensity-control problems.

PROOF. We write the proof in the univariate case. First we decompose $V(t, N_t)$ at the jumps of N_t:

$$V(t, N_t) = V(0, 0) + \sum_{0 < T_n \le t} [V(T_n, N_{T_n}) - V(T_{n-1}, N_{T_{n-1}})]$$

$$+ V(t, N_t) - V(\theta_t, N_{\theta_t}), \quad (2.9)$$

where θ_t is the last jump before t:

$$\theta_t = \sup\{T_n, n \ge 0 \mid T_n \le t\} = T_{N_t}. \quad (2.10)$$

For each $n \ge 1$, we have the further decomposition

$$V(T_n, N_{T_n}) - V(T_{n-1}, N_{T_{n-1}}) = V(T_n, N_{T_n}) - V(T_n, N_{T_{n-1}})$$
$$+ V(T_n, N_{T_{n-1}}) - V(T_{n-1}, N_{T_{n-1}}). \quad (2.11)$$

But

$$V(T_n, N_{T_{n-1}}) - V(T_{n-1}, N_{T_{n-1}}) = \int_{T_{n-1}}^{T_n} \frac{\partial V}{\partial s}(s, N_s)\, ds, \quad (2.12)$$

and also, since $N_{\theta_t} = N_t$,

$$V(t, N_t) - V(t, N_{\theta_t}) = \int_{\theta_t}^{t} \frac{\partial V}{\partial s}(s, N_s)\, ds. \quad (2.13)$$

Therefore

$$V(\theta_t, N_t) = V(0, 0) + \int_0^t \frac{\partial V}{\partial s}(s, N_s)\, ds$$

$$+ \sum_{0 < T_n \le t} (V(T_n, N_{T_n}) - V(T_n, N_{T_{n-1}})). \quad (2.14)$$

Now, since $N_{T_n} = N_{T_{n-1}} + 1$ and $N_{T_{n-1}} = N_{T_n-}$, we have

$$\sum_{0 < T_n \le t} [V(T_n, N_{T_n}) - V(T_n, N_{T_{n-1}})] = \int_0^t [V(s, N_{s-} + 1) - V(s, N_{s-})]\, dN_s. \quad (2.15)$$

Hence

$$V(t, N_t) = V(0, 0)$$

$$+ \int_0^t \left\{ \frac{\partial V}{\partial s}(s, N_s) + [V(s, N_s + 1) - V(s, N_s)]\lambda_s(u) \right\} ds$$

$$+ \int_0^t [V(s, N_{s-} + 1) - V(s, N_{s-})](dN_s - \lambda_s(u)\, du). \quad (2.16)$$

And therefore, since by hypothesis $V(T, N_T) = \Psi(T, N_T)$,

$$\int_0^T C(s, N_s, u_s)\, ds + \Psi(T, N_T) = V(0, 0)$$

$$+ \int_0^T \left\{ \frac{\partial V}{\partial s}(s, N_s) + [V(s, N_s + 1) - V(s, N_s)]\lambda(s, N_s, u) \right.$$

$$\left. + C(s, N_s, u_s) \right\} ds$$

$$+ \int_0^T [V(s, N_{s-} + 1) - V(s, N_{s-})](dN_s - \lambda_s(u)\, ds). \quad (2.17)$$

Since $V(s, N_{s-} + 1) - V(s, N_{s-})$ is a bounded $\mathscr{F}_t^N$-predictable process, it follows by II, T8 that

$$E_u\left[\int_0^T [V(s, N_{s-} + 1) - V(s, N_{s-})](dN_s - \lambda_s(u)\, ds)\right] = 0. \quad (2.18)$$

Hence, by integration of (2.17),

$$J(u) = V(0, 0)$$
$$+ E_u \left[\int_0^T \left\{ \frac{\partial V}{\partial s}(s, N_s) + [V(s, N_s + 1) - V(s, N_s)]\lambda(s, N_s, u_s) \right. \right.$$
$$\left. \left. + C(s, N_s, u_s) \right\} ds \right]. \quad (2.19)$$

By the hypothesis (2.6),

$$J(u) \geq V(0, 0), \quad (2.20)$$

and u^* defined by (2.7) and (2.8) is such that

$$J(u^*) = V(0, 0). \quad (2.21)$$

Hence

$$J(u^*) = \inf J(u) = V(0, 0). \quad (2.22)$$

$\square$

Such a general form of the dynamic programming conditions as given in T1 is not very useful in practice, because ω does not intervene as a mere parameter in the equations (2.6). It is linked in an essential way to the parameter t by the $\mathscr{F}_t^N$-progressiveness requirement for $(t, \omega) \to V(t, \omega, n)$. Only in the markovian case will the Hamilton–Jacobi conditions such as (2.6) be usable. Before we proceed to give sufficient conditions of existence of a solution to (2.6) in the markovian case, we will give the physical interpretation of $V(t, N_t)$.

Optimal Cost-to-Go or Value Process

We can perform the calculations in the proof of T1 starting from any t, $0 \leq t \leq T$, instead of 0, to obtain

$$\int_t^T C(s, N_s, u_s) \, ds + \Psi(T, N_T) = V(t, N_t)$$
$$+ \int_t^T \left\{ \frac{\partial V}{\partial s}(s, N_s) + [V(s, N_s + 1) - V(s, N_s)]\lambda(s, N_s, u_s) \right.$$
$$\left. + C(s, N_s, u_s) \right\} ds$$
$$+ m_T(u) - m_t(u), \quad (2.23)$$

where $m_t(u)$ is the $(P_u, \mathscr{F}_t^N)$-martingale $\int_0^t [V(s, N_{s-} + 1) - V(s, N_{s-})]$ $(dN_s - \lambda_s(u)\,ds)$. Hence

$$E_u\left[\int_t^T C(s, N_s, u_s)\,ds + \Psi(T, N_T)\Big|\mathscr{F}_t^N\right]$$
$$= V(t, N_t) + E_u\left[\int_t^T \left\{\frac{\partial V}{\partial s}(s, N_s) + [V(s, N_s + 1)\right.\right.$$
$$\left.\left. - V(s, N_s)]\lambda(s, N_s, u_s) + C(s, N_s, u_s)\right\}ds\Big|\mathscr{F}_t^N\right]. \quad (2.24)$$

Now, if u and $\tilde{u}$ are such that $u_s \equiv \tilde{u}_s$, $s \in [0, t]$ (this will be denoted by $\pi_t u = \pi_t \tilde{u}$), then P_u and $P_{\tilde{u}}$ coincide on $\mathscr{F}_t^N$ by the uniqueness theorem (III, T8). Hence expressions of the type

$$P_u \operatorname*{ess\,inf}_{\substack{\tilde{u} \in \mathscr{U} \\ \pi_t \tilde{u} = \pi_t u}} E_{\tilde{u}}[X|\mathscr{F}_t^N]$$

make sense, and from (2.24) and the definition of u^* it follows that

$$P_u \operatorname*{ess\,inf}_{\substack{\tilde{u} \in \mathscr{U} \\ \pi_t u = \pi_t \tilde{u}}} E_{\tilde{u}}\left[\int_t^T C(s, N_s, \tilde{u}_s)\,ds + \Psi(T, N_T)\Big|\mathscr{F}_t^N\right]$$
$$= V(t, N_t)$$
$$= E_{u^*}\left[\int_t^T C(s, N_s, u_s^*)\,ds + \Psi(T, N_T)\Big|\mathscr{F}_t^N\right]. \quad (2.25)$$

For this reason $V(t, N_t)$ is called the *optimal cost-to-go at time t given $\pi_t u$ and $\mathscr{F}_t^N$*, or more simply, *the optimal cost-to-go*.

Markovian Controls: Deterministic Hamilton–Jacobi Equations

C2 Corollary [3]. *Suppose that $\lambda(t, \omega, n, v)$, $C(t, \omega, n, v)$, and $\Psi(t, \omega, n)$ do not depend upon ω, and that there exists for each $n \in N_+^k$ a function $V(t, n)$ such that*

$$\frac{\partial V}{\partial t}(t, n) + \inf_{v \in U_t} \left\{\sum_{i=1}^k \lambda(i, t, n, v)[V(t, n + e_i) - V(t, n)] + C(t, n, v)\right\} = 0,$$
$$V(T, n) = \Psi(T, n). \quad (2.26)$$

Suppose also that there exists for each $n \in N_+^k$ a measurable R^l-valued function $u^*(t, n)$ such that

$$u^*(t, n) \in U_t, \qquad t \in [0, T], \quad (2.27)$$

2. Dynamic Programming for Intensity Controls: Complete-Observation Case

and

$$u^*(t, n) = \arg\min_{v \in U_t} \left\{ \sum_{i=1}^{k} \lambda(i, t, n, v)[V(t, n + e_i) - V(t, n) + C(t, n, v)] \right\},$$

$$t \in [0, T]. \quad (2.28)$$

Then u_t^* defined by

$$u_t^*(\omega) = u^*(t, N_t(\omega)) \quad (2.29)$$

is an optimal solution.

The equations (2.26) are called the *Hamilton–Jacobi equations* for markovian intensity control problems.

The above result is an immediate corollary of Theorem T1. In the situation described by C2, we say that the optimal control is markovian with respect to N_t; by "a control markovian with respect to N_t," we simply mean a control of the form $f(t, N_t)$ where $f(t, n)$ for each $n \in N_+^k$ is an R^l-valued measurable *deterministic* function. When the value J^* of a control problem is equal to the value J_M^* of the same problem restricted to markovian controls (i.e. $\mathcal{U}$ is replaced by $\mathcal{U}_M$, where $\mathcal{U}_M = \{u \in \mathcal{U} \mid u_t(\omega) = f(t, N_t)\}$), we say that the class of markovian controls is *complete*.

EXAMPLE (Hitting a Target with Maximum Probability). Let N_t be a univariate point process on $(\Omega, \mathcal{F})$, and a and b two nonnegative real numbers such that $0 < a \le b < \infty$. Let $\mathcal{U}$ be the set of measurable processes u_t adapted to $\mathcal{F}_t^N$ bounded by a and b:

$$a \le u_t(\omega) \le b, \quad \omega \in \Omega, \quad t \in [0, T]. \quad (2.30)$$

For each $u \in \mathcal{U}$ suppose that there exists a probability P_u on $(\Omega, \mathcal{F})$ such that N_t admits the $(P_u, \mathcal{F}_t^N)$-intensity $\lambda_t(u) = u_t$.

We seek $u^* \in \mathcal{U}$ which maximizes

$$J(u) = P_u[N_T = K], \quad (2.31)$$

where K is some nonnegative integer. One can think of the number K as a *target* which is to be attained at time T by modulating the intensity between two extremal values a and b.

Let us write down the system (2.26) specialized to $\lambda(t, n, v) = v$, $C(t, n, v) = 0$, and $\Psi(t, n) = 1$ if $n = K$, $\Psi(t, n) = 0$ if $n \ne K$. We obtain

$$\frac{\partial V}{\partial t}(t, n) = \sup_{v \in [a, b]} \{v[V(t, n + 1) - V(t, n)]\} = 0,$$

$$V(T, n) = 0 \quad \text{if } n \ne K, \quad (2.32)$$

$$V(T, K) = 1.$$

It is clear that on $N_t \geq K + 1$, the target is missed, and therefore the optimal condition cost-to-go is 0. Hence, we are tempted to make the educated guess $V(t, n) = 0$ for all $n \geq K + 1$, all $t \geq 0$. The system (2.32) then reduces to

$$\frac{\partial V}{\partial t}(t, n) + \sup_{v \in [a, b]} \{v(V(t, n + 1) - V(t, n)\} = 0, \qquad 0 \leq n < K,$$

$$V(T, n) = 0, \qquad 0 \leq n \leq K, \tag{2.33}$$

$$\frac{\partial V}{\partial t}(t, K) + \sup_{v \in [a, b]} \{-vV(t, K)\} = 0,$$

$$V(T, K) = 1.$$

We therefore have to solve a system of $K + 1$ differential equations with final conditions, and this is quite easy, since we have a lipschitzian system for which the classical theory of differential equations applies smoothly. The solution $V(t, n)$ of (2.32) therefore exists and is unique. Also the optimal control

$$u_t^* = a + (b - a)1(V(t, N_t + 1) \geq V(t, N_t)) \tag{2.34}$$

is *bang-bang* in the sense that it takes the boundary values $\{a, b\}$ of the constraint set $[a, b]$.

The above particular example was reduced to solving a finite set of ordinary differential equations. In the more general case of an infinite number of such equations, a similar result holds.

Existence of an Optimal Control via Hamilton–Jacobi Equations

In order to use Corollary C2, we must show the existence of a bounded solution of the system (2.26) and of a measurable vector $u^*(t) = (u^*(t, n), n \in N_+^k)$ that achieves the minimum in (2.28). Existence is guaranteed under various conditions, the simplest and most commonly used being the following one:

T3 Theorem. *Suppose that $U_t \equiv U$, a compact subset of R^l, the mappings $t, v \to \lambda(t, n, v)$ and $t, v \to C(t, n, v)$ are piecewise continuous and uniformly bounded in t, n, v, and the $\Psi(T, n)$'s are uniformly bounded in n. Then the system (2.26) admits a unique solution $(V(t, n), n \in N_+^k)$ such that the $V(t, n)$'s are uniformly bounded in t, n. Moreover there exists a measurable R^l-valued function $u^*(t, n)$ satisfying (2.28).*

PROOF. For the sake of notational simplicity, we present the proof in the case $k = 1$ (univariate point processes). Let l^∞ be the Banach space of real

2. Dynamic Programming for Intensity Controls: Complete-Observation Case

bounded sequences $x = (x_n, n \geq 0)$ with the norm $\|x\|_\infty = \sup_{n \geq 0} |x_n|$, and consider the following differential equation in l^∞:

$$\dot{x} = f(x, t), \qquad x(T) = y(T), \qquad (2.35)$$

where

$$\begin{aligned}
x(t) &= (x(t, 0), x(t, 1), \ldots), \\
y(t) &= (y(t, 0), y(t, 1), \ldots), \\
f(x, t) &= (f(0, x, t), f(1, x, t), \ldots), \qquad (2.36)\\
f(n, x, t) &= \inf_{v \in U} \{\lambda(t, n, v)(x_{n+1} - x_n) + C(t, n, v)\}, \\
y(t, n) &= \Psi(n, t).
\end{aligned}$$

The symbol $\dot{x}$ denotes differentiation with respect to t and relative to the sup norm of l^∞. In fact the differentiability of $t \to x(t)$ in the l^∞ sense implies the differentiability of $t \to x(t, n)$ for all $n \geq 0$ in the usual sense, and moreover

$$\dot{x}(t) = \left(\frac{dx(t, 0)}{dt}, \frac{dx(t, 1)}{dt}, \ldots\right). \qquad (2.37)$$

Under the conditions of boundedness for the intensity and the cost per unit time, it is not difficult to show that the mapping from R into R defined by

$$y \to \inf_{u \in U} \{\lambda(t, n, v)y + C(t, n, v)\} \qquad (2.38)$$

is lipschitzian and this for all $n \geq 0$, all $t \in [0, T]$. Now, the mapping $x \to Ax$ from l^∞ into l^∞ given by $(Ax)(n) = x(n + 1) - x(n)$ is lipschitzian, and therefore, by composition, $x \to f(x, t)$ is lipschitzian for all $t \geq 0$. The continuity of $t \to f(x, t)$ for all x follows from the continuity conditions imposed on λ and C. We can therefore apply the classical results on differential equations on Banach spaces that guarantee the existence of a unique solution of (2.26) in l^∞. The last assertion follows from the classical results on measurable selections.

Control of the Service Potential of a Queue

E1 Exercise. Let Q_t be a simple queue such that $Q_0 = 0$ defined on $(\Omega, \mathscr{F})$, and let $\mathscr{U}$ be the set of random processes $u = (u_t, t \in [0, T])$ of the form

$$u_t(\omega) = u(t, \omega, Q_t(\omega)), \qquad (2.39)$$

where for each $n \geq 0$, the mapping $(t, \omega) \to u(t, \omega, n)$ is $[0, c]$-valued and $\mathscr{P}(\mathscr{F}_t^Q)$-measurable. Let $(P_u, u \in \mathscr{U})$ be a set of probability measures on $(\Omega, \mathscr{F})$ such that for each $u \in \mathscr{U}$, Q_t has the $(P_u, \mathscr{F}_t^Q)$-parameters λ and u_t, where λ is a nonnegative number. Also define for each $u \in \mathscr{U}$

$$L(u) = \int_0^T Ku_t\, dt + \Phi(Q_T), \tag{2.40}$$

where K is a strictly positive number, and Φ is a nonnegative function on N_+. Solve the problem of minimization of $J(u) = E_u[L(u)]$ among all $u \in \mathscr{U}$.

Comment. c can be interpreted as the maximum service potential available, Kx is the cost per unit time of the service potential x, and $\Phi(Q_T)$ is the loss incurred for not having completed service or not having attended the Q_T customers remaining at the closure (terminal) time T.

Controlling the Kolmogorov Equations

E2 Exercise [4]. Let Q_t be a simple queue defined on $(\Omega, \mathscr{F})$, and let $\mathscr{U}$ be the set of R_+^2-valued processes $u = (v_t, w_t, t \in [0, T])$ that are adapted to $\mathscr{F}_t^Q$ and such that $v_t \in [a, b]$, $w_t \in [c, d]$ for some nonnegative numbers a, b, c, d, $0 \le a \le b$, $0 \le c \le d$. Let $(P_u, u \in \mathscr{U})$ be a set of probability measures on $(\Omega, \mathscr{F})$ such that for each $u \in \mathscr{U}$, Q_t has the $(P_u, \mathscr{F}_t^Q)$-parameters v_t and w_t, and $P_u[Q_0 = n] = p(n)$, $n \ge 0$ (initial distribution independent of $u \in \mathscr{U}$). Let $(J(u), u \in \mathscr{U})$ be a criterion of the form

$$J(u) = E_u \left[\int_0^T a(t, Q_t)v_t + b(t, Q_t)w_t + c(t, Q_t) \right] dt + \Psi(T, Q_T), \tag{2.41}$$

where for each $n \ge 0$ the functions $t \to a(t, n)$, $t \to b(t, n)$, $t \to c(t, n)$, $t \to \Psi(t, n)$ are nonnegative and continuous. Show that an optimal solution of the problem $\min_{u \in \mathscr{U}} J(u)$ is

$$u^* = (v_t^*, w_t^*, t \in [0, T]), \tag{2.42}$$

where $v_t^* \doteq v^*(t, Q_t)$, $w_t^* = w^*(t, Q_t)$, where in turn the $v^*(t, n)$, $w^*(t, n)$ are optimal solutions of the following problem:

$(\tilde{\mathscr{P}})$: minimize

$$\sum_{n=0}^\infty \left\{ \int_0^T [a(t, n)v(t, n) + b(t, n)w(t, n) + c(t, n)]p(n, t)\, dt + \Psi(T, n)p(n, T) \right\}$$

among all measurable mappings $(t \to v(t, n), t \to w(t, n), n \ge 0)$ of R_+ into R_+, subject to the following constraints:

$$p(n, t) = p(n) + \int_0^t [p(n-1, s)1(n > 0)v(s, n-1)$$
$$- p(n, s)(v(s, n) + w(s, n)1(n > 0))$$
$$+ p(n+1, s)w(s, n+1)]\, ds \quad \text{(state constraints)}, \tag{2.43}$$

$$\left.\begin{array}{l} v(t, n) \in [a, b] \\ w(t, n) \in [c, d] \end{array}\right\} \quad \text{(control constraints)}.$$

3. Input Regulation. A Case Study in Impulsive Control

In problems of impulsive control a decision d_n is to be taken at time τ_n for each $n \geq 1$, where $(\tau_n, n \geq 1)$ is a sequence of increasing times. The controller has the choice of both sequences $(\tau_n, n \geq 1)$ and $(d_n, n \geq 1)$, and this choice is adapted to the flow of information $(\mathscr{F}_t, t \geq 0)$ [an increasing family of sub-σ-fields of $\mathscr{F}$, where $(\Omega, \mathscr{F}, P)$ is the subjacent probability space] in the following way: for each $n \geq 1$, τ_n is an $\mathscr{F}_t$-stopping time and d_n is an $\mathscr{F}_{\tau_n}$-measurable random variable, taking its values in the decision space D.

Such problems have received attention from Bensoussan and Lions [2] when $\mathscr{F}_t$ is generated by a Wiener process or a diffusion. The corresponding dynamic-programming conditions are not of the Hamilton–Jacobi type, but take the form of "quasivariational inequalities."

In the case of point-process observations (when $\mathscr{F}_t$ is generated by a point process) quasivariational inequations also arise; however, when the sequence $(\tau_n, n \geq 1)$ is constrained to be contained in the sequence of jumps of the observation process, it is possible to restate the problem as one of control of the intensity of a point process for which the optimality conditions are of the Hamilton–Jacobi type.

Although the ideas in this section and the next can be developed in a general framework, we have preferred to present case studies. First, the input regulation problem. There one takes decisions at the jump times $(T_n, n \geq 1)$ of a point process, and for each n the decision consists in canceling or not canceling the corresponding point T_n in order to regulate the thinned process (the point process after cancellations) to a given rate. The line of argument for this case can be followed in a large class of control problems where one has to take decisions at the jumps of the observed point process: dynamic file allocation (§4), nonpreemptive dynamic priority assignment in queueing systems, routing in communications networks, etc.

The choice of the regulation problem to illustrate the method was motivated by a question concerning *random strategies* for which each cancellation is decided after tossing a coin; the bias $\tilde{u}_n$ of the coin depends upon the observation $\mathscr{F}_n$ at time T_n, and the sequence $(\tilde{u}_n, n \geq 1)$ is the control. Is there a difference between the three cases where $\mathscr{F}_n$ is the past at time T_n of (i) the original point process only, (ii) the thinned point process only, or (iii) both processes? When the point process before thinning is Poisson, it turns out that if one is interested in optimal controls, there is no difference. This follows from the completeness of the class of *pure strategies* ($\tilde{u}_n = 0$ or 1 for all $n \geq 1$) and the fact that for pure strategies, the three information patterns are equivalent. We now proceed to prove this assertion.

Input Regulation: Statement of the Problem

Let T_n, or N_t, be a point process given on some $(\Omega, \mathscr{F}, P)$. The T_n's can be interpreted as arrival times of tasks (or customers) in a processing unit (or service station).

Now let $(X_n, n \geq 1)$ be a sequence of $\{0, 1\}$-valued random variables, to be interpreted in the following manner: if $X_n = 1$, then the task (or customer) arriving at time T_n is admitted for processing (or service); otherwise it is dispatched somewhere else. Therefore, the flow of tasks (or customers) in the processing unit (or service station) is represented by the counting process Y_t defined by

$$Y_t = \sum_{n \geq 1} X_n 1(T_n \leq t). \tag{3.1}$$

The nth cancellation decision X_n is taken after tossing a coin: if the coin falls heads up then $X_n = 1$; otherwise 0. The coin is biased, and its bias varies with n and the available observations at time n. If $\mathscr{F}_n$ for each $n \geq 1$ is a sub-σ-field of $\mathscr{F}$ representing the observation at time n, the probability that $X_n = 1$ conditioned by $\mathscr{F}_n$ is $\tilde{u}_n$. Therefore for each n, $\tilde{u}_n$ has to be $\mathscr{F}_n$-measurable. Otherwise the choice of the sequence $\tilde{u} = (\tilde{u}_n, n \geq 1)$ is at the discretion of the controller. The set $\tilde{U}$ consisting of all sequences $(\tilde{u}_n, n \geq 1)$ of $[0, 1]$-valued random variables such that for each $n \geq 1$, $\tilde{u}_n$ is $\mathscr{F}_n$-measurable is called the set of admissible controls.

A control $\tilde{u}^* \in \tilde{U}$ is sought that minimizes the expectation of

$$\int_0^T (Y_t - rt)^2 \, dt, \tag{3.2}$$

where T and r are positive real constants; T is the terminal time, and r is a rate at which the input Y_t is to be regulated by cancellation over N_t, or thinning of N_t [the average value of the quantity in (3.2) gives a measure of the quality of the regulation]. This particular choice of a criterion is not crucial; the square criterion $(y - rt)^2$ could be replaced by a more general nonnegative $c(y, t)$.

To the following three information patterns:

(i) $\mathscr{F}_n = \sigma(T_1, \ldots, T_n, X_1, \ldots, X_{n-1})$,
(ii) $\mathscr{F}_n = \sigma(T_1, \ldots, T_n)$,
(iii) $\mathscr{F}_n = \sigma(T_1 X_1, \ldots, T_{n-1} X_{n-1}, T_n)$,

there correspond respectively the classes of admissible control $\tilde{U}_1$, $\tilde{U}_2$, and $\tilde{U}_3$. For each $1 \leq i \leq 3$, define $\tilde{U}_{p,i}$ to be the set of pure strategies in $\tilde{U}_i$, that is to say, the set of controls $\tilde{u}$ such that $\tilde{u} \in U_i$ and $\tilde{u}_n = 0$ or 1 for all $n \geq 1$.

It will be shown that in the case where N_t is Poisson, there exists a control $\tilde{u}$ that belongs to the intersection $\bigcap_{i=1}^3 \tilde{U}_{p,i}$ and is optimal for all classes of admissible controls $\tilde{U}_i$, $i = 1, 2, 3$. In other words the class of pure strategies

3. Input Regulation. A Case Study in Impulsive Control

is complete, and as far as optimality is concerned, the three above observation patterns are equivalent.

The plan of the proof is as follows:

(a) We show that an optimal control for the minimization problem corresponding to $\tilde{U}_1$ is in $\tilde{U}_{p,3}$.
(b) We show that $\tilde{U}_{p,1} \equiv \tilde{U}_{p,2} \equiv \tilde{U}_{p,3}$.

Before proving the announced result, we need to state the problem in more precise mathematical terms. In particular the probability structure corresponding to each control has to be made explicit. When this is done, we will *replace the original problem with one of control of the intensity* of a point process, as studied in §1 and §2. We then apply the method of dynamic programming, via martingales.

The Transformed Problem: Control of Intensity

Let Ω be the set of double sequences $\omega = (t_n, x_n, n \geq 0)$ where

$$t_0 = 0; \quad t_n < \infty \Rightarrow t_n < t_{n+1}, \quad n \geq 0; \quad (3.3)$$
$$x_0 = 0; \quad x_n = 0 \text{ or } 1, \quad n \geq 0.$$

Let $\mathscr{F}$ be the σ-field generated by the mappings $T_n: \omega \to t_n$ and $X_n: \omega \to x_n$, for all $n \geq 0$. Let N_t and Y_t be defined as above. Let $\tilde{u} = (\tilde{u}_n, n \geq 1)$ be a sequence of $[0,1]$-valued random variables such that for each $n \geq 1$, $\tilde{u}_n$ is $\sigma(X_0, \ldots, X_{n-1}, T_0, \ldots, T_n)$-measurable. The requirements

$$\tilde{P}_{\tilde{u}}[T_{n+1} - T_n \leq t \,|\, \sigma(X_0, \ldots, X_n, T_0, \ldots, T_n)]$$
$$= 1 - \exp\left\{-\int_{T_n}^{T_n+t} \lambda(s) \, ds\right\}, \quad (3.4)$$
$$\tilde{P}_{\tilde{u}}[X_n = 1 \,|\, \sigma(X_0, \ldots, X_{n-1}, T_0, \ldots, T_n)] = \tilde{u}_n,$$

where $t \to \lambda(t)$ is some nonnegative measurable locally integrable real-valued function, completely characterize a probability measure $\tilde{P}_{\tilde{u}}$ on $(\Omega, \mathscr{F})$, up to completion. We assume in the sequel that $(\Omega, \mathscr{F}, \tilde{P}_{\tilde{u}})$ is indeed complete. Define Z_t to be the bivariate point process (N_t, Y_t). Then:

R4 Result

(a) N_t is, with respect to $\tilde{P}_{\tilde{u}}$, a Poisson process with the intensity $\lambda(t)$, and for all $0 \leq s \leq t$, $N_t - N_s$ is $\tilde{P}_{\tilde{u}}$-independent of $\mathscr{F}_s^Z$.
(b) For some $\mathscr{F}_t^Z$-predictable $[0,1]$-valued process u_t, Y_t has the $(\tilde{P}_{\tilde{u}}, \mathscr{F}_t^Z)$-intensity $\lambda(t)u_t$, and moreover

$$u_{T_n} = \tilde{u}_n \quad \tilde{P}_{\tilde{u}}\text{-a.s.} \quad (3.5)$$

PROOF. (a): Recall (III, T2) that

$$\mathscr{F}^Z_{T_n} = \sigma(X_0, X_1, \ldots, X_n, T_0, \ldots, T_n),$$
$$\mathscr{F}^Z_{T_n-} = \sigma(X_0, X_1, \ldots, X_{n-1}, T_0, \ldots, T_n). \tag{3.6}$$

Therefore (3.3) reads

$$\tilde{P}_{\tilde{u}}[T_{n+1} - T_n \leq t | \mathscr{F}^Z_{T_n}] = 1 - \exp\left\{-\int_{T_n}^{T_n+t} \lambda(s)\,ds\right\}. \tag{3.7}$$

From III, T7 it follows that $\lambda(s)$ is the $(\tilde{P}_{\tilde{u}}, \mathscr{F}^Z_t)$-intensity of N_t. From the version of Watanabe's characterization theorem (II, T6), this suffices to ensure that conclusion (a) is true.

(b): For any nonnegative $\mathscr{F}^Z_t$-predictable process C_t and any $\tilde{u}$,

$$0 \leq \tilde{E}_{\tilde{u}}\left[\int_0^\infty C_s\,dY_s\right] \leq \tilde{E}_{\tilde{u}}\left[\int_0^\infty C_s\,dN_s\right]. \tag{3.8}$$

This implies that on $\mathscr{P}(\mathscr{F}^Z_t)$—the $\mathscr{F}^Z_t$-predictable σ-field on $(0, \infty) \times \Omega$—the measure $\tilde{P}_{\tilde{u}}(d\omega)\,dY_t(\omega)$ is absolutely continuous with respect to the measure $\tilde{P}_{\tilde{u}}(d\omega)\,dN_t(\omega)$, and that a version of the corresponding Radon-Nikodym derivative, $u_t(\omega)$, is $[0, 1]$-valued. Since, by definition of the intensity, $\tilde{P}_{\tilde{u}}(d\omega)\,dN_t(\omega) = \tilde{P}_{\tilde{u}}(d\omega)\lambda(t)\,dt$ on $\mathscr{P}(\mathscr{F}^Z_t)$, we have $\tilde{P}_{\tilde{u}}(d\omega)\,dY_t(\omega) = \tilde{P}_{\tilde{u}}(d\omega)u_t(\omega)\lambda(t)\,dt$ on $\mathscr{P}(\mathscr{F}^Z_t)$. In other words $u_t\lambda(t)$ is the $(\tilde{P}_{\tilde{u}}, \mathscr{F}^Z_t)$-intensity of Y_t. Note that the process u_t is $[0, 1]$-valued and $\mathscr{F}^Z_t$-predictable [as the Radon–Nikodym derivative of measures on $\mathscr{P}(\mathscr{F}^Z_t)$, $(t, \omega) \to u_t(\omega)$ is $\mathscr{P}(\mathscr{F}^Z_t)$-measurable]. Moreover, by II, T15:

$$\tilde{P}_{\tilde{u}}[X_n = 1 | \mathscr{F}^Z_{T_n-}] = u_{T_n}. \tag{3.9}$$

With the help of R4 we can replace the original problem $\tilde{\mathscr{P}}$, which consists of minimizing $\tilde{J}(\tilde{u}) = \tilde{E}_{\tilde{u}}[\int_0^T (Y_t - rt)^2\,dt]$ among all $\tilde{u} \in \tilde{U}$ (recall that $\tilde{U}$ can take six values: $\tilde{U}_i$, $\tilde{U}_{p,i}$, $1 \leq i \leq 3$, and therefore there are six problems $\tilde{\mathscr{P}}$: $\tilde{\mathscr{P}}_i$, $\tilde{\mathscr{P}}_{p,i}$, $1 \leq i \leq 3$), by an equivalent problem of control of the intensity of a point process:

Problem $\mathscr{P}$. Let U be the set of nonnegative $\mathscr{F}^Z_t$-predictable $[0, 1]$-valued processes u, and for each $u \in U$, let P_u be *the* probability measure on $(\Omega, \mathscr{F})$ that makes N_t a point process with the $(P_u, \mathscr{F}^Z_t)$-intensity $\lambda(t)$ and Y_t a point process with the $(P_u, \mathscr{F}^Z_t)$-intensity $\lambda(t)u_t$. Problem $\mathscr{P}$ consists in minimizing $J(u) = E_u[\int_0^T (Y_t - rt)^2\,dt]$. Here again we can define six problems $\mathscr{P}_i$, $\mathscr{P}_{p,i}$, $1 \leq i \leq 3$, and six sets of admissible control U_i, $U_{p,i}$, $1 \leq i \leq 3$, as we did above for $\tilde{\mathscr{P}}$ and $\tilde{U}$.

Since the goal is to show that the optimal control $\tilde{u}^*$ for $\tilde{\mathscr{P}}_1$ is in $\tilde{U}_{p,3}$ we will—for notational convenience—denote $\tilde{U}_1$ and $\tilde{\mathscr{P}}_1$ by $\tilde{U}$ and $\tilde{\mathscr{P}}$. Similarly U_1 and $\mathscr{P}_1$ will be denoted by U and $\mathscr{P}$.

3. Input Regulation. A Case Study in Impulsive Control   215

R5 Result. *If u^* is an optimal solution for $\mathcal{P}$, then $\tilde{u}^*$ defined by*

$$\tilde{u}_n^* = u_{T_n}^*, \quad n \geq 1, \tag{3.10}$$

is an optimal solution for $\tilde{\mathcal{P}}$.

PROOF. To any $u \in U$, we can associate $\tilde{u} \in \tilde{U}$ by

$$\tilde{u}_n = u_{T_n}, \quad n \geq 1, \tag{3.11}$$

and moreover $P_u = \tilde{P}_{\tilde{u}}$: indeed, $\tilde{P}_{\tilde{u}}$ is the unique probability measure on $(\Omega, \mathcal{F})$ such that (3.4) is verified, and P_u also verifies (3.4) (in the course of the proof of R4 we have seen that $P_u[X_n = 1 | \sigma(X_0, \ldots, X_n, T_0, \ldots, T_n)] = u_{T_n}$, which is equal to $\tilde{u}_n$ by the definition of $\tilde{u}$).

Similarly, to $\tilde{u} \in \tilde{U}$ one can associate, according to Result R4, $u \in U$ such that (3.11) holds, and moreover $P_u = \tilde{P}_{\tilde{u}}$, by the same arguments as above. The result is then clear, since $\tilde{J}(\tilde{u}) = J(u)$ when $P_u \equiv \tilde{P}_{\tilde{u}}$.   □

We now proceed to the solution of $\mathcal{P}$.

The Solution via Dynamic Programming

The following remark will be useful: if we let N be any fixed integer greater than rT, then clearly one can restrict attention to controls u such that $u_t(\omega) = 0$ on $\{Y_t(\omega) \geq N\}$. Indeed, if for some $t_0 \in [0, T]$ we have $Y_{t_0} \geq rT$, then for all $t \in [t_0, T]$ and all $k \geq 0$, $(Y_t + k - rt)^2 \geq (Y_t - rt)^2$. Let us denote by $\hat{U}$ the class of controls $U \cap \{u | u_t(\omega) = 0 \text{ on } \{Y_t(\omega) \geq N\}\}$.

R6 Result. *Suppose there exists a family $(t \to V(t, n), n \geq 0)$ of measurable mappings from $\mathbb{R}_+$ into $\mathbb{R}$ such that for all $n \geq 0$*

$$\frac{dV}{dt}(t, n) + \lambda(t) \mathbf{1}(V(t, n+1) - V(t, n) \leq 0)[V(t, n+1) - V(t, n)]$$
$$+ (n - rt)^2 = 0,$$

$$V(T, n) = 0. \tag{3.12}$$

Then u^* defined by

$$u_t^* = \mathbf{1}(V(t, Y_{t-} + 1) - V(t, Y_{t-}) \leq 0) \tag{3.13}$$

is optimal for $\mathcal{P}$. Moreover,

$$V(0, 0) = J(u^*) = \inf_{u \in U} J(u). \tag{3.14}$$

PROOF. We first observe that

$$\mathbf{1}(V(t, n+1) - V(t, n) \leq 0)[V(t, n+1) - V(t, n)]$$
$$= \inf_{u \in [0, 1]} \{u[V(t, n+1) - V(t, n)]\}, \tag{3.15}$$

so that (3.12) reads

$$\frac{dV}{dt}(t, n) + \inf_{u \in [0, 1]} \{\lambda(t)u[V(t, n + 1) - V(t, n)] + (n - rt)^2\} = 0. \quad (3.16)$$

Let us decompose $V(t, Y_t)$ as follows (see the proof of T1):

$$V(t, Y_t) = V(0, 0) + \sum_{0 < T_n \le t} (V(T_n, Y_{T_n}) - V(T_{n-1}, Y_{T_{n-1}}))$$
$$+ V(t, Y_t) - V(\theta_t, Y_{\theta_t}), \quad (3.17)$$

where $\theta_t = T_{N_t}$. Equivalently,

$$V(t, Y_t) = V(0, 0) + \sum_{0 < T_n \le t} (V(T_n, Y_{T_n}) - V(T_n, Y_{T_{n-1}}))$$
$$+ \sum_{0 < T_n \le t} (V(T_n, Y_{T_{n-1}}) - V(T_{n-1}, Y_{T_{n-1}}))$$
$$+ V(t, Y_t) - V(\theta_t, Y_{\theta_t}). \quad (3.18)$$

Now, observing that $Y_{T_{n-1}} = Y_{T_n-}$ and $Y_t = Y_{\theta_t}$, we have

$$\sum_{0 < T_n \le t} (V(T_n, Y_{T_{n-1}}) - V(T_{n-1}, Y_{T_{n-1}})) + V(t, Y_t)$$
$$- V(\theta_t, Y_{\theta_t}) = \int_0^t \frac{dV}{ds}(s, Y_s)\, ds. \quad (3.19)$$

Also

$$V(T_n, Y_{T_n}) - V(T_n, Y_{T_{n-1}}) = V(T_n, Y_{T_n-} + X_n) - V(T_n, Y_{T_n-})$$
$$= (V(T_n, Y_{T_n-} + 1) - V(T_n, Y_{T_n-}))X_n,$$

and therefore

$$E[(V(T_n, Y_{T_n}) - V(T_n, Y_{T_{n-1}}))1(T_n \le t)] =$$
$$E[(V(T_n, Y_{T_n-} + 1) - V(T_n, Y_{T_n-}))u_{T_n}1(T_n \le t)] \quad (3.20)$$

(where we have made use of the fact that T_n is $\mathscr{F}^Z_{T_n-}$-measurable and that $P_u[X_n = 1 | \mathscr{F}^Z_{T_n-}] = u_{T_n}$). Therefore, for any $u \in \tilde{U}$,

$$E\left[\sum_{0 < T_n \le t} (V(T_n, Y_{T_n}) - V(T_n, Y_{T_{n-1}}))\right]$$
$$= E\left[\int_0^t [V(s, Y_{s-} + 1) - V(s, Y_{s-})]u_s\, dN_s\right]$$
$$= E\left[\int_0^t [V(s, Y_{s-} + 1) - V(s, Y_{s-})]u_s \lambda(s)\, ds\right]. \quad (3.21)$$

3. Input Regulation. A Case Study in Impulsive Control   217

Combining (3.18), (3.19), and (3.21), we obtain

$$E_u[V(t, N_t)] = V(0, 0)$$
$$+ E_u\left[\int_0^t \left\{\frac{dV}{ds}(s, Y_s) + \lambda(s)u_s[V(s, Y_s + 1) - V(s, Y_s)]\right\} ds\right].$$
(3.22)

Adding $E_u[\int_0^t (Y_s - rs)^2 \, ds]$ to both sides of (3.22), we obtain, after letting $t = T$ and using the terminal condition (3.12),

$$J(u) = V(0, 0)$$
$$+ E_u\left[\int_0^T \left\{\frac{dV}{ds}(s, Y_s) + \lambda(s)u_s[V(s, Y_s + 1) - V(s, Y_s)]\right.\right.$$
$$\left.\left. + (Y_s - rs)^2\right\} ds\right].$$
(3.23)

By (3.12) and (3.23), $J(u) \geq V(0, 0)$ for all $u \in U$, and by the definition of u^*, $J(u^*) = V(0, 0)$. □

R7 Result. *There exists a solution of* (3.12).

PROOF. Let N be as in the remark at the beginning of the present section. Then clearly

$$V(t, n) = \int_t^T (n - rs)^2 \, ds, \qquad n \geq N,$$
(3.24)

satisfies, for all $n \geq N$,

$$\frac{dV}{dt}(t, n) + \lambda(t)1(V(t, n + 1) \leq V(t, n))[V(t, n + 1)$$
$$- V(t, n)] + (n - rt)^2 = 0,$$
(3.25)
$$V(T, n) = 0,$$

since for such $V(t, n)$, $1(V(t, n + 1) \leq V(t, n)) = 0$ for all $n \geq N$. Therefore what remains to be solved is the finite system of differential equations

$$\frac{dV}{dt}(t, n) + \lambda(t)1(V(t, n + 1) \leq V(t, n))[V(t, n + 1) - V(t, n)]$$
$$+ (n - rt)^2 = 0, \qquad 0 \leq n \leq N - 1),$$
(3.26)
$$V(T, n) = 0, \qquad 0 \leq n \leq N - 1,$$

where $V(t, N) = \int_t^T (N - rs)^2 \, ds$. The proof is then similar to that of the example after C2: hitting a target with maximum probability. □

We can now state the main result:

R8 Result (Completeness of Pure Strategies). *There exists a common optimal solution to $\tilde{\mathscr{P}}_i$ (or $\mathscr{P}_i$), $1 \leq i \leq 3$, corresponding to a pure strategy.*

PROOF. By Results R6 and R7 there exists an optimal solution to $\mathscr{P}_1$ which belongs to $U_{p,3}$. Equivalently, by T5 there exists an optimal solution of $\tilde{\mathscr{P}}_1$ that belongs to $\tilde{U}_{p,3}$. The conclusion follows from:

R9 Result. $\tilde{U}_{p,1} = \tilde{U}_{p,2} = \tilde{U}_{p,3}$.

PROOF. Clearly it suffices to show that $\tilde{U}_{p,1} \subset \tilde{U}_{p,2}$ and $\tilde{U}_{p,1} \subset \tilde{U}_{p,3}$. We only prove $\tilde{U}_{p,1} \subset \tilde{U}_{p,2}$, since $\tilde{U}_{p,1} \subset \tilde{U}_{p,3}$ is proven with exactly the same arguments (also, we only need to show that $\tilde{U}_{p,1} = \tilde{U}_{p,2} \supset \tilde{U}_{p,3}$, since we have shown the existence of an optimal solution of $\tilde{\mathscr{P}}_1$ that is in $\tilde{U}_{p,3}$).

Let $\tilde{u} \in \tilde{U}_{p,1}$. There exists a sequence $(f^{(n)}, n \geq 1)$ of measurable functions $f^{(n)}: R^{2n} \to \{0, 1\}$ such that

$$\tilde{u}_n = f^{(n)}(T_1, \ldots, T_n, X_0, \ldots, X_{n-1}) \quad \text{for all } n \geq 1. \tag{3.27}$$

Now since $\tilde{u}_n \in \{0, 1\}$ and $\tilde{u}_n = \tilde{P}_{\tilde{u}}[X_n = 1 | \mathscr{F}_n]$, we have $X_n = \tilde{u}_n$, $\tilde{P}_{\tilde{u}}$-a.s.; therefore

$$\tilde{u}_1 = f^{(1)}(T_1, X_0) = f^{(1)}(T_1, 0) = g^{(1)}(T_1)$$
$$\tilde{u}_2 = f^{(2)}(T_1, T_2, X_1, X_0) = f^{(2)}(T_1, T_2, g^{(1)}(T_1), 0) \tag{3.28}$$
$$= g^{(2)}(T_1, T_2) \quad \tilde{P}_{\tilde{u}}\text{-a.s.},$$

and so on, leading the general representation

$$\tilde{u}_n = g^{(n)}(T_1, T_2, \ldots, T_n) \quad \tilde{P}_{\tilde{u}}\text{-a.s.}, \tag{3.29}$$

where $g^{(n)}$ is a measurable mapping from R^n into $\{0, 1\}$. If we let $\tilde{u}'$ be defined by

$$\tilde{u}'_n = g^{(n)}(T_1, T_2, \ldots, T_n), \tag{3.30}$$

then firstly $\tilde{u}' \in \tilde{U}_{p,2}$, and secondly $\tilde{u}'_n = \tilde{u}_n$, $\tilde{P}_{\tilde{u}}$-a.s. And this implies that $\tilde{P}_{\tilde{u}} = \tilde{P}_{\tilde{u}'}$, at least if both $(\Omega, \mathscr{F}, \tilde{P}_{\tilde{u}})$ and $(\Omega, \mathscr{F}, \tilde{P}_{\tilde{u}})$ are completed, as was assumed.

E3 Exercise. (a) Show that the dynamic-programming equations (3.10), (3.11) are the same as those corresponding to the following intensity-control problem: N_t being a point process given on some measurable space $(\Omega, \mathscr{F})$,

$$\underset{u \in \mathscr{U}}{\text{minimize}} \quad E_u\left[\int_0^T (N_s - rs)^2 \, ds\right],$$

where the set of admissible controls $\mathscr{U}$ is the set of $[0, 1]$-valued measurable processes u_t adapted to $\mathscr{F}_t^N$, and for each $u \in \mathscr{U}$, P_u is a probability measure on $(\Omega, \mathscr{F})$ which gives the $(P_u, \mathscr{F}_t^N)$-intensity $\lambda(t)u_t$ to N_t.

(b) Show that for the same problem as above where $\lambda(t)$ is replaced by $\lambda(N_t)$ with $\lambda(n)$ bounded, the dynamic-programming equations are

$$\frac{dV}{dt}(t, n) + \inf_{u \in [0, 1]} \{\lambda(n)u[V(t, n + 1) - V(t, n)]\} + (n - rt)^2 = 0,$$

$$V(T, n) = 0.$$

(3.31)

E4 Exercise. Consider the input-regulation problem of the present section, replacing $\lambda(t)$ by $\lambda(N_t)$ where $\lambda(n)$ is bounded. Show that the optimal solution $\tilde{u}_n^*$ is given by

$$\tilde{u}_n^* = u_{T_n}^*,$$

(3.32)

where

$$u_t^* = \begin{cases} 1 & \text{if } V(t, N_{t-} + 1, Y_{t-} + 1) - V(t, N_{t-} + 1, Y_{t-}) \\ & \leq V(t, N_{t-}, Y_{t-}) - V(t, N_{t-} + 1, Y_{t-}), \\ 0 & \text{otherwise,} \end{cases}$$

(3.33)

and where V satisfies, for all $t \in [0, T]$ and all nonnegative integers n and y,

$$\frac{dV}{dt}(t, n, y) + \lambda(n) \inf_{u \in [0, 1]} \{u(V(t, n + 1, y + 1) + (1 - u)V(t, n + 1, y) - V(t, n, y))\}$$

$$+ (y - rt)^2 = 0,$$

$$V(T, n, y) = 0.$$

(3.34)

Comment. The point of Exercises E3 and E4 is to show that if the intensity of N_t in the original regulation problem depends upon N_t, then the regulation problem is not equivalent to the "obvious" intensity-control problem.

4. Attraction Controls

In this section we will see how to formalize a certain class of impulsive controls at the jumps of a basic point process; this will be done with the help of two typical examples of interest in operations research.

Optimal File Allocation

Statement of the Problem [7]. Two point processes $N_t(1)$ and $N_t(2)$ are given on a fixed probability space $(\Omega, \mathscr{F}, P)$. Each $N_t(i)$ has the $(P, \mathscr{F}_t^N(i))$-intensity $\lambda_t(i) = \lambda(i, t)$ (deterministic) and $\int_0^t \lambda^{(i)}(s)\, ds < \infty$, for all $t \geq 0$. Moreover $N_t(1)$ and $N_t(2)$ are independent, which implies that for each i, $N_t(i)$ has the $(P, \mathscr{G}_t)$-intensity $\lambda\,(i, t)$, where $\mathscr{G}_t = \mathscr{F}_t^N(1) \vee \mathscr{F}_t^N(2)$ and $N_t(1)$ and $N_t(2)$ have no common jumps, P-a.s. The interpretation of $N_t(i)$ is as follows: it is the counting process of requests at a given computer center located at location i for a file which is stored either at location 1 or at location 2. The problem is to optimally transfer this file back and forth between locations 1 and 2 in order to minimize some operating cost to be described.

The *control* is a process which we call the *attraction process*: it is a $\mathcal{G}_t$-adapted measurable process u_t taking its value in $\{0, 1\}$. If $u_t = 1$, then location 1 attracts the file, which means: if $u_t = 1$ and the file is at location 1 it remains there, whatever happens; if $u_t = 1$ and the file is at location 2, then the file is transferred from 2 to 1 if and only if there is a special request from location 1 $[\Delta N_t = 1]$. The symmetric interpretation is in force when $u_t = 0$: then location 2 attracts the file.

The attraction process u_t is denoted u, and the set of all such processes is called $\mathcal{U}$, the admissible set of controls. The *state of the file* is a right-continuous process Y_t taking its values in $\{0, 1\}$; if $Y_t = 1$, the file is at location 1; if $Y_t = 0$, the file is at location 2. This process changes values only at the jumps of $N_t(1)$ or $N_t(2)$, because the file switches from i to j only if a request emanates from j. The state equation is therefore

$$dY_t = -Y_{t-}(1 - u_t)\, dN_t(2) + (1 - Y_{t-})u_t\, dN_t(1). \qquad (4.1)$$

The *cost* corresponding to the control u_t is

$$J(u) = E\left[\int_0^T \{S_1 Y_t + S_2(1 - Y_t)\}\, dt\right]$$

$$+ E\left[\int_0^T \{C_{12} Y_{t-} u_t\, dN_t(2) + C_{21}(1 - Y_{t-})(1 - u_t)\, dN_t(1)\}\right]$$

$$+ E\left[\int_0^T \{T_{12} Y_{t-}(1 - u_t)\, dN_t(2) + T_{21}(1 - Y_{t-})u_t\, dN_t(1)\}\right], \qquad (4.2)$$

where the S's, C's, and T's are nonnegative real numbers. The S's are *storing costs*: S_i is the cost of storage at location i per unit of time. The C's are *communication costs*: C_{ij} is the cost of a communication from i to j; for instance the term $C_{12} Y_{t-} u_t \Delta N_t(2)$ is 0 or C_{12}, and it is C_{12} if and only if, prior to t, the file is in 1 (hence $Y_{t-} = 1$), it remains there at t (hence $u_t = 1$), and there is a request from 2 (hence $dN_t(2) = 1$). The T's are *transfer costs*: T_{ij} is the cost of transferring the file from i to j; for instance $T_{12} Y_{t-}(1 - u_t)\, dN_t(2)$ is 0 or T_{12}, and it is T_{12} if and only if, prior to t, the file is in 1 (hence $Y_{t-} = 1$), then at t it is in 2 (hence $1 - u_t = 1$), and as was said above, this transfer can occur only upon request emanating from 2 (hence $dN_t(2) = 1$).

The set of admissible controls $\mathcal{U}$ consists of $\{0, 1\}$-valued processes $u = (u_t, t \in [0, T])$ of the form

$$u_t(\omega) = u(t, \omega, N_t(1, \omega), N_t(2, \omega)), \qquad (4.3)$$

where the mappings $(t, \omega) \to u(t, \omega, n_1, n_2)$ are $\mathcal{P}(\mathcal{G}_t)$-measurable for all n_1, n_2 in N_+.

4. Attraction Controls

Solution. In order to find the optimal control $u^* \in \mathcal{U}$, we first try to obtain an equation for the optimal cost-to-go at time t, conditioned by $\mathcal{G}_t$ and by the past of the control $u \in \mathcal{U}$ up to the time t. We make the educated guess that it is of the form $V(t, Y_t)$ [recall that Y_t depends upon u, as is seen in (4.1)]. We obtain, using a now familiar decomposition technique and supposing that $t \to V(t, y)$ is differentiable for $y = 0$ or 1,

$$V(t, Y_t) = V(0, Y_0) + \int_0^t \frac{\partial V}{\partial s}(s, Y_s) \, ds$$

$$+ \int_0^t (V(s, Y_s) - V(s, Y_{s-}))(dN_s(1) + dN_s(2)). \quad (4.4)$$

Now, from Equation (4.1) we see that

$$\begin{aligned} dN_t(1) = 1 &\Rightarrow Y_t = Y_{t-} + (1 - Y_{t-})u_t, \\ dN_t(2) = 1 &\Rightarrow Y_t = u_t Y_{t-}. \end{aligned} \quad (4.5)$$

Observing that u_t and Y_t take their values in $\{0, 1\}$, we see that

$$(V(s, Y_s) - V(s, Y_{s-})) \, dN_s(1) = (V(s, 1) - V(s, 0))u_s(1 - Y_{s-}) \, dN_s(1),$$
$$(V(s, Y_s) - V(s, Y_{s-})) \, dN_s(2) = (V(s, 0) - V(s, 1))(1 - u_s)Y_{s-} \, dN_s(2). \quad (4.6)$$

Therefore, by the definition of intensity,

$$E\left[\int_0^t (V(s, Y_s) - V(s, Y_{s-})) \, dN_s(1)\right]$$

$$= E\left[\int_0^t (V(s, 1) - V(s, 0))u(s, N_{s-}(1) + 1, N_{s-}(2))(1 - Y_{s-}) \, dN_s(1)\right]$$

$$= E\left[\int_0^t (V(s, 1) - V(s, 0))u(s, N_s(1) + 1, N_s(2))(1 - Y_s)\lambda(1, s) \, dN_s(1)\right], \quad (4.7)$$

where we have used the fact that if $dN_t(1) = 1$, then $N_t(1) = N_{t-}(1) + 1$ and $N_t(2) = N_{t-}(2)$. Similarly

$$E\left[\int_0^t (V(s, Y_s) - V(s, Y_{s-})) \, dN_s(2)\right]$$

$$= E\left[\int_0^t (V(s, 0) - V(s, 1))(1 - u(s, N_s(1), N_s(2) + 1))\right.$$

$$\left. \times (1 - Y_s)\lambda(2, s) \, ds\right]. \quad (4.8)$$

The cost associated to $u \in \mathscr{U}$ can be expressed as

$$J(u) = E\left[\int_0^T (S_1 Y_s + S_2(1 - Y_s))\, ds\right.$$

$$+ \int_0^T \{C_{12} Y_s u(s, N_s(1), N_s(2) + 1)\lambda(2, s)$$

$$+ C_{21}(1 - Y_s)(1 - u(s, N_s(1) + 1, N_s(2)))\lambda(1, s)\}\, ds$$

$$+ \int_0^T \{T_{12} Y_s(1 - u(s, N_s(1), N_s(2) + 1))\lambda(2, s)$$

$$+ \left. T_{21}(1 - Y_s)u(s, N_s(1) + 1, N_s(2))\lambda(1, s)\}\, ds \right]. \quad (4.9)$$

Finally, if we make the hypothesis $V(T, 1) = V(T, 0) = 0$ [which is natural in view of the interpretation of $V(t, Y_t)$ as the optimal cost-to-go], then by combining the above equalities and observing that

$$\frac{\partial V}{\partial t}(t, Y_t) = \frac{\partial V}{\partial t}(t, 0)(1 - Y_t) + \frac{\partial V}{\partial t}(t, 1)Y_t,$$

we obtain

$$J(u) = V(0, Y_0) + E\left[\int_0^T \left\{\frac{\partial V}{\partial s}(s, 1)Y_s + \frac{\partial V}{\partial s}(s, 0)(1 - Y_s)\right.\right.$$

$$+ (V(s, 1) - V(s, 0))u_s(1)(1 - Y_s)\lambda(1, s)$$

$$+ (V(s, 0) - V(s, 1))(1 - u_s(2))Y_s\lambda(2, s) + S_1 Y_s + S_2(1 - Y_s)$$

$$+ C_{12} Y_s u_s(2)\lambda(2, s) + C_{21}(1 - Y_s)(1 - u_s(1))\lambda(1, s)$$

$$+ \left.\left. T_{12} Y_s(1 - u_s(2))\lambda(2, s) + T_{21}(1 - Y_s)u_s(1)\lambda(1, s)\right\} ds\right], \quad (4.10)$$

where

$$u_s(1) = u(s, N_s(1) + 1, N_s(2)),$$
$$u_s(2) = u(s, N_s(1), N_s(2) + 1). \quad (4.11)$$

The control u minimizing the quantity in braces in (4.10) can be chosen in the form $u(s, Y_{t-})$, in which case $u_s(1) = u_s(2) = u(s, Y_{s-})$. Therefore in (4.10), both $u_s(1)$ and $u_s(2)$ can be replaced by $u_s(s, Y_s)$ as far as we are seeking to minimize to 0 the quantity in braces. Now, using the fact that $Y_t = 0$ or 1, one can easily check the following:

4. Attraction Controls

R10 Result. *If there exists two differentiable real-valued functions $t \to V(t, 0)$ and $t \to V(t, 1)$ satisfying*

$$\frac{\partial V}{\partial s}(s, 1) + \inf_{u_1 \in \{0, 1\}} \{(V(s, 0) - V(s, 1))(1 - u_1)$$
$$+ C_{12}\lambda(2, s)u_1 + T_{12}\lambda(2, s)(1 - u_1)\} + S_1 = 0,$$

$$\frac{\partial V}{\partial s}(s, 0) + \inf_{u_0 \in \{0, 1\}} \{(V(s, 1) - V(s, 0))u_0 \qquad (4.12)$$
$$+ C_{21}\lambda(1, s)(1 - u_0) + T_{21}\lambda(1, s)u_0\} + S_2 = 0,$$

$$V(T, i) = 0, \quad i = 1, 2,$$

then the control $u^* \in \mathcal{U}$ given by

$$u_t^* = u^*(t, Y_t), \qquad (4.13)$$

where $u^*(t, 1)$ minimizes $\{\cdots\}$ in the first equation of (4.12) and $u^*(t, 0)$ minimizes the $\{\cdots\}$ in the second equation of (4.12), is optimal.

The above sketch of a proof is easily formalized, but since our goal in this chapter and especially in §3 is to illustrate the basic computational mechanisms, we leave such formalization to the reader.

E5 Exercise. Prove R10.

The solution of the optimal-allocation problem is easily implemented once $V(t, 1)$ and $V(t, 0)$ are obtained. But $V(t, 1)$ and $V(t, 0)$ are in turn not difficult to compute: the dimension of the differential system (4.12) is 2; it would be N if there were N stations sharing the same file. The general case is treated in Sismaïl [10].

Optimal Priority Allocation with no Preemption

In Example 3 of §1 concerning a problem of priority assignment in a queue, there was no switching cost involved: the costs relative to the transfer of the server from queue 1 to queue 2 and vice versa were supposed negligible. If one wishes to take such costs into account, the nature of the problem changes drastically and falls into the category of impulsive controls, for which the dynamic-programming equations are no longer of the Hamilton–Jacobi type. However, if the switching strategy is restricted so as to guarantee that the service of a customer will not be interrupted once started, the problem reduces to an intensity-control problem in a way quite similar to the dynamic file-allocation problem. More precisely:

Statement of the Problem. Two queues $Q_t(1)$ and $Q_t(2)$ with poissonian arrivals are controlled by switching a server from 1 to 2 and vice versa.

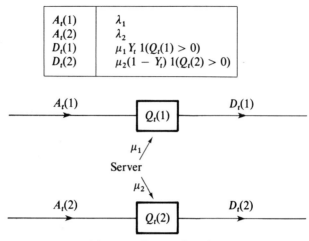

Figure 1 Server allocation.

When attending customers of queue 1, this server has a potential μ_1, and when serving customer of queue 2, its potential is μ_2. Mathematically, if we let $\mathscr{F}_t$ be the history common to both queues $[\mathscr{F}_t = \mathscr{F}_t^Q(1) \vee \mathscr{F}_t^Q(2)]$ and if we denote by Y_t the process that takes the value 1 if the server is assigned to queue 1, and the value 0 if he is assigned to queue 2, then the tableau of Figure 1 gives the $\mathscr{F}_t$-intensities of the various flows. It is supposed that the two queues do not have "hard interactions," in other words, $A_t(1)$, $A_t(2)$, $D_t(1)$, and $D_t(2)$ have no common jumps. Also Y_t must depend on $\mathscr{F}_t$. In fact Y_t is described as follows in terms of the control u_t (the attraction process): u_t takes the values 0 or 1; it is 1 if queue 1 attracts the server, 0 if queue 2 does. Switching of the server from queue 1 to queue 2 can be effected at time t only in the following situation:

(a) the server is attending queue 1 at time $t-$ (indeed, if one wants to switch from 1 to 2 at time t, the server *must* be in 1 at time $t-$), and therefore $Y_{t-} = 1$;
(b) queue 2 must attract the server, i.e. $u_t = 0$.

Also either one of the following conditions must hold:

(c') queue 1 is empty at time $t-$ and there is a new arrival in 2 at time t: $dA_t(2) = 1$;
(c") a customer leaves queue 1 at time t: $dD_t(1) = 1$.

The switching from 2 to 1 is governed by the symmetric rules.
With such rules one observes that:

(i) the server is idle only if both queues are empty;
(ii) the priority discipline is not preemptive: for instance, switching from 1 to 2 can occur only if either queue 1 is empty or a customer of queue 1 has just received complete service and is departing.

E6 Exercise. Find the evolution equation of Y_t in terms of $A_t(1)$, $D_t(1)$, $A_t(2)$, $D_t(2)$, and u_t.

The cost $J(u)$ relative to a strategy $u = (u_t, t \in [0, T])$ is decomposed into four parts:

(α) a *final cost*
$$E[M_1 Q_T(1) + M_2 Q_T(2)],$$

(β) a *sojourn cost*
$$E\left[\int_0^T (L_1 Q_t(1) + L_2 Q_t(2))\, dt\right],$$

(γ) a *hiring cost* for the server
$$E\left[\int_0^T (K_1 Y_t + K_2(1 - Y_t))\, dt\right],$$

(δ) a *switching cost*
$$E\left[\int_0^\cdot (Y_{t-} S_{12} - (1 - Y_{t-}) S_{21})\, dY_t\right].$$

Parts (α) and (β) are easily interpreted and were already considered in Example 3 of §1. Part (γ) was also considered in this example (with Y_t replacing u_t). As for part (δ), S_{ij} is the cost of switching from i to j; hence the cost S_{12} is incurred iff $Y_{t-} = 1$ and $dY_t = 1$; cost S_{21} is incurred iff $Y_{t-} = 0$ and $dY_t = -1$.

E7 Exercise. Find an expression for the total cost $J(u)$ [i.e., sum the four parts, replacing the integral with respect to dY_t in part (δ) by the appropriate integral with respect to Lebesgue measure dt].

Taking for the set of admissible controls $\mathcal{U}$ the $\{0, 1\}$-valued processes u_t of the form $u_t(\omega) = u(t, Y_{t-}(\omega))$, solving the problem of minimization of $J(u)$ among all $u \in \mathcal{U}$ should not be too difficult. It is however very computational as the reader will find if he tries his skill at it.

5. Existence via Likelihood Ratio

The existence of an optimal control is not always granted: it can happen that the infimum $\inf_{u \in \mathcal{U}} J(u)$ [or the supremum $\sup_{u \in \mathcal{U}} J(u)$] is not attained by an admissible control $u^* \in \mathcal{U}$. This should not be surprising: just think of the deterministic optimization problem $\min_{x \in U} x$, where U is the *open* interval $(0, 1)$; then $\min_{x \in U} x = 0$ but there is no x in U such that $x^* = \min_{x \in U} x$, the only such x^* being 0.

What is always granted however is the existence, for any fixed $\varepsilon > 0$, of at least one control u_ε^* in $\mathscr{U}$ such that (for the minimization problem for instance)

$$\inf_{u \in \mathscr{U}} J(u) \leq J(u_\varepsilon^*) < \inf_{u \in \mathscr{U}} J(u) + \varepsilon. \tag{5.1}$$

Such a control u_ε^* is called an ε-optimal control; clearly if ε is small enough, u_ε^* will be considered to be an acceptable approximate solution of the original minimization problem.

From the algorithmical point of view, one can argue that it is *only* possible to find an ε-optimal solution, for some $\varepsilon > 0$. From an analytical point of view, one can reply that to prove that a control $u \in \mathscr{U}$ is ε-optimal for an $\varepsilon > 0$ fixed in advance is as difficult, perhaps even more difficult, than to show that it is optimal. A happy compromise in this matter is to prove the existence *and* find the optimal control at the same time, as was done in T1, C2, and T3. But this is not always possible: recall that T3 is only concerned with a markovian problem, for which the Hamilton–Jacobi equations are deterministic.

Rather than developing arguments pro and con existence theory per se, we will present an existence theorem, following a method invented by Beneš [1] for Wiener-driven stochastic systems and based on compactness arguments (as usual in existence theory) of Radon–Nikodym derivatives. Since we only want to show the beauty of the method, we will present this theory in a simple case: most of the technical difficulties are thus avoided.

Statement of the Problem

Let N_t be a univariate point process defined on $(\Omega, \mathscr{F})$, and let $\mathscr{U}$ be the set of admissible controls consisting of those processes u_t that are $\mathscr{F}_t^N$-predictable and satisfy, for some positive real constant K,

$$0 \leq u_t \leq K. \tag{5.2}$$

The terminal time T is taken to be 1 (for notational convenience). The dynamics $(P_u, u \in \mathscr{U})$ have the following structure: there exists a probability measure P on $(\Omega, \mathscr{F})$ such that N_t is a $(P, \mathscr{F}_t^N)$-Poisson process with intensity 1, and for each $u \in \mathscr{U}$

$$\frac{dP_u}{dP} = L_1(u) = L(u), \tag{5.3}$$

where

$$L_t(u) = \left(\prod_{n \geq 1} u_{T_n} 1(T_n \leq t) \right) \exp \left\{ \int_0^t (1 - u_s)\, ds \right\}. \tag{5.4}$$

5. Existence via Likelihood Ratio

In other words, by VI, T3, N_t admits the $(P_u, \mathscr{F}_t^N)$-intensity u_t. Now let Φ be a random variable on $(\Omega, \mathscr{F})$ which is P-square-integrable:

$$E[|\Phi|^2] < \infty. \tag{5.5}$$

Then:

E8 Exercise. Show that Φ is P_u-integrable for all $u \in \mathscr{U}$.

Define

$$J(u) = E_u[\Phi], \qquad J^* = \inf_{u \in \mathscr{U}} J(u). \tag{5.6}$$

The result to be proven is:

T11 Theorem. *There exists at least one control $u^* \in \mathscr{U}$ such that $J(u^*) = J^*$.*

PROOF. Let $(u^{(n)}, n \geq 1)$ be a minimizing sequence of $\mathscr{U}$, that is to say, $u^{(n)}$ is in $\mathscr{U}$ for all $n \geq 1$ and $\lim_{n \uparrow \infty} J(u^{(n)}) = J^*$. The cost $J(u)$ can be expressed in terms of the probability P and of the Radon–Nikodym derivative $L_1(u)$ as $J(u) = E[L(u)\Phi]$. The random variable Φ is in $L^2(P)$ by hypothesis, and for all $u \in \mathscr{U}$, $L(u)$ is in $L^2(P)$ (by VI, T4, since u_t is bounded). If the family $(L(u), u \in \mathscr{U})$ is shown to be weakly compact in $L^2(P)$, then T11 will be proven: indeed, from the minimizing sequence $(u^{(n)}, n \geq 1)$ one can then extract a subsequence $(u^{(n')}, n' \geq 1)$ such that $L(u^{(n')}) \to L(u^*)$, $L^2(P)$-weakly, for some $u^* \in \mathscr{U}$, which implies $\lim_{n \uparrow \infty} J(u^{(n')}) = J(u^*)$, and therefore $J(u^*) = J^*$, since $(u^{(n')}, n' \geq 0)$ is also a minimizing sequence. The control u^* is therefore optimal, being admissible ($\in \mathscr{U}$) and achieving the infimum of the cost.

In order to show that $(L(u), u \in \mathscr{U})$ is weakly compact in $L^2(P)$ it is sufficient to show that it is convex and strongly closed. This will be done in the two following lemmas. □

L12 Lemma. *The family $(L(u), u \in \mathscr{U})$ is convex.*

PROOF. We have to show that if $u(1)$ and $u(2)$ are two admissible controls and λ_1, λ_2 are two positive real numbers such that $\lambda_1 + \lambda_2 = 1$, then $L = \lambda_1 L(u(1)) + \lambda_2 L(u(2))$ is equal to $L(u)$ for some $u \in \mathscr{U}$. Recall that for $i = 1, 2$

$$L_t(u(i)) = 1 + \int_0^t L_{s-}(u(i))(u_s(i) - 1)(dN_s - ds), \tag{5.7}$$

so that

$$L_t = 1 + \int_0^t \left(\sum_{i=1}^2 \lambda_i L_{s-}(u(i))(u_s(i) - 1) \right)(dN_s - ds). \tag{5.8}$$

Let τ be the following $\mathscr{F}_t^N$-stopping time:

$$\tau = \inf\{t \mid L_{t-} = 0\} \wedge 1, \tag{5.9}$$

and define

$$u_t = \frac{\lambda_1 L_{t-}(u(1))}{L_{t-}} u_t(1) + \frac{\lambda_2 L_{t-}(u(2))}{L_{t-}} u_t(2) \tag{5.10}$$

where $0/0 = 0$ by convention. Recall that $L_{t-} = 0$ implies that $L_{t-}(u(1)) = L_{t-}(u(2)) = 0$, since λ_1 and λ_2 are strictly nonnegative; also from the product form of $L_t(u)$, it is clear that $L_t(u(i)) = 0$ for all $t \geq \tau$, $i = 1, 2$, so that for all $t \geq \tau$, u_t is null.

From these remarks, Equation (5.8), and the definition of u, it follows that

$$L_t = 1 + \int_0^t L_{s-}(u_s - 1)(dN_s - ds). \tag{5.11}$$

Also $u \in \mathcal{U}$, since it is $\mathscr{F}_t^N$-predictable, and the constraint (5.2) is respected [as can be directly checked with the definition (5.10) of u]. Therefore by VI, T3, $L_t = L_t(u)$. □

L13 Lemma. *The family $(L(u), u \in \mathcal{U})$ is strongly closed in $L^2(P)$.*

PROOF. To be proven: if $(L(u^{(n)}), n \geq 1)$ is a sequence of $\mathcal{U}$ that converges in $L^2(P)$ to some random variable ρ, then $\rho = L(u)$ for some $u \in \mathcal{U}$. Taking a subsequence if necessary, we can assume that

$$\rho = \lim_{n \uparrow \infty} L(u^{(n)}) \quad P\text{-a.s.} \tag{5.12}$$

The random variable ρ is in $L_2(P, \mathscr{F}_1^N)$, and therefore, by III, T12, it can be represented as a stochastic integral:

$$\rho = 1 + \int_0^1 \psi_s(dN_s - ds), \tag{5.13}$$

where ψ_t is an $\mathscr{F}_t^N$-predictable process such that

$$E\left[\int_0^1 |\psi_s|^2 \, ds\right] < \infty. \tag{5.14}$$

Let us define

$$\rho_t = 1 + \int_0^t \psi_s(dN_s - ds) \tag{5.15}$$

(so that $\rho = \rho_1$). By Jensen's inequality, since $\rho_t - L_t(u^{(n)})$ is a martingale,

$$\int_0^1 E[|\rho_{t-} - L_{t-}(u^{(n)})|^2] \, dt \leq \int_0^1 E[|\rho - L(u^{(n)})|^2] \, dt \tag{5.16}$$

(see the proof of I, E12 for the Jensen argument). Since ρ is the $L_2(P)$-limit of $L(u^{(n)})$, it follows from (5.16) that[2]

$$\rho_{t-} = \lim_{n \uparrow \infty} L_{t-}(u^{(n)}) \quad dt \times dP\text{- or } dN_t \times dP\text{-a.e.} \tag{5.17}$$

[at least for a subsequence of $(u^{(n)}, n \geq 1)$, still denoted $(u^{(n)}, n \geq 1)$]. By the isometry (III, T13),

$$E[|\rho - L(u^{(n)})|^2] = E\left[\int_0^1 |L_{s-}(u^{(n)})(u_s^{(n)} - 1) - \psi_s|^2 \, ds\right]. \tag{5.18}$$

By the definition of ρ, the right-hand sides of (5.16) and (5.18) tend to 0. Therefore, taking subsequences if necessary,

$$\psi_t = \lim L_{t-}(u^{(n)})(u_t^{(n)} - 1) \quad dt \times dP\text{- or } dN_t \times dP\text{-a.e.} \tag{5.19}$$

From (5.17) and (5.19) we deduce that on $\{(t, \omega) | \rho_{t-}(\omega) > 0\}$

$$\lim_{n \uparrow \infty} (u_t^{(n)} - 1) = \frac{\psi_t}{\rho_{t-}} \quad dt \times dP\text{- or } dN_t \times dP\text{-a.e.} \tag{5.20}$$

Defining

$$u_t = \frac{\psi_t}{\rho_{t-}} 1(\rho_{t-} > 0) + 1 \tag{5.21}$$

or equivalently

$$\rho_{t-}(u_t - 1) = \psi_t, \tag{5.22}$$

we see from (5.15) that

$$\rho_t = 1 + \int_0^t \rho_{s-}(u_s - 1)(dN_s - ds) = L_t(u). \tag{5.23}$$

By (5.20), $u \in \mathcal{U}$, being bounded by K.

Remark. The result can be improved in several directions: indeed, the above proof would hold word by word if at each time t, u_t were constrained to a compact convex set U_t in such a way as to ensure that $L(u)$ was in $L^2(P)$ for all $u \in \mathcal{U}$. Also Φ could be replaced by $\Phi(u)$ in such a way that $\Phi(u)$ was in $L_2(P)$ for all $u \in \mathcal{U}$, and moreover that $(\Phi(u), u \in \mathcal{U})$ was in $L^2(P)$-relatively compact [this would be guaranteed by $\Phi(u) \leq K$ for all $u \in \mathcal{U}$, for instance].

[2] Recall that on $\mathcal{P}(\mathcal{F}_t^N)$, $dt \times dP = dN_t \times dP$ (N_t is a Poisson process with intensity 1).

References

[1] Beneš, V. (1971) Existence of optimal stochastic control laws, SIAM J. Control **9**, pp. 446–475.
[2] Bensoussan, L. and Lions, J. L. (1975) Nouvelles méthodes en contrôle impulsionnel, J. Appl. Math. Optimization **1**, pp. 298–312.

[3] Boel, K. and Varaiya, P. (1977) Optimal control of jump processes, SIAM J. Control **15**, pp. 92–119.
[4] Brémaud, P. (1976) Bang-bang controls of point processes, Adv. Appl. Probab. **8**, pp. 385–394.
[5] Brémaud, P. (1979) Optimal thinning of a point process, SIAM J. Control and Optimization **17**, 2.
[6] Davis, M. H. A. and Varaiya, P. (1973) Dynamic programming conditions for partially observed stochastic systems, SIAM J. Control and Optimization **11**, pp. 226–261.
[7] Martins-Netto, A. and Wong, E. (1976) A martingale approach to queues, Math. Prog. Study **6**, pp. 97–110.
[8] Rishel, R. (1976) Optimal control of a Poisson source, *Proc. JACC Conference at Purdue*, pp. 531–535.
[9] Segall, A. (1976) Dynamical file assignment in computer networks, Part I, IEEE Transactions **AC-21**, 2, pp. 161–173.
[10] Sismaïl, K. (1981) Représentation intégrale des systèmes markoviens de files d'attente, Thèse de 3ème cycle, U. de Paris IX.

Solutions to Exercises, Chapter VII

E1. We make the educated guess that the optimal cost-to-go has the form $V(t, Q_t)$. The development of §2 can be imitated word for word. For instance we find that

$$V(t, Q_t) = V(0, 0) + \int_0^t \frac{\partial V}{\partial s}(s, Q_s)\, ds$$

$$+ \int_0^t \{\lambda[V(s, Q_s + 1) - V(s, Q_s)] + u_t[V(s, Q_s - 1) - V(s, Q_s)]\}\, ds$$

$$+ \text{martingale},\qquad(1)$$

and then

$$E\int_0^t K u_s\, ds + V(T, Q_T) = V(0, 0)$$

$$+ E\int_0^t \left\{\frac{\partial v}{\partial s}(s, Q_s) + \lambda[V(s, Q_s + 1) - V(s, Q_s)]\right.$$

$$\left. + u_s[V(s, Q_s - 1)1(Q_s > 0) - V(s, Q_s)] + K u_s\right\} ds.\qquad(2)$$

If V is chosen such that

$$V(T, n) = \Phi(n),\qquad(3)$$

then (2) and (3) yield

$$J(u) = E\int_0^t \left\{\frac{\partial V}{\partial s}(s, Q_s) + \lambda[V(s, Q_s + 1) - V(s, Q_s)]\right.$$

$$\left. + u_s[V(s, Q_s - 1) - V(s, Q_s)] + K u_s\right\} ds.\qquad(4)$$

Solutions to Exercises, Chapter VII

Hence we are led to solve the system of differential equations

$$\frac{\partial V}{\partial t}(t, n) + \lambda[V(t, n + 1) - V(t, n)]$$

$$+ \inf_{u \in [0, c]} \{u[V(t, n - 1)1(n > 0) - V(t, n)] + Ku\} = 0, \quad n \geq 0, \quad (5)$$

$$V(T, n) = \Phi(n), \quad n \geq 0.$$

Here again the optimal control is bang-bang, since u^* is given by

$$u^*(t, \omega) = \tilde{u}^*(t, Q_t(\omega)), \quad (6)$$

where

$$\tilde{u}^*(t, n) = \begin{cases} 0 & \text{if } V(t, n - 1)1(n > 0) - V(t, n) + K \geq 0, \\ c & \text{if } V(t, n - 1)1(n > 0) - V(t, n) + K < 0. \end{cases} \quad (7)$$

E2. In view of II, E11, for any $u = (v_t, w_t) \in \mathcal{U}$, $P(Q_t = n) = p(n, t)$ satisfies (2.43) where $\tilde{u} = (v(t, n), w(t, n), n > 0)$ is defined by

$$E[v_t 1(Q_t = n)] = v(t, n)p(n, t),$$
$$E[w_t 1(Q_t = n)] = w(t, n)p(n, t), \quad (1)$$

and $J(u)$ can be expressed as follows:

$$J(u) = \sum_{n=0}^{\infty} \left(\int_0^t (a(t, n)v(t, n) + b(t, n)w(t, n) + c(t, n))p(n, t) \, dt + \Psi(T, n)p(n, T) \right). \quad (2)$$

The rest of the proof follows easily; see [4].

E3. Just apply C2.

E4. We develop the same type of calculations as in the proof of R6 to obtain

$$V(t, N_t, Y_t) = V(0, 0, 0) + \int_0^t \frac{\partial V}{\partial s}(s, N_s, Y_s) \, ds$$

$$+ \sum_{n \geq 1} (V(T_n, N_{T_n}, Y_{T_n}) - V(T_n, N_{T_{n-1}}, Y_{T_{n-1}}))1(T_n \leq t). \quad (1)$$

But

$$(V(T_n, V_{T_n}, Y_{T_n}) - V(T_n, N_{T_{n-1}}, Y_{T_{n-1}}))1(T_n \leq t)$$
$$= X_n(V(T_n, N_{T_n-} + 1, Y_{T_n-}, Y_{T_n-}))1(T_n \leq t)$$
$$+ (1 - X_n)(V(T_n, N_{T_n-} + 1, Y_{T_n-}) - V(T_n, N_{T_n-}, Y_{T_n-}))1(T_n \leq t). \quad (2)$$

The proof then parallels the proof of R6 and leads to the equality

$$V(t, N_t, Y_t) = V(0, 0, 0) + \int_0^t \frac{\partial V}{\partial s}(s, N_s, Y_s) \, ds$$

$$+ \int_0^t \lambda(N_s)\{u_s V(s, N_s + 1, Y_s + 1)$$

$$+ (1 - u_s)V(s, N_s + 1, Y_s) - V(s, N_s, Y_s)\} \, ds, \tag{3}$$

which is the key to the solution.

E5. Let $V(t, 1)$ and $V(t, 0)$ satisfy the conditions of R10. For each $u \in \mathscr{U}$, $J(u)$ has the form given in (4.10). For the control u^* defined by (4.13),

$$J(u^*) = V(0, Y_0), \tag{1}$$

whereas for any other control $u \in \mathscr{U}$, the condition (4.12) implies

$$J(u) \geq V(0, Y_0). \tag{2}$$

Therefore u^* is optimal, and moreover

$$V(0, Y_0) = \inf_{u \in \mathscr{U}} J(u). \tag{3}$$

Remark. $V(t, Y_t)$ can be interpreted as the optimal cost-to-go at time t, given $\mathscr{F}_t^N(1) \vee \mathscr{F}_t^N(2) \vee \sigma(Y_0)$ (same line of argument as in §2 after the proof of T1).

E6. Solution:

$$dY_t = -Y_{t-}(1 - u_t)[1(Q_{t-}(1) > 0) \, dA_t(2) + dD_t(1)]$$
$$+ (1 - Y_{t-})u_t[1(Q_{t-}(2) > 0) \, dA_t(1) + dD_t(2)] \tag{1}$$

(the first term of the right-hand side corresponds to the transition from location 1 to 2; the second term, to the transition from 2 to 1).

E7. Only the switching cost needs to be reexpressed; in view of E6 and the definition of stochastic intensity,

$$E\left[\int_0^T (Y_{t-}S_{12} - (1 - Y_{t-})S_{21}) \, dY_t\right]$$

$$= E\left[\int_0^T \{(1 - Y_s)u_s 1(Q_s(2) > 0)(\lambda_1 + \mu_2) \right.$$

$$\left. + Y_s(1 - u_s)1(Q_s(1) > 0)(\lambda_2 + \mu_1)\} \, ds\right]. \tag{1}$$

E8. By VI, T3, $L(u)$ is P-square-integrable, since u_t is bounded. Now $E_u[|\Phi|] = E[L(u)|\Phi|]$; it follows (from the Schwarz inequality and the assumption that Φ is P-square-integrable) that the latter quantity is finite.

CHAPTER VIII

Marked Point Processes

1. Counting Measure and Intensity Kernels
2. Martingale Representation and Filtering
3. Radon–Nikodym Derivatives
4. Towards a General Theory of Intensity
References
Solutions to Exercises, Chapter VIII

1. Counting Measure and Intensity Kernels

Let N_t be a point process. The sequence T_n of its jump times is interpreted as follows: T_n is the nth occurrence of a given physical phenomenon. If for each $n \geq 1$ the occurrence T_n is described by the value Z_n of some attribute (say Z_n is the number of customers in the nth batch of arrivals, or Z_n is the amount of service required by the nth customer entering a service facility), then we obtain a double sequence $(T_n, Z_n, n \geq 1)$ called a marked point process, Z_n being the "mark" corresponding to T_n. Sometimes a marked point process is called a *space-time point process* (see [11]).

D1 Definition. Let there be defined on some measurable space $(\Omega, \mathscr{F})$:

(i) a point process T_n (or N_t),
(ii) a sequence $(Z_n, n \geq 1)$ of E-valued random variables.

The double sequence $(T_n, Z_n, n \geq 1)$ is called an *E-marked point process*. The measurable space $(E, \mathscr{E})$ on which the sequence $(Z_n, n \geq 1)$ takes its values is the *mark space*.

By convention, $Z_0 = \delta$, where δ is a point outside E.

EXAMPLE 1 (Point Process). E consists of a single point. We then identify $(T_n, n \geq 1)$ and $(T_n, Z_n, n \geq 1)$.

EXAMPLE 2 (Multivariate Point Process). E consists of k points: $z_1, z_2, \ldots, z_k$. If we define, for each $1 \leq i \leq k$,

$$N_t(i) = \sum_{n \geq 1} 1(Z_n = z_i)1(T_n \leq t), \qquad (1.1)$$

then we obtain a k-variate point process $(N_t(1), \ldots, N_t(k))$.

EXAMPLE 3 (Queueing Process). Here we give a new description of a queueing process by means of a marked point process $(T_n, Z_n, n \geq 1)$ where the mark space $E = R_+ - \{0\}$; T_n is the arrival time of the nth customer, and Z_n is the amount of service required. See E3 and E4 for applications of this desscription.

Going back to the general case, we associate to each $A \in \mathscr{E}$ the counting process $N_t(A)$ defined by

$$N_t(A) = \sum_{n \geq 1} 1(Z_n \in A)1(T_n \leq t). \qquad (1.2)$$

In particular, $N_t(E) = N_t$. The internal history $\mathscr{F}_t^p$ of $(T_n, Z_n, n \geq 1)$ is defined by

$$\mathscr{F}_t^p = \sigma(N_s(A); 0 \leq s \leq t, A \in \mathscr{E}). \qquad (1.3)$$

Note that each T_n is an $\mathscr{F}_t^p$-stopping time. The superscript p in $\mathscr{F}_t^p$ stands for the *counting measure* $p(dt \times dz)$:

$$p(\omega, (0, t] \times A) = N_t(\omega, A), \qquad t \geq 0, \quad A \in \mathscr{E}. \qquad (1.4)$$

It is a transition measure from $(\Omega, \mathscr{F})$ into $((0, \infty) \times E, (0, \infty) \otimes \mathscr{E})$.

Clearly, for a given $\omega \in \Omega$, $p(\omega, dt \times dz)$ is σ-finite if and only if the realization $T_n(\omega)$ is nonexplosive.

The following notation is similar to that used for univariate point processes:

$$\int_0^\infty \int_E H(s, z)p(ds \times dz) = \sum_{n=1}^\infty H(T_n, Z_n)1(T_n \leq \infty),$$

$$\int_0^t \int_E H(s, \omega, z)p(\omega, ds \times dz) = \sum_{n=1}^\infty H(T_n(\omega), \omega, Z_n(\omega))1(T_n(\omega) \leq t).$$

We recall a convention already used in the previous chapters: that the symbol $\int_a^b$ is to be interpreted as $\int_{(a,b]}$ if $b < \infty$, and $\int_{(a,b)}$ if $b = \infty$.

The counting measure $p(dt \times dz)$ and the double sequence $(T_n, Z_n, n \geq 1)$ will be identified, both being called the E-marked point process (or marked point process for the sake of euphony).

1. Counting Measure and Intensity Kernels

Let $\mathscr{F}_t$ be a history of $p(dt \times dz)$, i.e. a history such that

$$\mathscr{F}_t \supset \mathscr{F}_t^p, \quad t \geq 0.$$

On $(0, \infty) \times \Omega \times E$ we define the σ-field

$$\tilde{\mathscr{P}}(\mathscr{F}_t) = \mathscr{P}(\mathscr{F}_t) \otimes \mathscr{E}, \quad (1.5)$$

where $\mathscr{P}(\mathscr{F}_t)$ is the predictable σ-field on $(0, \infty) \times \Omega$ (I, §3).

Any mapping $H: (0, \infty) \times \Omega \times E \to R$ which is $\tilde{\mathscr{P}}(\mathscr{F}_t)$-measurable is called an $\mathscr{F}_t$-predictable process indexed by E. Clearly $\tilde{\mathscr{P}}(\mathscr{F}_t)$ is generated by the following mappings H:

$$H(t, \omega, z) = C_t(\omega) 1_A(z), \quad (1.6)$$

where C_t is a $\mathscr{F}_t$-predictable process and $A \in \mathscr{E}$.

D2 Definition. Let $p(dt \times dz)$ be a E-marked point process with a history $\mathscr{F}_t$. Suppose that for each $A \in \mathscr{E}$, $N_t(A)$ admits the $(P, \mathscr{F}_t)$-predictable intensity $\lambda_t(A)$, where $\lambda_t(\omega, dz)$ is a transition measure from $(\Omega \times [0, \infty), \mathscr{F} \otimes \mathscr{B}_+)$ into $(E, \mathscr{E})$. We then say that $p(dt \times dz)$ admits the $(P, \mathscr{F}_t)$-intensity kernel $\lambda_t(dz)$.

T3 Theorem (Projection Theorem). *Let $p(dt \times dz)$ be a E-marked point process with the $(P, \mathscr{F}_t)$-intensity kernel $\lambda_t(dz)$. Then for each nonnegative $\mathscr{F}_t$-predictable E-marked process H*

$$E\left[\int_0^\infty \int_E H(s, z) p(ds \times dz)\right] = E\left[\int_0^\infty \int_E H(s, z) \lambda_s(dz) \, ds\right]. \quad (1.7)$$

PROOF. (1.7) is true for H of the form (1.6), by the definition of $\lambda_t(A)$ as the $(P, \mathscr{F}_t)$-intensity of $N_t(A)$. The extension to general nonnegative H's follows from the monotone class theorem. □

C4 Corollary (Integration Theorem). *Let $p(dt \times dz)$ be a E-marked point process with the $(P, \mathscr{F}_t)$-intensity kernel $\lambda_t(dz)$. Let H be a $\mathscr{F}_t$-predictable E-indexed process such that, for all $t \geq 0$, we have*

$$E\left[\int_0^t \int_E |H(s, z)| \lambda_s(dz) \, ds\right] < \infty,$$

$$\left[\int_0^t \int_E |H(s, z)| \lambda_s(dz) \, ds < \infty \quad P\text{-a.s.,}\right] \quad (1.8)$$

then, defining $q(ds \times dz) = p(ds \times dz) - \lambda_s(dz) \, ds$,

$$\int_0^t \int_E H(s, z) q(ds \times dz) \text{ is a } (P, \mathscr{F}_t)\text{-martingale [local martingale]}. \quad (1.9)$$

Remark. C4 is the analogue of II, T8.

PROOF. Same line of argument as in the proof of II, T8, i.e., apply T2 with H replaced by H' defined by

$$H'(t, \omega, z) = H(t, \omega, z)1_A(\omega)1(a \le t \le b)1(t \le T_n(\omega)),$$

where $0 \le a \le b$, $A \in \mathscr{F}_a$, and T_n is an appropriate sequence. □

E1 Exercise. What appropriate sequence?

E2 Exercise. Show that if $\int_0^t \int_E |H(s, z)|\lambda_s(dz)\,ds < \infty$, P-a.s., $t \ge 0$, then $\int_0^t \int_E |H(s, z)|p(ds \times dz) < \infty$, P-a.s., $t \ge 0$, and that outside a P-null set, $\int_0^t \int_E H(s, z)\,q(ds \times dz)$ is then well defined for all $t \ge 0$.

D5 Definition. Let $p(dt \times dz)$ be a E-marked point process with $(P, \mathscr{F}_t)$-intensity kernel $\lambda_t(dz)$ of the form

$$\lambda_t(dz) = \lambda_t \Phi_t(dz), \qquad (1.10)$$

where λ_t is a nonnegative $\mathscr{F}_t$-predictable process and $\Phi_t(\omega, dz)$ is a probability transition kernel from $(\Omega \times [0, \infty), \mathscr{F} \otimes \mathscr{B}_+)$ into $(E, \mathscr{E})$. The pair $(\lambda_t, \Phi_t(dz))$ is called the $(P, \mathscr{F}_t)$-*local characteristics* of $p(dt \times dz)$.

Comment. Since $\Phi_t(dz)$ is a probability, $\Phi_t(E) = 1$, so that

$$\lambda_t \equiv \lambda_t(E) \qquad (1.11)$$

is the $(P, \mathscr{F}_t)$-intensity of the underlying point process $N_t = N_t(E)$. The kernel $\Phi_t(dz)$ can be interpreted along the lines of II, T15:

T6 Theorem. *Let $p(dt \times dz)$ be a E-marked point process with the $(P, \mathscr{F}_t)$-local characteristics $(\lambda_t, \Phi_t(dz))$. If the history $\mathscr{F}_t$ has the special form*

$$\mathscr{F}_t = \mathscr{F}_0 \vee \mathscr{F}_t^p, \qquad (1.12)$$

then for all $n \ge 1$ and all $A \in \mathscr{E}$,

$$\Phi_{T_n}(A) = P[Z_n \in A | \mathscr{F}_{T_n-}] \quad P\text{-a.s. on } \{T_n < \infty\}. \qquad (1.13)$$

PROOF. First observe that for all $A \in \mathscr{E}$ and all $n \ge 1$, $\Phi_{T_n}(A)1(T_n < \infty)$ is $\mathscr{F}_{T_n}$-measurable. Indeed, by (1.10), $\Phi_{T_n}(A)\lambda_{T_n} = \lambda_{T_n}(A)$, and $\lambda_{T_n} > 0$ on $\{T_n < \infty\}$ by II, T12. Therefore $\Phi_{T_n}(A)1(T_n < \infty) = (\lambda_{T_n}(A)/\lambda_{T_n})1(T_n < \infty)$, and the result follows by I, E10. The proof is then a reproduction of the proof of II, T15: let B be a set of $\mathscr{F}_{T_n-}$; by A2, T31 there exists an $\mathscr{F}_t$-predictable process X_t such $X_{T_n} = 1_B$ on $\{T_n < \infty\}$ and X_t is null outside

1. Counting Measure and Intensity Kernels   237

$[T_n, T_{n+1})$. Now let H be defined by $H(t, z) = X_t 1_A(z)$. Then

$$E[1_B \lambda_{T_n} \Phi_{T_n}(A) 1(T_n < \infty)] = E\left[\int_0^\infty \int_E H(s, z) \lambda_s p(ds \times dz)\right]$$

$$= E\left[\int_0^\infty \lambda_s X_s \, dN_s(A)\right]$$

$$= E\left[\int_0^\infty \lambda_s X_s \lambda_s(A) \, ds\right]$$

$$= E\left[\int_0^\infty \lambda_s(A) X_s \, dN_s^*\right]$$

$$= E[1_B \lambda_{T_n}(A) 1(T_n < \infty)],$$

where all equalities follow by definition of the quantities and symbols involved. The set B being arbitrary in $\mathcal{F}_{T_n-}$, the announced equality is proven. □

E3 Exercise (Marked Poisson Processes). Suppose that $p(dt \times dz)$ admits the $(P, \mathcal{F}_t)$-local characteristics $(\lambda(t), F(dz))$ where $\lambda(t)$ is a locally integrable deterministic function and $F(dz)$ is a probability distribution on $(E, \mathcal{E})$. Show that whenever A and B of $\mathcal{E}$ are disjoint, $N_t(A)$ and $N_t(B)$ are independent Poisson processes.

E4 Exercise. The mathematical setting is as in E3 with $E = R_+ - \{0\}$. The interpretation is in terms of queueing: Z_n is the amount of service required by the nth customer. Define the clearing time W_t at time t by

$$t + W_t = \begin{cases} \inf\{s/s \geq t, D_s - D_t = Q_t\} & \text{if } \{\cdots\} \neq \emptyset, \\ +\infty & \text{otherwise,} \end{cases} \quad (1.14)$$

where Q_t is the number of customers present at time t and D_t is the departure process. Assume that there exists a unique server taking care of one server at a time, providing service at unit rate, and never idle if the queue is not empty. Find a semimartingale representation for W_t and for $E[e^{iuW_t}]$. Assume that the average service time $E(Z) = \int_0^\infty z F(dz)$ is finite and that the queue is in equilibrium with respect to the clearing time W_t, i.e. $E[e^{iuW_t}] = E[e^{iuW_0}]$. Find an expression for $E[e^{iuW_t}]$.

E5 Exercise (Backlog Equilibrium). Let $(T_n, Z_n, n \geq 1)$ be an $R_+ - \{0\}$-marked point process with the same queueing interpretation as in E4. Z_n is the amount of service required by the nth customer arriving at a processing center. Let b_t be a nonnegative process adapted to $\mathcal{F}_t^p$; $\int_a^b b_t \, dt$ is the amount of service provided by the center between times a and b. Let Q_t be the number of customers in the center at time t, waiting for service or having received partial service.

Let W_t be the backlog at time t, i.e. the total remaining service to be provided to the Q_t customers present in the center. Assume that Z_n is independent of T_n, and that each Z_n has mean $1/\mu$.

Find an expression for $E[W_t]$ in terms of $E[W_0]$, $E[N_t]$, $E[b_t]$, and μ. Suppose that N_t is weakly stationary in the sense that $E[N_t] = \lambda t$ for some $\lambda > 0$, and suppose that "backlog equilibrium" has been reached, i.e. $E[W_t] = E[W_0]$. Show that in such a situation

$$E[b_t] = \rho,$$

where ρ is the traffic intensity λ/μ. In particular, if $b_t = C1(Q_t > 0)$, i.e., there is one single server with working capacity C, then

$$P[Q_t = 0] = 1 - \frac{\rho}{C}.$$

E6 Exercise. Same situation as in E3. Let $\tilde{A}$ and $\tilde{B}$ be two disjoint sets of $(\mathcal{B}_+ \otimes \mathcal{E})$ and define $N(\tilde{A}) = \int \int_{\tilde{A}} p(dt \times dz)$, $N(\tilde{B}) = \int \int_{\tilde{B}} p(dt \times dz)$. Show that $N(\tilde{A})$ and $N(\tilde{B})$ are independent Poisson random variables with parameter $\int \int_{\tilde{A}} \lambda(t) F(dz)$ and $\int \int_{B} \lambda(t) F(dz)$ respectively (assume these quantities finite). Show that the point process T'_n defined by $T'_n = T_n + Z_n$ is Poisson, and compute its intensity, assuming that F admits a density f.

Remark. The last result of E6 is due to Mirasol {108} and stated as follows: the output of an $M/G/\infty$ queue is Poisson.

2. Martingale Representation and Filtering

Integral Representation of Martingales

The results of Chapter III can readily be extended to marked point processes. Since the proofs are exactly the same, they will be omitted.

T7 Theorem [6]. *Let $(T_n, Z_n, n \geq 1)$ or $p(dt \times dz)$ be an E-marked point process with the internal history $\mathcal{F}_t^p$, and let $\mathcal{F}_t$ be a history of the form*

$$\mathcal{F}_t = \mathcal{F}_0 \vee \mathcal{F}_t^p. \tag{2.1}$$

Suppose that for each $n \geq 1$, there exists a regular conditional distribution of (S_{n+1}, Z_{n+1}) given $\mathcal{F}_{T_n}$ of the form

$$P[S_{n+1} \in A, Z_{n+1} \in C | \mathcal{F}_{T_n}] = \int_A g^{(n+1)}(s, C) \, ds, \qquad A \in \mathcal{B}_+, \quad C \in \mathcal{E}, \tag{2.2}$$

where $g^{(n+1)}(\omega, s, C)$ is a finite kernel from $(\Omega \times [0, \infty), \mathcal{F}_{T_n} \otimes \mathcal{B}_+)$ into $(E, \mathcal{E})$, that is to say:

(α) for fixed (ω, s), $C \to g^{(n+1)}(\omega, s, C)$ is a finite measure on $(E, \mathcal{E})$.
(β) for fixed $C \in \mathcal{E}$, the mapping $(\omega, s) \to g^{(n+1)}(\omega, s, C)$ is $\mathcal{F}_{T_n} \otimes \mathcal{B}_+$-measurable.

Then $p(dt \times dz)$ admits the $\mathcal{F}_t$-local characteristics $(\lambda_t, \Phi_t(dz))$ defined by

$$\lambda_t(C) = \frac{g^{(n+1)}(t - T_n, C)}{1 - \int_0^{t-T_n} g^{(n+1)}(x, E)\,dx} \quad \text{on } (T_n, T_{n+1}],$$

$$\lambda_t = \lambda_t(E), \tag{2.3}$$

$$\Phi_t(C) = \frac{\lambda_t(C)}{\lambda_t(E)}.$$

In other words, for each $n \geq 1$, each $C \in \mathcal{E}$,

$$N_{t \wedge T_n}(C) - \int_0^{t \wedge T_n} \lambda_s(C)\,ds \text{ is a } (P, \mathcal{F}_t)\text{-martingale}. \tag{2.4}$$

We suppose now that $(\Omega, \mathcal{F}, \mathcal{F}_t, P)$ have been modified as explained in III, §5, and we will keep the same notation for the modified objects. As was mentioned in III, §5 and is shown in A2, §3, there is no essential difficulty in doing so.

T8 Theorem [1, 6]. *Let the conditions of T6 prevail after the proper modification of $(\Omega, \mathcal{F}, \mathcal{F}_t, P)$ just mentioned. Then any $(P, \mathcal{F}_t)$-martingale M_t admits the stochastic integral representation*

$$M_t = M_0 + \int_0^t \int_E H(s, z)q(ds \times dz) \quad P\text{-a.s.}, \quad t \geq 0, \tag{2.5}$$

where

$$q(dt \times dz) = p(dt \times dz) - \lambda_t(dz)\,dt \tag{2.6}$$

and H is an E-indexed $\mathcal{F}_t$-predictable process such that

$$\int_0^t \int_E |H(s, z)|\lambda_s(dz)\,ds < \infty \quad P\text{-a.s.}, \quad t \geq 0. \tag{2.7}$$

The E-indexed $\mathcal{F}_t$-predictable process H in the above representation is essentially unique with respect to the measure $P(d\omega)\lambda_t(\omega, dz)\,dt$ on $(\Omega \times [0, \infty) \times E, \mathcal{P}(\mathcal{F}_t) \otimes \mathcal{E})$.

Moreover, if M_t is square-integrable, H satisfies a stronger condition than (2.7), namely

$$E\left[\int_0^t \int_E |H(s, z)|^2 \lambda_s(dz)\,ds\right] < \infty, \quad t \geq 0. \tag{2.8}$$

PROOF. Same as for T9, T15 and T17 of Chapter III. See Itmi [12] for a detailed proof.

Filtering with Marked-Point-Process Observations

Here again, the theory is exactly the same as for multivariate point process observations, and we will therefore be content with a summary of the results.

Let Z_t be a real-valued stochastic process defined on $(\Omega, \mathscr{F}, P)$ with the following semimartingale representation:

$$Z_t = Z_0 + \int_0^t f_s \, ds + m_t^d + m_t^c, \tag{2.9}$$

where $\mathscr{F}_t$ is some history defined on $(\Omega, \mathscr{F})$, and

(α) f_t is an $\mathscr{F}_t$-progressive process such that

$$E\left[\int_0^t |f_s| \, ds\right] < \infty, \, t \geq 0, \tag{2.10}$$

(β) m_t^d is an $\mathscr{F}_t$-martingale of integrable variation over finite intervals,
(γ) m_t^c is a continuous $\mathscr{F}_t$-martingale,
(δ) $Z_t - m_t^c$ is a bounded process.

Now let $(T_n, Z_n, n \geq 1)$ or $p(dt \times dz)$ be an E-marked point process with the $\mathscr{F}_t$-local characteristics $(\lambda_t, \Phi_t(dz))$ and the $\mathscr{G}_0 \vee \mathscr{F}_t^p$-local characteristics $(\hat{\lambda}_t, \hat{\Phi}_t(dz))$, where $\mathscr{G}_0 \subset \mathscr{F}_0$.

T9 Theorem [5, 2]. *Let τ be a finite $\mathscr{G}_0 \vee \mathscr{F}_t^p$-stopping time. Then the following holds:*

$$E[Z_\tau | \mathscr{G}_0 \vee \mathscr{F}_\tau^p] = \hat{Z}_\tau, \tag{2.11}$$

where

$$\hat{Z}_t = E[Z_0 | \mathscr{G}_0] + \int_0^t \hat{f}_s \, ds$$

$$+ \int_0^t \int_E K(s, z)(p(ds \times dz) - \hat{\lambda}_s \hat{\Phi}_s(dz) \, ds). \tag{2.12}$$

Here $\hat{f}_t$ is a $\mathscr{G}_0 \vee \mathscr{F}_t^p$-progressive version of $E[f_t | \mathscr{G}_0 \vee \mathscr{F}_t^p]$ in the sense given in IV, §2, and $K(t, z)$ is a $\mathscr{G}_0 \vee \mathscr{F}_t^p$-predictable process indexed by E, that is $P(d\omega)p(dt \times dz)$-essentially uniquely, defined by

$$K(t, z) = \Psi_1(t, z) - \Psi_2(t, z) + \Psi_3(t, z), \tag{2.13}$$

where the $\Psi_i(t, z)$ are $\mathscr{G}_0 \vee \mathscr{F}_t^p$-predictable processes indexed by E and are $P(d\omega)p(dt \times dz)$-essentially uniquely, defined by the following equalities holding for all $t \geq 0$ and all bounded $\mathcal{G}_0 \vee \mathcal{F}_t^p$-predictable processes $C(t, z)$ indexed by E:

$$E\left[\int_0^t \int_E \Psi_1(s, z)C(s, z)\hat{\lambda}_s(dz)\, ds\right] = E\left[\int_0^t \int_E Z_s C(s, z)\lambda_s(dz)\, ds\right],$$

$$\Psi_2(s, z) = \hat{Z}_{s-} \tag{2.14}$$

$$E\left[\int_0^t \int_E \Psi_3(s, z)C(s, z)\hat{\lambda}_s(dz)\, ds\right] = E\left[\int_0^t \Delta Z_s C(s, z)p(dt \times dz)\right].$$

3. Radon–Nikodym Derivatives

T10 Theorem [1, 6]. *Let $p(dt \times dz)$ be an E-marked point process with the $(P, \mathcal{F}_t)$-local characteristics $(\lambda_t, \Phi_t(dz))$. Let μ_t be a nonnegative $\mathcal{F}_t$-predictable process, and let $h(t, z)$ be an $\mathcal{F}_t$-predictable E-indexed nonnegative process such that*

$$\int_0^t \mu_s \lambda_s\, ds < \infty \quad P\text{-a.s.,} \quad t \geq 0,$$

$$\int_E h(t, z)\Phi_t(dz) = 1 \quad P\text{-a.s.,} \quad t \geq 0. \tag{3.1}$$

Define for each $t \geq 0$

$$L_t = L_0 \left(\prod_{n \geq 1} \mu_{T_n} h(T_n, Z_n) 1(T_n \leq t)\right)$$

$$\times \exp\left\{\int_0^t \int_E (1 - \mu_s h(s, z))\lambda_s \Phi_s(dz)\, ds\right\}, \tag{3.2}$$

where L_0 is a nonnegative $\mathcal{F}_0$-measurable random variable such that $E[L_0] = 1$ (as usual, the product $\prod_{n=1}$ is taken to be 1 if $T_1 > t$). Then:

(a) *L_t is a nonnegative $(P, \mathcal{F}_t)$-local martingale and a nonnegative $(P, \mathcal{F}_t)$-supermartingale.*

(b) *If $E[L_1] = 1$, L_t is a $(P, \mathcal{F}_t)$-martingale over $[0, 1]$. Defining the probability $\tilde{P}$ by $d\tilde{P}/dP = L_1$, $p(dt \times dz)$ admits over $[0, 1]$ the $(\tilde{P}, \mathcal{F}_t)$-local characteristics $(\mu_t \lambda_t, h(t, z)\Phi_t(dz))$.*

PROOF. A sketch of proof will be sufficient, since we will use the same arguments as in the proofs of VI, T2 and VI, T3.

(a) Rewrite L_t using the exponential formula (Appendix A4, T4):

$$L_t = L_0 + \int_0^t \int_E (\mu_s h(s, z) - 1)L_{s-}\, q(ds \times dz), \tag{3.3}$$

where $q(dt \times dz) = p(dt \times dz) - \lambda_t \Phi_t(dz) \, dt$. Stop at

$$S_n = \text{``inf}\left\{t \mid L_{t-} + \int_0^t \mu_s \lambda_s \, ds \geq n\right\} \text{ or } \infty\text{''} \tag{3.4}$$

and apply C4 to obtain that $L_{t \wedge S_n}$ is a $(P, \mathscr{F}_t)$-martingale. Now $S_n \uparrow \infty$, P-a.s., by the condition (3.1); it follows that L_t is a $(P, \mathscr{F}_t)$-local martingale, and, being nonnegative, it is a $(P, \mathscr{F}_t)$-supermartingale (recall I, E8).

(b) If $E[L_0] = E[L_1]$ and if L_t is a supermartingale, it is a martingale over $[0, 1]$ by I, E6. To be proven now:

$$\tilde{E}\left[\int_0^1 \int_E H(s, z) p(ds \times dz)\right] = \tilde{E}\left[\int_0^1 \int_E H(s, z) \mu_s \lambda_s h(s, z) \Phi_s(dz)\right] \tag{3.5}$$

for any nonnegative $\mathscr{F}_t$-predictable E-indexed process H. This is done through the following sequence of equalities:

$$\tilde{E}\left[\int_0^1 \int_E H(s, z) p(ds \times dz)\right]$$

$$= E\left[L_1 \int_0^1 \int_E H(s, z) p(ds \times dz)\right]$$

(by the definition of $\tilde{P}$: $L_1 = d\tilde{P}/dP$)

$$= E\left[\int_0^1 \int_E L_s H(s, z) p(ds \times dz)\right] \quad \text{(by A2, T19)}$$

$$= E\left[\int_0^1 \int_E L_{s-} H(s, z) \mu_s h(s, z) p(ds \times dz)\right]$$

(since on $\{T_n < \infty\}$, $L_{T_n} = L_{T_n-} \mu_{T_n} h(T_n, Z_n)$)

$$= E\left[\int_0^1 \int_E L_s H(s, z) \mu_s h(s, z) \lambda_s \Phi_s(dz) \, ds\right] \quad \text{(by T3)}$$

$$= E\left[L_1 \int_0^1 \int_E H(s, z) \mu_s \lambda_s h(s, z) \Phi_s(dz) \, ds\right]$$

(by Appendix A2, T19 again)

$$= \tilde{E}\left[\int_0^1 \int_E H(s, z) \mu_s \lambda_s h(s, z) \Phi_s(dz) \, ds\right] \quad \text{(by the definition of } \tilde{P}\text{)}. \tag{3.6}$$

$\square$

We now give a sufficient condition for $E[L_1] = 1$.

T11 Theorem [1]. *Let the conditions (3.1) of T9 prevail, and suppose in addition that there exists a deterministic increasing real-valued function* $B(t)$

3. Radon–Nikodym Derivatives

and finite constants K_1, K_2, α, *where* $\alpha > 1$, *such that for all* $t \in [0, 1]$,

$$\int_E (\mu_s h(t, z))^\alpha \Phi_t(dz) \leq K_1 + K_2\left(N_t + \int_0^t \lambda_s \, ds\right) \quad \text{P-a.s.,} \tag{3.7}$$

$$\int_0^t \lambda_s \, ds \leq B(t) \quad \text{P-a.s.} \tag{3.8}$$

Suppose also that for all $0 < M < \infty$

$$E[L_0 \exp\{MN_1\}] < \infty. \tag{3.9}$$

Then $E[L_1] = 1$.

EXAMPLE. Take $\lambda_t = \lambda(t)$, where $\lambda(t)$ is locally integrable and deterministic, and $\Phi_t(\omega, dz) = F(dz)$. [Thus, under P, $p(dt \times dz)$ is a marked Poisson process; see E3.] The conditions (3.8) and (3.9) are then automatically satisfied. For the condition (3.9) for instance, N_1 being independent of $\mathscr{F}_0$, $E[L_0 \exp\{MN_1\}] = E[L_0 E[\exp\{MN_1\}]] = E[\exp\{MN_1\}] < \infty$, the last inequality being a consequence of the fact that N_1 is a Poisson random variable.

PROOF. Define for $1 < \gamma < \alpha^{1/2}$ (such a choice of γ is possible, since $\alpha > 1$):

$$f_t(\gamma) = L_0^{1/\gamma}\left(\prod_{n=1} \tilde{h}(T_n, Z_n)^\gamma 1(T_n \leq t)\right)$$

$$\times \exp\left\{\int_0^t \int_E \left(\frac{1}{\gamma} - \frac{\tilde{h}(s, z)^{\gamma^2}}{\gamma}\right)\lambda_s(dz) \, ds\right\},$$

$$g_t(\gamma) = L_0^{(\gamma-1)/\gamma}\exp\int_0^t \int_E \left(-\gamma(\tilde{h}(s, z) - 1) - \frac{1}{\gamma} + \frac{\tilde{h}(s, z)^{\gamma^2}}{\gamma}\right)\lambda_s(dz) \, ds,$$

where $\tilde{h}(s, z) = \mu_s h(s, z)$. Hence

$$L_t^\gamma = f_t(\gamma)g_t(\gamma).$$

But $(f_t(\gamma))^\gamma$ is equal to L_t of (3.2) if we set $\mu_s h(s, z) = \tilde{h}(s, z)^{\gamma^2}$. Since $\gamma^2 \leq \alpha$, it follows from T10 and (3.7) that

$$E[f_t(\gamma)^\gamma] \leq 1.$$

By Hölder's inequality,

$$E[L_t^\gamma] = E[f_t(\gamma)^\gamma]^{1/\gamma} E[g_t(\gamma)^{\gamma/(\gamma-1)}]^{(\gamma-1)/\gamma},$$

and therefore

$$E[L_t^\gamma] \leq E[g_t(\gamma)^{\gamma/(\gamma-1)}]^{(\gamma-1)/\gamma}.$$

But, since $h \geq 0$ implies $-(h - 1) \leq 1$, it follows from the definition of $g_t(\gamma)$ that

$$g_t(\gamma)^{\gamma/(\gamma-1)} \leq L_0 \exp\left\{\int_0^t \int_E \left(\gamma - \frac{1}{\gamma} + \frac{\tilde{h}(s, z)^{\gamma^2}}{\gamma}\right)\lambda_s(dz) \, ds\right\},$$

so that, using the assumptions of T10, and denoting $\int_0^t \lambda_s \, ds$ by Λ_t,

$$g_t(\gamma)^{\gamma/(\gamma-1)} \leq L_0 \exp\left\{\frac{\gamma^2 - 1}{\gamma} B(s) + \int_0^t (K_1 + K_2(N_s + \Lambda_s))\Lambda_s \, ds\right\}$$

$$\leq L_0 \exp\left\{\left(\frac{\gamma^2 - 1}{\gamma} + K_1 + K_2 B(s) + K_2 N_s\right)B(s) \, ds\right\},$$

and therefore, if $t \leq 1$:

$$g_t(\gamma)^{\gamma/(\gamma-1)} \leq L_0 \exp \beta \exp\{K_2 B(1) N_1\}$$

for some constant β. Thus, for all $0 \leq t \leq 1$, $E[L_t^\gamma]$ is bounded by a constant C.

The conditions satisfied by $\mu_s h(s, z)$ are *a fortiori* satisfied by $\mu_s h(s, z) \times 1(s \leq S_n)$, where S_n is defined by (3.4). Hence

$$E[L_{1 \wedge S_n}^\gamma] \leq C.$$

In particular by Appendix A1, T25 the family $(L_{1 \wedge S_n}, n \geq 1)$ is uniformly integrable and therefore, by Appendix A1, T26,

$$\lim_{n \uparrow \infty} E[L_{1 \wedge S_n}] = E\left[\lim_{n \uparrow \infty} L_{1 \wedge S_n}\right].$$

But $E[L_{1 \wedge S_n}] = 1$ for all $n \geq 1$ and $S_n \uparrow \infty$, P-a.s.; therefore $E[L_1] = 1$, q.e.d. □

4. Towards a General Theory of Intensity

Let N_t be a point process adapted to a history $\mathscr{F}_t$. We will assume that $(\Omega, \mathscr{F}, P, \mathscr{F}_t)$ have been properly completed, i.e. that $(\Omega, \mathscr{F}, P)$ is a complete probability space and $\mathscr{F}_0$ contains all P-negligible events of $\mathscr{F}_\infty$. Moreover we suppose that $\mathscr{F}_t$ is right-continuous. We then say that the "usual conditions" are verified.

Even in the case where N_t is nonexplosive, the existence of an $\mathscr{F}_t$-intensity λ_t for N_t is not granted, as the following very simple counterexample shows: Let Ω consist of a single point, a function $(n(t), t \geq 0)$ where $n(t) = 1(t \geq c)$ for some strictly positive real number c. There is just one choice for $\mathscr{F}, \mathscr{F}_t$, and P, and any candidate $\mathscr{F}_t$ for the status of an $\mathscr{F}_t$-intensity is a locally integrable deterministic function $\lambda(t)$, which must satisfy

$$n(t) - n(s) = \int_s^t \lambda(u) \, du. \tag{4.1}$$

A contradiction arises at point c.

4. Towards a General Theory of Intensity

Other counterexamples, less trivial, can be constructed. So the mathematical question is: what happens when there is no intensity?

T12 Theorem. *Let N_t be a point process on $(\Omega, \mathscr{F}, P)$ with the history $\mathscr{F}_t$, and assume that the "usual conditions" are verified. There exists a unique (up to P-undistinguishability) right-continuous $\mathscr{F}_t$-predictable nondecreasing process A_t with $A_0 \equiv 0$ such that for all nonnegative $\mathscr{F}_t$-predictable processes C_t,*

$$E\left[\int_0^{T_n} C_s \, dN_s\right] = E\left[\int_0^{T_n} C_s \, dA_s\right]. \qquad (4.2)$$

A_t *is the "dual $\mathscr{F}_t$-predictable projection" of N_t.*

Note that we do not assume N_t to be nonexplosive. In the case where N_t is indeed nonexplosive, if A_t is absolutely continuous with respect to the Lebesgue measure in the sense that

$$A_t = \int_0^t \lambda_s \, ds \qquad (4.3)$$

for some $\mathscr{F}_t$-progressive nonnegative process λ_t, then we are back to the situation considered in all previous chapters.

Theorem T12 is an immediate corollary of a fundamental result of the general theory of stochastic processes, the dual predictable projection theorem of Dellacherie and Meyer (see {48}, V T28) which we will state without proof.

T13 Theorem (Dual Predictable Projection [4]). *Let B_t be a right-continuous nondecreasing process with $B_0 \equiv 0$ and $E[B_\infty] < \infty$. Let $\mathscr{F}_t$ be some history of B_t, and assume that the "usual conditions" hold. Then there exists a unique (up to P-indistinguishability) right-continuous $\mathscr{F}_t$-predictable nondecreasing process $\tilde{B}_t$ with $\tilde{B}_0 \equiv 0$ and $E[\tilde{B}_\infty] < \infty$ such that*

$$E\left[\int_0^\infty C_s \, dB_s\right] = E\left[\int_0^\infty C_s \, d\tilde{B}_s\right] \qquad (4.4)$$

for all nonnegative $\mathscr{F}_t$-predictable processes C_t; $\tilde{B}_t$ is called the dual $\mathscr{F}_t$-predictable projection of B_t.

PROOF OF T12. Take $B_t = 1(t \geq T_n)$ in T13 and obtain the existence of $\tilde{B}_t$, denoted $B_t^{(n)}$. Then set $A_t = \sum_{n \geq 1} \tilde{B}_t^{(n)}$. □

Theorem T12 can be extended to the case of marked point processes. However, we need some additional assumptions on the space of marks $(E, \mathscr{E})$.

Assumption (L). E is a Borel subset of a compact metric space and $\mathscr{E}$ consists of the Borelians of E.

In particular, $\mathscr{E}$ is separable (i.e. generated by a countable family of sets in $\mathscr{E}$). Assumption (L) will allow us to use a result of disintegration of measures.

Let $p(dt \times dz)$ be an E-marked point process defined on $(\Omega, \mathscr{F}, P)$, and let $\mathscr{F}_t$ be a history of $p(dt \times dz)$. Assume that the "usual conditions" are verified for $(\Omega, \mathscr{F}, P, \mathscr{F}_t)$, and suppose that Assumption (L) holds for $(E, \mathscr{E})$.

Let M_p be the measure on $\mathscr{P}(\mathscr{F}_t) \otimes \mathscr{E}$ associated with $p(dt \times dz)$ by the formula

$$M_p(H) = E\left[\int_0^\infty \int_E H(s, z) p(ds \times dz)\right] \tag{4.5}$$

for all nonnegative $\mathscr{F}_t$-predictable E-indexed processes H. Let m_p be the measure on $\mathscr{P}(\mathscr{F}_t)$ obtained by "projection" of M_p on $\mathscr{P}(\mathscr{F}_t)$, i.e.

$$m_p(C) = M_p(C1_E), \tag{4.6}$$

where C_t is any nonnegative $\mathscr{F}_t$-predictable process, and $H = C1_E$ is defined by $H(t, z) = C_t 1_E(z)$.

Assume in a first step that M_p is σ-finite. It can then be disintegrated [here we need Assumption (L)]:

$$M_p(d\omega \times dt \times dz) = m_p(d\omega \times dt)\Phi_t(\omega, dz), \tag{4.7}$$

where $\Phi_t(\omega, dz)$ is a transition measure from $((0, \infty) \times \Omega, \mathscr{P}(\mathscr{F}_t))$ into $(E, \mathscr{E})$. Now, $m_p(d\omega \times dt) = P(d\omega) \, dN_t(\omega)$. Indeed,

$$m_p(C) = E\left[\int_0^\infty C_t \, dN_t\right]. \tag{4.8}$$

Therefore by T13,

$$m_p(d\omega \times dt) = P(d\omega) \, dA_t(\omega) \tag{4.9}$$

for a unique right-continuous $\mathscr{F}_t$-predictable increasing process A_t such that $A_0 = 0$. Finally,

$$M_p(d\omega \times dt \times dz) = P(d\omega) \, dA_t(\omega)\Phi_t(\omega, dz). \tag{4.10}$$

That is to say,

$$E\left[\int_0^\infty \int_E H(s, z) p(ds \times dz)\right] = E\left[\int_0^\infty \int_E H(s, z)\Phi_s(dz) \, dA_s\right] \tag{4.11}$$

for all nonnegative $\mathscr{F}_t$-predictable E-indexed processes H.

In general, without the σ-finiteness assumption for M_p:

T14 Theorem [6]. *Let $(\Omega, \mathscr{F}, P, \mathscr{F}_t)$ verify the usual conditions, where $\mathscr{F}_t$ is a history of an E-marked point process $p(dt \times dz)$. Suppose that the mark space $(E, \mathscr{E})$ satisfies Assumption (L). Then there exists*

(α) *a unique (up to P-indistinguishability) right-continuous increasing $\mathscr{F}_t$-predictable process A_t with $A_0 \equiv 0$,*

(β) *a transition measure* $\Phi_t(\omega, dz)$ *form* $((0, \infty) \times \Omega, \mathscr{P}(\mathscr{F}_t))$ *into* $(E, \mathscr{E})$, *such that, for all* $n \geq 1$,

$$E\left[\int_0^{T_n} \int_E H(s, z) p(ds \times dz)\right] = E\left[\int_0^{T_n} \int_E H(s, z) \Phi_s(dz) \, dA_s\right] \quad (4.12)$$

for all nonnegative $\mathscr{F}_t$-*predictable E-indexed processes* H.

PROOF. Replace $p(dt \times dz)$ by $p^{(n)}(dt \times dz)$ in the proof preceding the statement of T14, with

$$p^{(n)}(\omega, dt \times dz) = 1(t \leq T_n(\omega)) p(\omega, dt \times dz) \quad (4.13)$$

and observe that $M_{p^{(n)}}$ is indeed σ-finite (it is in fact finite). □

Remark. If N_t is P-nonexplosive, M_p is σ-finite, so that (4.11) holds.

Remark. $\Phi_t(dz)$ is unique in the following sense: if $\Phi'_t(dz)$ has the same properties as $\Phi_t(dz)$ in T14, then for any $C \in \mathscr{E}$, there exists a P-null set $\mathscr{N}(C)$ such that outside $\mathscr{N}(C)$

$$\int_0^t \int_C \Phi_s(\omega, dz) \, dA_s(\omega) = \int_0^t \int_C \Phi'_s(\omega, dz) \, dA_s(\omega), \quad t \geq 0. \quad (4.14)$$

This follows from the uniqueness result of T13 and the fact that $\int_0^t \int_C \Phi_s(dz) \, dA_s$ and $\int_0^t \int_C \Phi'_s(dz) \, dA_s$ are two versions of the dual $\mathscr{F}_t$-predictable projection of $N_t(C)$.

Since $\mathscr{E}$ is separable, one can find a "universal" P-null event $\mathscr{N}$ such that if $\omega \notin \mathscr{N}$, (4.14) holds for any $C \in \mathscr{E}$.

T14 is a generalization of T3. It has analogous consequences; for instance:

C15 Corollary. *Let the conditions of T14 prevail, and suppose in addition that the quantities on either side of (4.12) are finite. Then*

$$\int_0^{t \wedge T_n} \int_E H(s, z) q(ds \times dz) \text{ is a } (P, \mathscr{F}_t)\text{-martingale}, \quad (4.15)$$

where $q(ds \times dz) = p(ds \times dz) - \Phi_s(dz) \, dA_s$.

PROOF. Same as for C4. □

T16 Theorem. *Let the conditions of T14 prevail. Suppose in addition that*

$$\mathscr{F}_t = \mathscr{F}_0 \vee \mathscr{F}_t^p. \quad (4.16)$$

Then the transition measure $\Phi_t(dz)$ *satisfies, for all* $n \geq 1$,

$$\Phi_{T_n}(A) = P[Z_n \in A / \mathscr{F}_{T_n-}] \quad P\text{-a.s.}, \quad \text{on } \{T_n < \infty\}. \quad (4.17)$$

PROOF. Same as for T14. □

The pair $(A_t, \Phi_t(dz))$ is called the *generalized* $(P, \mathscr{F}_t)$-*local characteristics* of $p(dt \times dz)$. We will see that they are transformed in an absolutely continuous way by absolutely continuous changes of probability.

T17 Theorem [6]. *Let the conditions and notation of T14 prevail, and suppose moreover that N_t is P-nonexplosive. Let $\tilde{P}$ be a probability measure on $(\Omega, \mathscr{F})$ such that*

$$\tilde{P} \ll P \quad on \ (\Omega, \mathscr{F}_1). \tag{4.18}$$

Let $(\tilde{A}_t, \tilde{\Phi}_t(dz))$ be the generalized $(\tilde{P}, \mathscr{F}_t)$-local characteristics of $p(dt \times dz)$. There exists a nonnegative $\mathscr{F}_t$-predictable E-indexed process $h(t, z)$ and a nonnegative $\mathscr{F}_t$-predictable process μ_t such that

$$\tilde{A}_t = \int_0^t \mu_s \, dA_s,$$

$$\tilde{\Phi}_t(dz) = h(t, z)\Phi_t(dz). \tag{4.19}$$

PROOF. We will need in the course of the proof the following extension of (A2, T19):

L18 Lemma [4]. *Let B_t be a right-continuous $\mathscr{F}_t$-predictable nondecreasing process such that $B_0 \equiv 0$ and $B_1 < \infty$, P.a.s., and let L_t be a nonnegative $\mathscr{F}_t$-martingale, right-continuous. Then*

$$E[L_1 B_1] = E\left[\int_0^1 L_{s-} \, dB_s\right]. \tag{4.20}$$

The proof of L18 can be found in [4, V. T27]. For the proof of T17 we proceed as follows: Let L_t be the Radon–Nikodym derivative of $\tilde{P}$ with respect to P on $(\Omega, \mathscr{F}_t)$, for $t \in [0, 1]$. We can assume that L_t is right-continuous, by Appendix A2, T20.

Define, for all nonnegative $\mathscr{F}_t$-predictable E-indexed processes H,

$$\tilde{M}_p^1(H) = \tilde{E}\left[\int_0^1 \int_E H(s, z)\tilde{\Phi}_s(dz) \, d\tilde{A}_s\right], \tag{4.21}$$

$$\tilde{M}^1(H) = \tilde{E}\left[\int_0^1 \int_E H(s, z)\Phi_s(dz) \, dA_s\right]. \tag{4.22}$$

We will show that $\tilde{M}_p^1 \ll \tilde{M}^1$, and this will guarantee the existence of a nonnegative $\mathscr{F}_t$-predictable E-indexed process $U(t, z)$ such that

$$\tilde{\Phi}_s(dz) \, d\tilde{A}_s = U(s, z)\Phi_s(dz) \, dA_s \quad \tilde{P}\text{-a.s.}, \tag{4.23}$$

from which the existence of μ_t and $h(t, z)$ is readily obtained.

4. Towards a General Theory of Intensity

In order to show the absolute continuity of $\tilde{M}_p^1$ with respect to M_p^1, first observe that

$$\tilde{M}_p(H1(L_- = 0)) = \tilde{E}\left[\int_0^1 \int_E H(s, z)1(L_{s-} = 0)\tilde{\Phi}(dz)\, d\tilde{A}_s\right]$$

$$= E\left[L_1 \int_0^1 \int_E H(s, z)1(L_{s-} = 0)\tilde{\Phi}_s(dz)\, d\tilde{A}_s\right]$$

$$= E\left[\int_0^1 \int_E H(s, z)L_{s-}1(L_{s-} = 0)\tilde{\Phi}_s(dz)\, d\tilde{A}_s\right] = 0 \quad (4.24)$$

(where L18 has been used). Therefore

$$\tilde{M}_p^1(H) = \tilde{M}_p^1(H1(L_- > 0)) = \tilde{E}\left[\int_0^1 \int_E H(s, z)1(L_{s-} > 0)\tilde{\Phi}_s(dz)\, d\tilde{A}_s\right]$$

$$= \tilde{E}\left[\int_0^1 \int_E H(s, z)1(L_{s-} > 0)p(ds \times dz)\right]$$

$$= E\left[L_1 \int_0^1 \int_E H(s, z)1(L_{s-} > 0)p(ds \times dz)\right]$$

$$= E\left[\int_0^1 \int_E H(s, z)L_s 1(L_{s-} > 0)p(ds \times dz)\right]$$

(4.25)

(where A2, T19 has been used). Invoking A2, T19 again,

$$\tilde{M}^1(H) = \tilde{E}\left[\int_0^1 \int_E H(s, z)\Phi_s(dz)\, dA_s\right]$$

$$= E\left[L_1 \int_0^1 \int_E H(s, z)\Phi_s(dz)\, dA_s\right]$$

$$= E\left[\int_0^1 \int_E H(s, z)L_{s-}\Phi_s(dz)\, dA_s\right]$$

$$= E\left[\int_0^1 \int_E H(s, z)L_{s-}p(ds \times dz)\right]. \quad (4.26)$$

The absolute continuity to be proven follows from the last expressions of $\tilde{M}_p^1$ and $\tilde{M}^1$ in (4.25) and (4.26) respectively. □

Remark. T17 shows that the form of T10 is the most general to be expected for an absolutely continuous change of probability measure $P \to \tilde{P}$ starting with a point process with $(P, \mathscr{F}_t)$-local characteristics $(\lambda_t, \Phi_t(dz))$.

We will stop at this point, although the general theory of intensity can be further developed. We refer to [6] for such developments, and, for the reader with an interest in the general theory of stochastic processes, to {78}, where marked point processes with multiple explosions are considered.

References

[1] Boel, R., Varaiya, P., and Wong, E. (1975) Martingales on jump processes; Part I: Representation results; Part II: Applications; SIAM J. Control **13**, pp. 999–1061.
[2] Brémaud, P. (1976) La méthode des semi-martingales en filtrage lorsque l'observation est un processus ponctuel marqué, *Séminaire Proba X*, Lect. Notes in Math. **511**, Springer, Berlin, pp. 1–18.
[3] Davis, M. H. A. (1976) The representation of martingales of jump processes, SIAM J. Control **14**, pp. 623–638.
[4] Dellacherie, C. (1972) *Capacités et Processus stochastiques*, Springer, Berlin.
[5] Grigelionis, B. (1973) On the non-linear filtering theory and absolute continuity of measures corresponding to stochastic processes, *Proc. Japan–SSSR Symposium*, Lect. Notes in Math. **330**, Springer.
[6] Jacod, J. (1975) Multivariate point processes: predictable projection, Radon–Nikodym derivative, representation of martingales, Z. für W. **31**, pp. 235–253.
[7] Kailath, T. and Segall, A. (1975) Radon–Nikodym derivatives with respect to measures induced by discontinuous independent increment processes, Ann. Probab. **3**, pp. 449–464.
[8] Kunita, H. and Watanabe, S. (1967) On square integrable martingales, Nagoya Math. J. **30**, pp. 209–245.
[9] Rubin, I. (1974) Regular jump processes and their detection, IEEE Trans. **IT-20**, pp. 617–624.
[10] Snyder, D. L. (1973) Information processing for observed jump processes, Information and Control **22**, pp. 69–78.
[11] Snyder, D. L. (1975) *Random Point Processes*, Wiley, New York.
[12] Itmi, M. (1981) Histoire interne des processus ponctuels, Thèse 3ème cycle, U. de Roven.

SOLUTIONS TO EXERCISES, CHAPTER VIII

E1. $S_n = \text{``inf}\left\{t \,\Big|\, \int_0^t \int_E |H(s,z)|\lambda_s \Phi_s(dz)\, ds \geq n\right\}$ or ∞."

E2. Let S_n be as in E1. Then by C4

$$E\left[\int_0^{S_n}\int_E |H(s,z)|p(ds \times dz)\right] = E\left[\int_0^{S_n}\int_E |H(s,z)|\lambda_s(dz)\,ds\right] < n \leq \infty, \quad (1)$$

so that $\int_0^{S_n}\int_E |H(s,z)|p(ds \times dz) < \infty$, P-a.s. The result follows, since $S_n \uparrow \infty$, P-a.s.

E3. If $A \cap B = \emptyset$, then $N_t(A)$ and $N_t(B)$ have no common jumps. Their $\mathscr{F}_t$-intensities are $F(A)\lambda(t)$ and $F(B)\lambda(t)$ respectively. The announced result then follows from Watanabe's multichannel theorem (II, T6).

E4. W_t is the amount of required service in excess of the amount of completed service, at time t. But if W_0 is the initial load, we have

$$\begin{cases} \text{required service:} & W_0 + \sum_{n \geq 1} Z_n 1(T_n \leq t), \\ \text{completed service:} & \int_0^t 1(Q_s > 0) \, ds. \end{cases} \quad (1)$$

Thus

$$W_t = W_0 + \int_0^t \int_0^\infty z p(ds \times dz) - \int_0^t 1(Q_s > 0) \, ds, \quad (2)$$

or, after compensation,

$$W_t = W_0 + \int_0^t \int_0^\infty z \lambda F(dz) \, ds + \int_0^t \int_0^\infty z q(dz \times ds) - \int_0^t 1(Q_s > 0) \, ds. \quad (3)$$

Equivalently,

$$W_t = W_0 + \lambda E[Z] t - \int_0^t 1(Q_s > 0) \, ds + \int_0^t \int_0^\infty z q(dz \times ds). \quad (4)$$

By C4, $\int_0^t \int_0^\infty z q(ds \times dz)$ is an $\mathscr{F}_t$-martingale.

E5. By arguments similar to those used in the solution of E4,

$$W_t = W_0 + \sum_{n \geq 1} Z_n 1(T_n \leq t) - \int_0^t b_s \, ds, \quad (1)$$

so that, using the independence assumption,

$$E[W_t] = E[W_0] + \sum_{n \geq 1} E[Z_n] E[1(T_n \leq t)] - \int_0^t E[b_s] \, ds. \quad (2)$$

But $E[Z_n] = E[Z]$ and $\sum_{n \geq 1} E[1(T_n \leq t)] = E[N_t]$. Therefore

$$E[W_t] = E[W_0] + \frac{E[N_t]}{\mu} - \int_0^t E[b_s] \, ds. \quad (3)$$

If $E[W_t] = E[W_0]$ and $E[N_t] = \lambda t$, for all $t \geq 0$, we must have

$$\frac{\lambda}{\mu} = E[b_t]. \quad (4)$$

In particular, if $b_t = C1(Q_t > 0)$,

$$P[Q_t = 0] = 1 - \rho/C. \quad (5)$$

E6. For any $\tilde{C} \in [0, \infty) \times E$, we have $N(\tilde{C}) = \int_0^\infty \int_E 1_{\tilde{C}}(s, z) p(ds \times dz)$, so that $E[e^{iuN(\tilde{A}) + ivN(\tilde{B})}] = E[Z_\infty]$, where

$$Z_t = \exp\left\{ iu \int_0^t \int_E 1_{\tilde{A}}(s, z) p(ds \times dz) + iv \int_0^t \int_E 1_{\tilde{B}}(s, z) p(ds \times dz) \right\}. \quad (1)$$

Now, at a jump T_n of N_t,

$$Z_{T_n} = Z_{T_n -}((e^{iu} - 1) 1_{\tilde{A}}(T_n, Z_n) + (e^{iv} - 1) 1_{\tilde{B}}(T_n, Z_n)), \quad (2)$$

where we have taken into account the hypothesis $\tilde{A} \cap \tilde{B} = \emptyset$. In other symbols, since Z_t is a step process,

$$Z_t = 1 + \int_0^t \int_E Z_{s-}((e^{iu} - 1)1_{\tilde{A}}(s, z) + (e^{iv} - 1)1_{\tilde{B}}(s, z))p(ds \times dz), \tag{3}$$

so that, by T3,

$$E[Z_t] = 1 + \int_0^t E[Z_s]\left((e^{iu} - 1)\tilde{\lambda}(s)\int_E 1_{\tilde{A}}(s, z)F(dz)\right.$$
$$\left. + (e^{iv} - 1)\lambda(s)\int_E 1_{\tilde{B}}(s, z)F(dz)\right) ds. \tag{4}$$

From this we obtain

$$E[Z_t] = \exp\left\{(e^{iu} - 1)\int_0^t \int_E \lambda(s) 1_{\tilde{A}}(s, z)\, dF(z)\, ds\right\}$$
$$\times \exp\left\{(e^{iv} - 1)\int_0^t \int_E \lambda(s) 1_{\tilde{B}}(s, z)\, dF(z)\, ds\right\}. \tag{5}$$

It suffices now to let t go to ∞ to obtain the first part of the result stated in E6.

At jumps, $W_{T_n} = W_{T_n-} + Z_n$; therefore

$$\exp\{iuW_{T_n}\} - \exp\{iuW_{T_n-}\} = \exp\{iuW_{T_n-}\}(e^{iuZ_n} - 1). \tag{6}$$

Between jumps,

$$de^{iuW_t} = -iue^{iuW_t}1(Q_t > 0)\, dt. \tag{7}$$

Hence

$$e^{iuW_t} = e^{iuW_0} + \int_0^t \int_0^\infty e^{iuW_{s-}}(e^{iuz} - 1)p(ds \times dz) - iu\int_0^t 1(Q_s > 0)\, ds, \tag{8}$$

or, equivalently,

$$e^{iuW_t} = e^{iuW_0} + \int_0^t \int_0^\infty e^{iuW_{s-}}(e^{iuz} - 1)\lambda F(dz)\, ds$$
$$- iu\int_0^t 1(Q_s > 0)\, ds + \int_0^t \int_0^\infty e^{iuW_{s-}}(e^{iuz} - 1)q(ds \times dz), \tag{9}$$

where the last term of the right-hand side is, by C4, a martingale m_t.

Observing that $Q_t = 0$ implies $e^{iuW_t} = 1$, and rewriting $\int_0^\infty (e^{iuz} - 1)F(dz)$ as $E[e^{iuZ} - 1]$, we have

$$e^{iuW_t} = e^{iuW_0} + \left(\int_0^t e^{iuW_s}\, ds\right)(\lambda E[e^{iuZ} - 1] - iu) + iu\int_0^t 1(Q_s = 0)\, ds + m_t. \tag{10}$$

By integration of (10), and after introducing the notation

$$\Phi_t(u) = E[e^{iuW_t}], \quad \Psi(u) = E[e^{iuZ}], \tag{11}$$

we obtain

$$\Phi_t(u) = \Phi_0(u) + (\lambda(\Psi(u) - 1) - iu) \int_0^t \Phi_s(u)\, ds + iu \int_0^t P[Q_s = 0]\, ds. \quad (12)$$

Assuming $\Phi_t(u) = \Phi_0(u) (= \Phi(u))$ and in particular $P[Q_t = 0] = P[W_t = 0] = p_0$, we find from (12)

$$\Phi(u) = \frac{iup_0}{iu - \lambda(\Psi(u) - 1)}. \quad (13)$$

Now by integration of (4),

$$\lambda E[Z] = 1 - p_0, \quad (14)$$

so that finally,

$$\Phi(u) = \frac{iu(1 - \lambda E[Z])}{iu - \lambda(\Psi(u) - 1)}. \quad (15)$$

The number $N'(A)$ of points T'_n in any bounded set $A \in \mathcal{B}_+$ is the number of points (T_n, Z_n) in the set $\tilde{A} = \{(t, z) \mid t + z \in A\}$. Since A is bounded, $\int_0^t \int_A \lambda(s) F(dz)\, ds < \infty$.

Take $A \in \mathcal{B}_+$, $B \in \mathcal{B}_+$, bounded and disjoint; the corresponding $\tilde{A}$ and $\tilde{B}$ satisfy the general conditions at the beginning of the exercise. Thus $N'(A)$ and $N'(B)$ are independent Poisson random variables; thus N'_t is Poisson. Since

$$E[N'_t] = \int_0^t \lambda(s) \int_0^{t-s} F(dz)\, ds = \int_0^t \lambda(s) F(t - s), \quad (16)$$

N'_t admits an intensity $(d/dt)E[N'_t]$, which is

$$\int_0^t \lambda(s) f(t - s)\, ds. \quad (17)$$

APPENDIX A1

Background in Probability and Stochastic Processes

1. Introduction
2. Monotone Class Theorem
3. Random Variables
4. Expectations
5. Conditioning and Independence
6. Convergence
7. Stochastic Processes
8. Markov Processes
References

1. Introduction

This appendix is written as a memory refresher, not as an introduction to probability theory. It contains all the elementary results in probability and stochastic processes which are used, sometimes without further reference, in the main body of the book.

A few proofs are included, especially those relative to the monotone class theorem and conditional expectations, the two basic tools of constant use in the book.

The more specialized topics on stochastic processes are covered in Appendix A2.

Probability theory deals with idealized mathematical models of physical systems driven by chance. Experiments, or *trials*, are performed, and all possible results, or *outcomes*, are grouped in a set Ω, the *sample space*, of generic element $\omega \in \Omega$. If tossing a coin, Ω will consist of two elements, "heads" and "tails"; in observing the level of water in a reservoir between times t_1 and t_2, Ω will be conveniently chosen to be the set of nonnegative functions $\omega = (\omega_t, t \in [t_1, t_2])$.

Outcomes are in turn grouped into subsets of Ω. To each subset A of Ω corresponds a characteristic property $\mathscr{P}_A$,[1] and only those subsets with an

---
[1] Such a characteristic property always exists: for instance $\mathscr{P}_A$ is the property of belonging to A. Of course a modeler working on physical situations will find a less tautological way of phrasing $\mathscr{P}_A$.

"interesting" property (from the point of view of the modeler) are called *events*. Thus, in general, the family $\mathscr{F}$ of events is a strict subset of $\mathscr{P}(\Omega)$, the family of all subsets of Ω. For instance, let the set Ω consist of all the outcomes of the experiments of drawing balls at random from an urn, and suppose that the balls are black or red, and/or big or small. If color only is important, the modeler will want the sets $A = \{\omega | \text{outcome is "red"}\}$ and $B = \{\omega | \text{outcome is "black"}\}$ to be in $\mathscr{F}$, and he will not care about the set $C = \{\omega | \text{outcome is "big"}\}$.

In summary, a mathematical model in probability theory must consist of a sample space Ω and a family $\mathscr{F}$ of subsets of Ω, the family of "interesting" subsets or events. Of course, it must also feature a measure of likelihood, the *probability* of the events of $\mathscr{F}$. This measure is a function P mapping $\mathscr{F}$ into $[0, 1]$. The choice of $[0, 1]$ as image space P is somewhat arbitrary, but mathematically simple and intuitively appealing.

If $P(A) = 0$, then one says that A is an *impossible* event; if $P(A) = 1$, A is said to be *certain*; the closer (PA) is to 1, the more likely the outcome ω has the property $\mathscr{P}_A$ characterizing the event A.

The family $\mathscr{F}$ and the mapping P must satisfy some requirements; these will be imposed in an axiomatic way corresponding to the intuitive notion of probability. For instance if A and B are events (i.e. are in $\mathscr{F}$), then one will want the intersection $A \cap B$ to be in $\mathscr{F}$. Indeed, an outcome ω in $A \cap B$ possesses the property "$\mathscr{P}_A$ and $\mathscr{P}_B$," and if $\mathscr{P}_A$ and $\mathscr{P}_B$ are "interesting" properties, then, most likely, the conjunction "$\mathscr{P}_A$ and $\mathscr{P}_B$" will be an "interesting" property. Also if A and B are events of $\mathscr{F}$ such that $A \subset B$, inclusion $A \subset B$ implies that if $\mathscr{P}_A$ is satisfied, then necessarily $\mathscr{P}_B$ is satisfied; therefore quite naturally the likelihood $P(B)$ should be greater than $P(A)$.

The above requirements are *natural* ones. However, there are other less natural requirements that one would like to impose on $(\Omega, \mathscr{F}, P)$ in order to build a tractable *mathematical* model for random phenomena, and we will now present them in a precise axiomatic way. For this we need a short review of set theory, culminating with one of the fundamental tools of probability theory: the indispensable *monotone class theorem*.

2. Monotone Class Theorem

Operations on Sets

Let Ω be a set and $\mathscr{P}(\Omega)$ be the collection of all subsets of Ω, including the empty set $\varnothing$. We denote by ω the generic element of Ω. The *union* of A and B in $\mathscr{P}(\Omega)$ is denoted by $A \cup B$, the *intersection* by $A \cap B$, the *complement* of A by A^c or $\overline{A}$. When A and B are *disjoint* (that is to say $A \cap B = \varnothing$), the union $A \cup B$ is also denoted $A + B$ and is called the *sum* of A and B. When A is included in B ($A \subset B$), the *strict difference* $B - A$ is the set of elements ω

2. Monotone Class Theorem

of Ω belonging to B *and* not belonging to A. For arbitrary subsets A and B of Ω their *symmetric difference*, $A \triangle B$, is the set $(A \cup B) - (A \cap B)$. For any collection $(A_i, i \in I)$ of subsets of Ω, the union of all its terms is denoted $\bigcup_{i \in I} A_i$; similarly, for the intersection one writes $\bigcap_{i \in I} A_i$. The notation $\sum_{i \in I} A_i$ (sum of the A_i's, $i \in I$) implies that the subsets A_i, $i \in I$, are pairwise disjoint (i.e. $i \neq j \Rightarrow A_i \cap A_j = \emptyset$) and represents the union $\bigcup_{i \in I} A_i$. Sometimes $\bigcup_{i \in I} A_i$ is denoted by $\sup_{i \in I} A_i$, and $\bigcap_{i \in I} A_i$ by $\inf_{i \in I} A_i$. Let $(A_n, n \geq 1)$ be a countable sequence in $\mathcal{P}(\Omega)$. It is said to be *increasing* if $A_{n+1} \supset A_n$, $n \geq 1$; *decreasing* if $A_{n+1} \subset A_n$, $n \geq 1$. Let $(A_n, n \geq 1)$ be an increasing sequence in $\mathcal{P}(\Omega)$; the union $A = \bigcup_{n \geq 1} A_n$ is called the *increasing limit* of $(A_n, n \geq 1)$; the two following notations are equivalent: $A = \lim \uparrow A_n$ and $A_n \uparrow A$. Similarly, for any *decreasing* sequence $(A_n, n \geq 1)$ in $\mathcal{P}(\Omega)$, $A = \lim \downarrow A_n$ and $A_n \downarrow A$ are notations used for the *decreasing limit* $A = \bigcap_{n \geq 1} A_n$.

Let $(A_n, n \geq 1)$ be an arbitrary countable sequence in $\mathcal{P}(\Omega)$. The sequence $(B_n, n \geq 1)$ defined by $B_n = \bigcup_{m \geq n} A_m$ is decreasing, and its decreasing limit $\bigcap_{n \geq 1} B_n$ is denoted by $\overline{\lim} \, A_n$ or $\limsup A_n$, so that by definition $\overline{\lim} \, A_n = \bigcap_{n \geq 1} \bigcup_{m \geq n} A_m = \lim \downarrow \sup_{m \geq n} A_m$. Similarly, the sequence $(C_n, n \geq 0)$ defined by $C_n = \bigcap_{m \geq n} A_m = \inf_{m \geq n} A_m$ is increasing, and its increasing limit is denoted by $\underline{\lim} \, A_n$ or $\liminf_n A_n$, so that, by definition $\underline{\lim} \, A_n = \bigcup_{n \geq 1} \bigcap_{m \geq n} = \lim \uparrow \inf_{m \geq n} A_m$. When $\underline{\lim}_n A_n = \overline{\lim}_n A_n$, one says that the sequence $(A_n, n \geq 1)$ admits a limit, denoted $\lim A_n$, and defined by $\lim A_n = \underline{\lim} A_n = \overline{\lim} A_n$. It is easy to see that in general $\underline{\lim} A_n \subset \overline{\lim} A_n$. In the sequel we will omit the subscript n in expressions like $\underline{\lim}_n A_n$, writing instead $\underline{\lim} A_n$. The notions of lim sup and lim inf have an intuitive meaning in terms of outcomes and events, as the following exercise will show:

E1 Exercise. Show that:

$$\omega \in \limsup A_n \Leftrightarrow \text{for all } n_0 \geq 1, \text{ there exists } n \geq n_0 \text{ such that } \omega \in A_n. \tag{2.1}$$

Let $\mathcal{P}_n$ be the characteristic property of A_n. Let ω be the outcome of a trial for which we are checking the properties $\mathcal{P}_n$, $n \geq 1$. If $\omega \in \limsup A_n$, then we are not sure that all the properties $\mathcal{P}_n$ are verified, but by (2.1) we are sure that an infinity of these properties are verified by ω. We could then say "$\mathcal{P}_n$ occurs infinitely often." Probabilists prefer to use for $\limsup A_n$ the notation $\{\omega | A_n \text{ i.o.}\}$, where "i.o." is an abbreviation of "infinitely occurs" or "infinitely often."

E2 Exercise. Show that

$$\omega \in \liminf A_n \Leftrightarrow \omega \in A_n \text{ for all but a finite number of } n\text{'s}. \tag{2.2}$$

In other words, starting from rank n_0, all properties $\mathscr{P}_n$ are satisfied by an outcome ω which belongs to lim inf A_n.

The notions of lim sup and lim inf are well known in real analysis, and their link with the corresponding concepts in set theory is very simple: Let A be any subset of Ω. The *indicator function* of A is the mapping $1_A : \Omega \to \{0, 1\}$ defined by $1_A(\omega) = 1$ if $\omega \in A$, $1_A(\omega) = 0$ if $\omega \in A^c$.

E3 Exercise. Show that

$$1_{\varliminf A_n}(\omega) = \varliminf 1_{A_n}(\omega), \qquad 1_{\varlimsup A_n}(\omega) = \varlimsup 1_{A_n}(\omega). \tag{2.3}$$

Let $\mathscr{C}$ be a subfamily of $\mathscr{P}(\Omega)$. One says that $\mathscr{C}$ is *stable by finite [countable, arbitrary] intersection* if the intersection of any finite [countable, arbitrary] collection of elements in $\mathscr{C}$ still belongs to $\mathscr{C}$. Stability by finite [countable, arbitrary] union, complementation, strict difference, etc., can be defined in an analogous manner. $\mathscr{C}$ is said to be *stable by sequential increasing [decreasing] limit* if the limit of any increasing [decreasing] sequence in $\mathscr{C}$ is still in $\mathscr{C}$. If $\mathscr{C}$ is stable by sequential increasing limit *and* by sequential decreasing limit, it is said to be *stable by sequential monotone limit*.

Let $(\mathscr{C}_i, i \in I)$ be an arbitrary collection of subfamilies of $\mathscr{P}(\Omega)$. The intersection $\bigcap_{i \in I} \mathscr{C}_i$ is defined to be the family of all subsets of Ω that belong to all $\mathscr{C}_i, i \in I$. The union $\bigcup_{i \in I} \mathscr{C}_i$ is the family of all subsets of Ω that are in $\mathscr{C}_i$ for at least one index $i \in I$.

Families of Sets

The following structures on subfamilies of $\mathscr{P}(\Omega)$ are notable:

π-*system*. $\mathscr{S} \subset \mathscr{P}(\Omega)$ is called a π-system iff it is stable by finite intersection.

Semialgebra. $\mathscr{S} \subset \mathscr{P}(\Omega)$ is called a semialgebra iff:

(a) $\Omega, \emptyset \in \mathscr{S}$,
(b) $\mathscr{S}$ is stable by finite intersection,
(c) if $S \in \mathscr{S}$, then S^c is the union of a finite collection of pairwise disjoint sets of $\mathscr{S}$.

d-system. $\mathscr{S} \subset \mathscr{P}(\Omega)$ is called a *d*-system iff:

(a) $\Omega, \phi \in \mathscr{S}$,
(b) $\mathscr{S}$ is stable by strict difference,
(c) $\mathscr{S}$ is stable by sequential increasing limit.

Algebra. $\mathscr{A} \subset \mathscr{P}(\Omega)$ is called an algebra iff:

(a) $\Omega, \emptyset \in \mathscr{A}$,
(b) $\mathscr{A}$ is stable by complementation,
(c) $\mathscr{A}$ is stable by finite intersection.

2. Monotone Class Theorem

σ-algebra or σ-field. $\mathscr{F} \subset \mathscr{P}(\Omega)$ is called a σ-algebra, or σ-field, iff it is an algebra and moreover it is stable by countable intersection.

E4 Exercise. Verify that if $\mathscr{A}[\mathscr{F}]$ is an algebra [σ-algebra], then it is stable by finite (countable) union.

E5 Exercise. Show that if $\mathscr{S} \subset \mathscr{P}(\Omega)$ is a d-system and a π-system, it is a σ-algebra.

E6 Exercise. Verify that if $(\mathscr{S}_i; i \in I)$ is an arbitrary collection of d-systems [algebras, σ-algebras] defined on Ω, the intersection $\bigcap_{i \in I} \mathscr{S}_i$ is also a d-system [algebra, σ-algebra].

Before proceeding further, we will mention that the structure chosen for the family $\mathscr{F}$ of "interesting" events in the axiomatic presentation of probability theory is that of a σ-algebra. There are technical reasons for such a choice that are best understood through practice; however, this choice can also be motivated to some extent by intuitive arguments such as the following one relative to stability by countable intersection. Let Ω be the set of infinite real sequences $\omega = (x_1, x_2, \ldots)$. Clearly the property "$x_n \geq a$" where $a \in R$ is an interesting property, and one would like to see the corresponding event $A_n = \{\omega | x_n \geq a\}$ belong to $\mathscr{F}$. Such an event is, intuitively speaking, an *elementary* event, and from such elementary events one can construct more complex ones, such as the one corresponding to the property "$\overline{\lim} x_n < a$," i.e. $C = \{\omega | \overline{\lim} x_n < a\}$. Clearly $\overline{C} = \overline{\lim} A_n = \{\omega | A_n \text{ i.o.}\}$. Now C is obtained from the A_n's by countable unions and countable intersections; therefore, in order to have it belong to $\mathscr{F}$, it is necessary and sufficient to require that $\mathscr{F}$ be a σ-algebra and contain the A_n's.

The above example serves to illustrate another important point. Frequently, one defines subsets of Ω, forming a family $\mathscr{C}$ that we call very imprecisely *elementary* events [for instance, the class $\mathscr{C} = (\{\omega | x_n \in [a, b]\}, n \geq 1, a \leq b, a, b \in R)$]. For reasons of the type invoked above, one insists on $\mathscr{F}$ being a σ-algebra. Quite naturally, one will define $\mathscr{F}$ to be the "smallest σ-algebra containing $\mathscr{C}$."

More generally, let $\mathscr{C}$ be a subfamily of $\mathscr{P}(\Omega)$, and consider the collection of all the d-systems [algebras, σ-algebras] defined on Ω and containing $\mathscr{C}$. This collection is not empty, because it counts $\mathscr{P}(\Omega)$ as a member. By Exercise E6, the intersection of all these d-systems [algebras, σ-algebras] is a d-system [algebra, σ-algebra], clearly the smallest one containing $\mathscr{C}$. It is called the *d-system [algebra, σ-algebra] generated by* $\mathscr{C}$ and is denoted by $d(\mathscr{C})$ $[\mathscr{A}(\mathscr{C}), \sigma(\mathscr{C})]$.

E7 Exercise. Verify that for any $\mathscr{C} \subset \mathscr{P}(\Omega)$, $d(\mathscr{C}) \subset \sigma(\mathscr{C})$. Observe that if $\mathscr{C}$ is a d-system [algebra, σ-algebra], then $\mathscr{C} = d(\mathscr{C})$ $[\mathscr{C} = \mathscr{A}(\mathscr{C}), \mathscr{C} = \sigma(\mathscr{C})]$. Verify that if $\mathscr{C}_1 \subset \mathscr{C}_2 \subset \mathscr{P}(\Omega)$, then $d(\mathscr{C}_1) \subset d(\mathscr{C}_2)$ $[\mathscr{A}(\mathscr{C}_1) \subset \mathscr{A}(\mathscr{C}_2), \sigma(\mathscr{C}_1) \subset \sigma(\mathscr{C}_2)]$.

Let $\mathscr{F}_1$ and $\mathscr{F}_2$ be two algebras [σ-algebras] defined on Ω. One can give very simple counterexamples for which $\mathscr{F}_1 \cup \mathscr{F}_2$ is not an algebra [a σ-algebra]. The σ-algebra generated by $\mathscr{F}_1$ and $\mathscr{F}_2$ is denoted $\mathscr{F}_1 \vee \mathscr{F}_2$, i.e. $\mathscr{F}_1 \vee \mathscr{F}_2 = \sigma(\mathscr{F}_1 \cup \mathscr{F}_2)$. Denote by $\mathscr{C}$ the class of subsets of Ω of the form $A \cap B$, $A \in \mathscr{F}_1$, $B \in \mathscr{F}_2$. Then clearly $\mathscr{F}_1 \cup \mathscr{F}_2 \subset \mathscr{C} \subset \mathscr{F}_1 \vee \mathscr{F}_2$; by Exercise E7, $\sigma(\mathscr{F}_1 \cup \mathscr{F}_2) \subset \sigma(\mathscr{C}) \subset \sigma(\mathscr{F}_1 \vee \mathscr{F}_2)$, and therefore $\sigma(\mathscr{C}) = \mathscr{F}_1 \vee \mathscr{F}_2$, since $\sigma(\mathscr{F}_1 \vee \mathscr{F}_2) = \mathscr{F}_1 \vee \mathscr{F}_2$. In other words, $\mathscr{F}_1 \vee \mathscr{F}_2$ is generated by the intersections of events of $\mathscr{F}_1$ with events of $\mathscr{F}_2$.

Borel σ-fields. Let $\Omega = E$ be a topological space (usually $E = R$ or $E = R^n$) the topology of which is defined by the collection $\mathscr{G}$ of its open sets. By definition, the σ-field generated by $\mathscr{G}$ is called the Borel σ-field of E, and it is denoted $\mathscr{B}(E)$. If $E \equiv R^n$, then $\mathscr{B}(E) = \mathscr{B}(R^n)$ is denoted by $\mathscr{B}^n$, and $\mathscr{B}(R) = \mathscr{B}^1$ is denoted by $\mathscr{B}$. If $E = R_+$, then $\mathscr{B}(E) = \mathscr{B}(R_+)$ is denoted $\mathscr{B}^+$, and $\overline{\mathscr{B}}$ represents $\mathscr{B}(\overline{R})$, where $\overline{R} = [-\infty, +\infty]$ is the extended real line; the topologies chosen are the euclidian topologies for R^n and R_+, and the extended euclidian topology for $\overline{R}$. For any subset A of a topological space E we define $\mathscr{B}(A) = \{C \mid C = A \cap B, B \in \mathscr{B}(E)\}$.

In R, the open sets may be defined in terms of the open intervals. Thus, it is natural to seek a definition of $\mathscr{B}$ in terms of intervals.

E8 Exercise. Let $\mathscr{S}$ be the class of all intervals (open, semiopen, closed) of R. Verify that $\mathscr{S}$ is a semialgebra. Show that $\mathscr{S}$ generates the algebra $\mathscr{A}$ of finite sums of disjoint intervals. Therefore $\sigma(\mathscr{S}) = \sigma(\mathscr{A})$. Show that $\sigma(\mathscr{S}) = \sigma(\mathscr{A}) = \mathscr{B}$. (Hint: Use the fact that every interval and every open set of R is a countable union or a countable intersection of open intervals.) Extend the above to R^n, replacing "intervals" by "rectangles."[2]

The Monotone Class Theorem

The above definition of $\mathscr{B}$ in terms of intervals is appealing, the intervals being "elementary" subsets of R. Generally, a σ-field $\mathscr{F}$ will be defined as the smallest σ-field generated by a class $\mathscr{C}$ of elementary events (see for instance the definition of the predictable σ-fields in Chapter 1). Sometimes, one has to show that a certain property is true for all elements $A \in \mathscr{F} = \sigma(\mathscr{C})$, and this might seem impossible in view of the abstract definition of $\mathscr{F}$ as $\sigma(\mathscr{C})$. However, if $\mathscr{C}$ is a π-system, and if one can show that the property in question is verified for all A in $d(\mathscr{C})$, then it is also verified for all A in $\sigma(\mathscr{C})$. Indeed:

T1 Theorem (Monotone Class Theorem). *Let $\mathscr{S}$ be a π-system defined on Ω. Then $d(\mathscr{S}) = \sigma(\mathscr{S})$.*

---

[2] A rectangle of R^n is a set of the form $(x_i \in \mathscr{I}_i, 1 \leq i \leq n)$ where $\mathscr{I}_i$ is an interval of R for all $1 \leq i \leq n$.

PROOF. Since $d(\mathcal{S}) \subset \sigma(\mathcal{S})$ (Exercise E7), it suffices to show that $d(\mathcal{S})$ is a σ-algebra. For this, it is enough to show that $d(\mathcal{S})$ is a π-system (Exercise E5). To do this, one proceeds as follows. Define

$$\mathcal{D}_1 = \{B \in d(\mathcal{S}) | B \cap A \in d(\mathcal{S}) \text{ for all } A \in \mathcal{S}\}. \tag{2.4}$$

One verifies that $\mathcal{D}_1$ is a d-system, and that $\mathcal{D}_1 \supset \mathcal{S}$, since $\mathcal{S}$ is a π-system. Therefore $\mathcal{D}_1 \supset d(\mathcal{S})$ (Exercise E7). By the definition of $\mathcal{D}_1$, $\mathcal{D}_1 \subset d(\mathcal{S})$, so that $\mathcal{D}_1 = d(\mathcal{S})$. Define now

$$\mathcal{D}_2 = \{B \in d(\mathcal{S}) | B \cap A \in d(\mathcal{S}) \text{ for all } A \in d(\mathcal{S})\}. \tag{2.5}$$

One verifies that $\mathcal{D}_2$ is a d-system. Also, if $A \in \mathcal{S}$, then $B \cap A \in d(\mathcal{S})$ for all $B \in \mathcal{D}_1 = d(\mathcal{S})$, and therefore $\mathcal{S} \subset \mathcal{D}_2$. Therefore $d(\mathcal{S}) \subset d(\mathcal{D}_2) = \mathcal{D}_2$ (Exercise E7). Also, by the definition of $\mathcal{D}_2$, $\mathcal{D}_2 \subset d(\mathcal{S})$, so that finally $\mathcal{D}_2 = d(\mathcal{S})$, which expresses that $d(\mathcal{S})$ is a π-system. □

3. Random Variables

Operations on Random Variables

With an outcome $\omega \in \Omega$, one may associate an element $X(\omega)$ of a space E. When $\Omega = \{\text{heads, tails}\}$, one may define the mapping $X: \Omega \to E$ by $X(\omega) = +1$ if $\omega = $ tails, $X(\omega) = -1$ if $\omega = $ heads. Here $E = \{-1, 1\}$ is a very simple set. In general, E will be more complex and a family $\mathcal{E} \subset \mathcal{P}(E)$ representing the interesting events may be selected, just as $\mathcal{F}$ was selected in $\mathcal{P}(\Omega)$. Let C be an interesting event of E, i.e. $C \in \mathcal{E}$; it is then natural to require that the corresponding set of outcomes ω—that is to say $\{\omega | X(\omega) \in C\}$, denoted $X^{-1}(C)$—belongs to $\mathcal{F}$, because it is an interesting event of Ω. In general, things work this way: $(E, \mathcal{E})$ is given, X is given, and the interesting events of Ω are determined by X and $\mathcal{E}$; the minimum requirement on $\mathcal{F}$ is $X^{-1}(C) \in \mathcal{F}$ for each $C \in \mathcal{E}$. The mapping X is then said to be $\mathcal{E}/\mathcal{F}$-measurable.

We now turn to more precise definitions. Let Ω and E be two sets, and let $X: \Omega \to E$ be a mapping from Ω into E. For any subset C of E, $X^{-1}(C)$ represents, by definition, the following subset of Ω: $\{\omega | X(\omega) \in C\}$. We recall that the mapping $X^{-1}: \mathcal{P}(E) \to \mathcal{P}(\Omega)$ preserves the set operations of complementation, union, and intersection, that is to say, for any $A \subset E$ and any family $(A_i, i \in I)$ of subsets of E,

$$X^{-1}(A^c) = (X^{-1}(A))^c,$$

$$X^{-1}\left(\bigcup_{i \in I} A_i\right) = \bigcup_{i \in I} X^{-1}(A_i), \tag{3.1}$$

$$X^{-1}\left(\bigcap_{i \in I} A_i\right) = \bigcap_{i \in I} X^{-1}(A_i).$$

From this it follows that if $\mathscr{E}$ is a π-system [algebra, σ-algebra] on E, then $X^{-1}(\mathscr{E})$ is a π-system [algebra, σ-algebra] on Ω.

Let $(\Omega, \mathscr{F})$ and $(E, \mathscr{E})$ be two measurable spaces, and let $X: \Omega \to E$ be a mapping such that

$$X^{-1}(C) \in \mathscr{F} \quad \text{for all } C \in \mathscr{E}. \tag{3.2}$$

In symbols: $X^{-1}(\mathscr{E}) \subset \mathscr{F}$. Then X is called a *measurable mapping* from $(\Omega, \mathscr{F})$ to $(E, \mathscr{E})$, and this is sometimes symbolized by $X: (\Omega, \mathscr{F}) \to (E, \mathscr{E})$ or $X \in \mathscr{E}/\mathscr{F}$. Equivalently we will write $X \in \mathscr{E}$ when there is no possible confusion about the choice of $\mathscr{F}$, and if the choice of $(\Omega, \mathscr{F})$ and $\mathscr{E}$ is clear from the context, we will say that X is an E-valued random variable.

Let $(\Omega, \mathscr{F})$, $(E, \mathscr{E})$, and $(U, \mathscr{U})$ be three measurable spaces, and X and Y be measurable mappings in $\mathscr{E}/\mathscr{F}$ and $\mathscr{U}/\mathscr{E}$ respectively. Define Z to be the composition of X by Y, i.e., $Z = Y \circ X$. From the definition of measurability it follows immediately that Z is a mapping of $\mathscr{U}/\mathscr{F}$. In other words: *measurability is a transitive property*.

In the case where $(E, \mathscr{E}) = (R, \mathscr{B})$ one speaks of the *real random variable X*, and if $(E, \mathscr{E}) = (\bar{R}, \bar{\mathscr{B}})$ of the *numerical random variable X*. The abbreviation "r.v." will stand for "random variable."

In some instances, one wishes to prove that X is a measurable mapping from $(\Omega, \mathscr{F})$ into $(E, \mathscr{E})$, but one is able to verify directly that $X^{-1}(C) \in \mathscr{F}$ only for subsets C of E that belong to a family $\mathscr{C} \subset \mathscr{P}(E)$ generating $\mathscr{E}$. The following result will imply that this is enough to ensure that $X \in \mathscr{E}/\mathscr{F}$.

T2 Theorem. *Let X be a mapping from Ω into E, and let $\mathscr{C} \subset \mathscr{P}(E)$. Then $\sigma(X^{-1}(\mathscr{C})) = X^{-1}(\sigma(\mathscr{C}))$.*

PROOF. Since $X^{-1}(\mathscr{C}) \subset X^{-1}(\sigma(\mathscr{C}))$ and $X^{-1}(\sigma(\mathscr{C}))$ is a σ-algebra, we have the inclusion $\sigma(X^{-1}(\mathscr{C})) \subset X^{-1}(\sigma(\mathscr{C}))$. Now let $\mathscr{F}' = \{C' | X^{-1}(C') \in \sigma(X^{-1}(\mathscr{C}))\}$. One verifies that $\mathscr{F}'$ is a σ-algebra on E which contains $\mathscr{C}$. Therefore $\mathscr{F}' \supset \sigma(\mathscr{C})$ and $X^{-1}(\sigma(\mathscr{C})) \subset X^{-1}(\mathscr{C}) \subset \sigma(X^{-1}(\mathscr{C}))$, the latter inclusion following from the definition of $\mathscr{F}'$.

E9 Exercise. Prove the result announced above: *in order for X to belong to $\mathscr{E}/\mathscr{F}$, it is enough that $X^{-1}(C) \in \mathscr{F}$ for all $C \in \mathscr{C}$, where $\mathscr{C} \subset \mathscr{P}(E)$ and $\sigma(\mathscr{C}) \supset \mathscr{E}$.*

Application. A continuous function $X: R \to R$ is $\mathscr{B}/\mathscr{B}$-measurable. Indeed, by definition of continuity, $X^{-1}(A)$ is an open set of R whenever A is an open set of R. Therefore, by Exercise E9, X is measurable, since the open sets of R generate $\mathscr{B}$.

The next result concerns numerical and real r.v.'s and their approximation by *step real r.v.s'* of $\mathscr{B}/\mathscr{F}$, that is to say, mappings $X: \Omega \to R$ of the form

$$X(\omega) = \sum_{i \in J} x_i 1_{A_i}(\omega), \tag{3.3}$$

where J is a finite index set, $(A_i, i \in J)$ is a partition of Ω consisting of elements of $\mathscr{F}$, and $(x_i, i \in J)$ is a set of real numbers.

3. Random Variables   263

T3 Theorem. *Let X be a mapping from Ω onto $\bar{R}$. For X to be a nonnegative numerical random variable defined on $(\Omega, \mathscr{F})$ it is necessary and sufficient that it be the pointwise limit of an increasing sequence of nonnegative step real random variables.*

PROOF. Let X be a nonnegative r.v. from $(\Omega, \mathscr{F})$ into $(\bar{R}, \mathscr{B})$. Define, for each $n \geq 0$, the mapping $X_n : \Omega \to R$ by

$$X_n(\omega) = \sum_{q=1}^{n2^n} \frac{q-1}{2^n} 1(q - 1 \leq X(\omega)2^n < q) + n1(X(\omega) \geq n). \quad (3.4)$$

Then, clearly X_n is a step real r.v. and $X_n(\omega) \uparrow X(\omega)$ for all $\omega \in \Omega$.

Conversely, let $(X_n, n \geq 0)$ be an increasing sequence of nonnegative step real r.v.'s, and let X be its pointwise limit. By E9, in order to verify that X is a numerical r.v., it is sufficient to verify that $\{\omega | X(\omega) \leq x\} \in \mathscr{F}$ for all $x \in \bar{R}$, since the class of intervals $((-\infty, x], x \in R)$ generates $\bar{\mathscr{B}}$. This verification follows from

$$\bigcap_{n \geq 1} \{\omega | X_n(\omega) \leq x\} = \{\omega | X(\omega) \leq x\} \quad (3.5)$$

and the fact that $\{\omega | X_n(\omega) \leq x\} \in \mathscr{F}$ for all $n \geq 0$ by the measurability of the X_n's. □

E10 Exercise. Let X and Y be numerical r.v.'s defined on $(\Omega, \mathscr{F})$, and $c \in R$. Then cX is a numerical r.v., and $X + Y$ [XY, X/Y] also, as long as for all $\omega \in \Omega$ we have $X(\omega) + Y(\omega) \neq +\infty - \infty$ [$X(\omega)Y(\omega) \neq 0 \cdot \infty$, $X(\omega)/Y(\omega) \neq \infty/\infty$ or $0/0$]. Also $\{\omega | X(\omega) = Y(\omega)\}$ is a measurable set.

E11 Exercise. Let $(X_n, n \geq 0)$ be a sequence of numerical r.v.'s. Then $\sup X_n$, $\inf X_n$, $\limsup X_n$, $\liminf X_n$ are numerical r.v.'s. Also the set of convergence of $(X_n, n \geq 0)$, i.e. $\{\omega | \liminf X_n(\omega) = \limsup X_n(\omega)\}$, is a measurable set, and $\lim X_n$, the pointwise limit of $(X_n, n \geq 0)$, is (when it exists) a numerical r.v.

A Theorem of Verification of Measurability

The two following results are consequences of the monotone class theorem. The first one is useful in verifying the measurability of a set of real-valued functions.

T4 Theorem (Verification Theorem). *Let Ω be a set and $\mathscr{S}$ a π-system on Ω. Let $\mathscr{H}$ be a vector space of R-valued functions such that:*

(a) $1 \in \mathscr{H}$, $1_A \in \mathscr{H}$ whenever $A \in \mathscr{S}$;
(b) *if $(X_n, n \geq 0)$ is an increasing sequence of nonnnegative functions of $\mathscr{H}$ such that $\sup X_n$ is finite [bounded], then $\sup X_n \in \mathscr{H}$.*

Then $\mathscr{H}$ contains all R-valued [bounded] mappings measurable with respect to $\sigma(\mathscr{S})$.

PROOF. Define $\mathscr{D} = \{A | 1_A \in \mathscr{H}\}$. Then, by (a), $\mathscr{S} \subset \mathscr{D}$. Also $\mathscr{D}$ is a d-system, since: (i) $\Omega \in \mathscr{D}$ ($1 \in \mathscr{H}$), $\varnothing \in \mathscr{H}$ ($1 - 1 \equiv 0 \in \mathscr{H}$, which is satisfied because $1 \in \mathscr{H}$ and $\mathscr{H}$ is a vector space), (ii) $\mathscr{D}$ is stable by strict difference, that is to say, $A \subset B$, $A \in \mathscr{D}$, $B \in \mathscr{D} \Rightarrow B - A \in \mathscr{D}$ ($1_{B-A} = 1_B - 1_A \in \mathscr{H}$, which is verified because $1_A \in \mathscr{H}$, $1_B \in \mathscr{H}$, and $\mathscr{H}$ is a vector space), (iii) $\mathscr{D}$ is stable by increasing sequential limit, that is to say, if $A_n \uparrow A$ and $A_n \in \mathscr{D}$, then $A \in \mathscr{D}(1_{\cup A_n} = \sup 1_{A_n} \in \mathscr{H}$, which is verified by (b), since $1_{A_n} \in \mathscr{H}$ for all $n \geq 0$). Therefore $d(\mathscr{D}) = \mathscr{D} \supset d(\mathscr{S}) = \sigma(\mathscr{S})$ (by MCT). In other words, $\mathscr{H}$ contains all the mappings ($1_A, A \in \sigma(\mathscr{S})$). Now let X be an R-valued [bounded] $\sigma(\mathscr{S})$-measurable r.v. Then X^+ and X^- are increasing limits of R-valued [bounded] step r.v. measurable with respect to $\sigma(\mathscr{S})$—each of these step r.v. belonging to $\mathscr{H}$ because $1_A \in \mathscr{H}$ for all $A \in \sigma(\mathscr{S})$ and $\mathscr{H}$ is a vector space. One concludes by using hypothesis (b) again. □

The Countable Dependency Theorem

Let Ω be a set and $(E, \mathscr{E})$ a measurable space. Let X be a mapping from Ω into E. We define $\sigma(X)$, *the σ-algebra generated by X*, by $\sigma(X) = X^{-1}(\mathscr{E})$. Clearly $\sigma(X)$ is the smallest σ-field $\mathscr{G}$ on Ω that makes the application X $\mathscr{E}/\mathscr{G}$-measurable, since for all such $\mathscr{G}$, $X^{-1}(\mathscr{E}) \subset \mathscr{G}$ by the definition of measurability.

Now let $\{(E_i, \mathscr{E}_i), i \in I\}$ be a family of measurable spaces, and $\{X_i, i \in I\}$ be a family of mappings $X_i : \Omega \to E_i$, $i \in I$. We define $\sigma(X_i, i \in I)$, *the σ-algebra generated by the family $\{X_i, i \in I\}$*, by $\sigma(X_i, i \in I) = \bigvee_{i \in I} X_i^{-1}(\mathscr{E}_i)$. Let $\mathscr{G}$ be a σ-algebra on $(\Omega, \mathscr{F})$ such that for all $i \in I$, X_i is $\mathscr{E}_i/\mathscr{G}$-measurable; then $\mathscr{G} \supset X^{-1}(\mathscr{E}_i)$ for all $i \in I$ (by the definition of measurability), and therefore $\mathscr{G} \supset \bigcup_{i \in I} X_i^{-1}(\mathscr{E}_i)$. Now this implies $\mathscr{G} = \sigma(\mathscr{G}) \supset \sigma(\bigcup_{i \in I} X_i^{-1}(\mathscr{E}_i)) = \bigvee_{i \in I} X_i^{-1}(\mathscr{E}_i) = \sigma(X_i, i \in I)$. Therefore $\sigma(X_i, i \in I)$ is the smallest σ-algebra $\mathscr{G}$ on Ω that makes X_i $\mathscr{E}_i/\mathscr{G}$-measurable for all $i \in I$.

E12 Exercise. Let S_i be, for each $i \in I$, a π-system of E_i that generates $\mathscr{E}_i$. Show that $\{\bigcap_{i \in J} X_i^{-1}(A_i) | A_i \in \mathscr{S}_i, i \in J$ finite subset of $I\}$ is a π-system of Ω that generates $\sigma(X_i, i \in I)$.

T5 Theorem. *Let Ω, E_i, $i \in I$, be sets, and for each $i \in I$ let $\mathscr{S}_i$ be a π-system on E_i. Let X_i, $i \in I$, be a family of applications $X_i : \Omega \to E_i$, $i \in I$. Let $\mathscr{H}$ be a vector space of R-valued functions defined on Ω such that*

(a) $1 \in \mathscr{H}$;
(b) *if $(Z_n, n \geq 0)$ is an increasing sequence of nonnegative functions of $\mathscr{H}$ such that $\sup Z_n$ is finite (respectively, bounded), then $\sup Z_n \in \mathscr{H}$;*
(c) *$\mathscr{H}$ contains all products of the form $\prod_{i \in J} 1_{A_i}(X_i)$, where J is a finite subset of I, and $A_i \in \mathscr{S}_i$ for all $i \in I$.*

Then $\mathscr{H}$ contains all R-valued (respectively, bounded) mappings measurable with respect to $\sigma(X_i, i \in I)$.

3. Random Variables

PROOF. Theorem T5 is a restatement of Theorem T4 with

$$\mathscr{S} = \left\{ \bigcap_{i \in J} X_i^{-1}(A_i) \, | \, A_i \in \mathscr{S}_i, i \in J, J \text{ a finite subset of } I \right\}.$$

Indeed, by E12, $\mathscr{S}$ is a π-system. Also, clearly, if $A \in \mathscr{S}$ then

$$A = \bigcap_{i \in J} X_i^{-1}(A_i), \, 1_A = \prod_{i \in J} 1_{A_i}(X_i). \qquad \square$$

T6 Theorem (A Characterization of Measurability). *Let $(\Omega, \mathscr{F})$ and $(E, \mathscr{E})$ be two measurable spaces, and Y a measurable mapping from $(\Omega, \mathscr{F})$ into $(E, \mathscr{E})$. Let X be a real random variable defined on $(\Omega, \mathscr{F})$ which is $\sigma(X)$-measurable. Then there exists a real random variable g defined on $(E, \mathscr{E})$ such that $X = g(Y)$.*

PROOF. Consider the set $\mathscr{H}$ of all the real random variables of the form $g(Y)$ where g is a real random variable defined on $(E, \mathscr{E})$. Clearly $\mathscr{H}$ is a vector space and contains the constants.

Let $(g_n(Y), n \geq 0)$ be an increasing sequence of nonnegative elements of $\mathscr{H}$ such that $Z = \sup_n g_n(Y)$ is finite. Define

$$A = \{y \in E \, | \sup g_n(y) < \infty\}. \tag{3.6}$$

Then A is in $\mathscr{E}$ and $Y(\Omega) \subset A$. Define $g : E \to R$ by

$$g(y) = \begin{cases} \sup g_n(y) & \text{on } y \in A, \\ 0 & \text{otherwise}. \end{cases} \tag{3.7}$$

Then g is a real random variable defined on $(E, \mathscr{E})$, and $Z(\omega) = g(Y(\omega))$. If $C \in \sigma(Y)$, then $C = Y^{-1}(B)$ for some $B \in \mathscr{E}$, and so $1_C(\omega) = 1_B(Y(\omega))$. Therefore $\mathscr{H}$ contains all the real-valued $\sigma(Y)$-measurable random variables X, according to T5.

If X is a numerical r.v., then

$$X' = \arctan X \tag{3.8}$$

(with the convention $\arctan \pm \infty = \pm \pi/2$) is real-valued, and by the above, $X'(\omega) = g'(Y(\omega))$ for some real-valued random variable g' defined on $(E, \mathscr{E})$. It can be assumed that $g'(E) \subset [-\pi/2, +\pi/2]$, since $X'(\omega) \in [-\pi/2, +\pi/2]$ for all ω. Therefore

$$X(\omega) = \tan g'(Y(\omega)). \tag{3.9}$$

The proof is completed by taking $g = \tan g'$ (with $\tan \pm \pi/2 = \pm \infty$). $\square$

T7 Theorem (Countable-Dependency Theorem). *Let Ω be some set on which is defined a family $(\mathscr{F}_i, i \in I)$ of σ-fields. Let $\mathscr{F} = \sigma(\mathscr{F}_i, i \in I)$. Then for any event A of $\mathscr{F}$, there exists a countable subset J of I such that A is $\sigma(\mathscr{F}_i, i \in J)$-measurable.*

PROOF. Consider the class $\mathscr{C}$ of events A satisfying the property in the statement of T7. Clearly $\mathscr{C}$ is a σ-algebra, and it contains each $\mathscr{F}_i$, $i \in I$. Hence

$$\mathscr{C} = \sigma(\mathscr{C}) \subset \sigma(\mathscr{F}_i, i \in I). \quad \square \tag{3.10}$$

4. Expectations

Probability and Measure

It is now time to introduce the main ingredient of probability theory: a *probability measure*.

D8 Definition. Let $(\Omega, \mathscr{F})$ be a measurable space together with a mapping $\mu: \mathscr{F} \to [0, +\infty]$ such that $\mu(\varnothing) = 0$ and

$$\mu\left(\sum_{i \in I} A_i\right) = \sum_{i \in I} \mu(A_i) \tag{4.1}$$

for any countable index set I, and A_i's in $\mathscr{F}$. μ is called a *measure* on $(\Omega, \mathscr{F})$, and $(\Omega, \mathscr{F}, \mu)$ a *measure space*.

The property (4.1) is σ-*additivity*. If (4.1) were restricted to finite index sets I, one would speak of *additivity*. When a property $\mathscr{P}$ is satisfied for all $\omega \in \Omega$, except for a set $N \in \mathscr{F}$ such that $\mu(N) = 0$, one says that $\mathscr{P}$ holds μ-*almost everywhere* (μ-a.e.). If $\sup_{A \in \mathscr{F}} \mu(A) < \infty$, μ is said to be *bounded*. If there exists a sequence $(C_n, n \geq 1)$ of events of $\mathscr{F}$ such that $C_n \uparrow \Omega$ and $\mu(C_n) < \infty$ for all $n \geq 1$, then μ is said to be σ-*finite*.

A *probability* P on $(\Omega, \mathscr{F})$ is a measure such that $P(\Omega) = 1$. When P is a probability, "P-a.e." is also written "P-a.s." (a.s. = almost surely).

An easy consequence of σ-additivity is *sub-σ-additivity*, i.e., for all sequences $(A_n, n \geq 1)$ of $\mathscr{F}$,

$$\mu\left(\bigcup_{n=1}^{\infty} A_n\right) \leq \sum_{n=1}^{\infty} \mu(A_n). \tag{4.2}$$

Also μ is sequentially continuous:

$$A_n \in \mathscr{F} \text{ for all } n \geq 1, \exists \lim A_n \;\Rightarrow\; \lim \mu(A_n) = \mu(\lim A_n). \tag{4.3}$$

We do not prove the above properties, nor most of the results given in this section; refer to any standard text book on measure theory.

Integrals

Let $(\Omega, \mathscr{F}, \mu)$ be a measure space. For any nonnegative numerical random variable X on $(\Omega, \mathscr{F})$, one can always unambiguously define a quantity denoted $\int_\Omega X \, d\mu$, $\int X \, d\mu$, $\int X(\omega)\mu(d\omega)$, or $\mu(X)$ and called the integral of X

4. Expectations

with respect to μ. This integral is a nonnegative real number, possibly infinite, and is equal to the limit as n tends to infinity of

$$\sum_{q=0}^{n2^n-1} \frac{q}{2^n} \mu\left(\left\{\omega \left| \frac{q}{2^n} \le X(\omega) < \frac{q+1}{2^n}\right.\right\}\right) + n\mu(\{\omega \mid X(\omega) \ge n\}). \quad (4.4)$$

For a numerical random variable that is not necessarily nonnegative three cases are to be considered:

Case 1: $\int X^+ \, d\mu < \infty$ and $\int X^- \, d\mu < \infty$. X is then said to be μ-integrable and $\int X \, d\mu$ is defined by

$$\int X \, d\mu = \int X^+ \, d\mu - \int X^- \, d\mu. \quad (4.5)$$

Then

$$\int |X| \, d\mu = \int X^+ \, d\mu + \int X^- \, d\mu < \infty. \quad (4.6)$$

Case 2: Either $\int X^+ \, d\mu < \infty$ or $\int X^- \, d\mu < \infty$. X is then said to be μ-summable and $\int X \, d\mu$ is defined by (4.2) (which has a meaning, since the right-hand side is not an indeterminate form $\infty - \infty$ or $-\infty + \infty$). Here also $\int |X| \, d\mu = \int X^+ \, d\mu + \int X^- \, d\mu$. But then $\int |X| \, d\mu$ is infinite.

Case 3: Both $\int X^+ \, d\mu$ and $\int X^- \, d\mu$ are infinite. In this case $\int X \, d\mu$ is not defined.

For any numerical random variable X defined on $(\Omega, \mathscr{F})$ and any set $A \in \mathscr{F}$ such that $X1_A$ is μ-summable, one defines $\int_A X \, d\mu$ by

$$\int_A X \, d\mu = \int 1_A X \, d\mu. \quad (4.7)$$

We now list the main properties of the integral:

For all $A \in \mathscr{F}$, $\int 1_A \, d\mu = \int_A d\mu = \mu(A).$ \quad (4.8)

If $X = 0$, μ-a.e., then X is μ-integrable and $\int X \, d\mu = 0$. \quad (4.9)

If X is μ-integrable, then $|X| < \infty$, μ-a.e. \quad (4.10)

If X and Y are μ-integrable and if for all $A \in \mathscr{F}$,

$$\int_A X \, d\mu = \int_A Y \, d\mu, \text{ then } X = Y, \mu\text{-a.e.} \quad (4.11)$$

(Monotonicity) If $X \le Y$, μ-a.e., and if X and Y are μ-summable, then $\int X \, d\mu \le \int Y \, d\mu.$ \quad (4.12)

If $|X| \leq Y$, μ-a.e., and Y is μ-integrable, then X is μ-integrable and
$$\int |X|\, d\mu \leq \int Y\, d\mu. \qquad (4.13)$$

(*Linearity*) Let a and b be real numbers, and X and Y be numerical r.v.'s. Then $\int (aX + bY)\, d\mu = a \int X\, d\mu + b \int X\, d\mu$ whenever all the quantities in the above equality have a meaning $\qquad (4.14)$

If X is a nonnegative random variable, then $v_X(A) = \int_A X\, d\mu$,
$A \in \mathscr{F}$, defines a measure v_X on $(\Omega, \mathscr{F})$. $\qquad (4.15)$

The next properties concern the *interchange of integration and pointwise limit*.

(*Monotone convergence*) Let $(X_n, n \geq 1)$ be a sequence of μ-a.e. nonnegative random variables such that
$$X_n \uparrow X, \mu\text{-a.e. Then } \lim \int X_n\, d\mu = \int X\, d\mu. \qquad (4.16)$$

(*Fatou's lemma*) Let $(X_n, n \geq 1)$ be a sequence of numerical random variables that are μ-a.e. uniformly bounded from below [from above]. Then
$$\int (\underline{\lim}\, X_n)\, d\mu \leq \underline{\lim} \int X_n\, d\mu \left[\overline{\lim} \int X_n\, d\mu \leq \int (\overline{\lim}\, X_n)\, d\mu \right]. \qquad (4.17)$$

(*Lebesgue dominated-convergence theorem*) Let $(X_n, n \geq 1)$ be a convergent sequence of numerical random variables, which are dominated in absolute value by an integrable random variable Y; then $\lim \int X_n\, d\mu = \int X\, d\mu.$ $\qquad (4.18)$

Expectations

In probability theory the integral is called *expectation*. Thus if X is a numerical random variable defined on a measurable space $(\Omega, \mathscr{F})$ and if P is a probability measure on $(\Omega, \mathscr{F})$, the expectation of X with respect to P is defined to be $\int X\, dP$ whenever X is P-summable. It is denoted $E_P[X]$, and when there is no doubt about which P is meant, the subscript P is omitted: $E[X]$. When P is a probability measure on $(\Omega, \mathscr{F})$, $(\Omega, \mathscr{F}, P)$ is called a *probability space*. If X is a P-summable random variable, the expectation

4. Expectations

$E[X]$ is also called the *mean of* X. If X^2 and X are P-integrable (actually X^2 P-integrable $\Rightarrow X$ P-integrable, as we will recall in a few lines), then the quantity $E[(X - E[X])^2]$ is called the *variance of* X. For two μ-summable numerical random variables X and Y, one defines $\int (X + iY)\,d\mu = \int X\,d\mu + i\int Y\,d\mu$. When $\mu = P$, and X is a P-almost everywhere finite r.v., the *characteristic function of* X is the mapping $\phi_X: R \to C$ defined by

$$\phi_X(u) = E[e^{iuX}], \qquad u \in R. \tag{4.19}$$

The P-distribution F of a numerical random variable X is defined by

$$F(x) = P[X \leq x]. \tag{4.20}$$

EXAMPLES

Poisson random variable. Let X be a random variable defined on $(\Omega, \mathscr{F}, P)$ and taking its values in $N_+ = \{0, 1, \ldots\}$. If for all $n \geq 0$

$$P[X = n] = e^{-\lambda}\frac{\lambda^n}{n!} \qquad (\text{convention}: 0! = 1), \tag{4.21}$$

where λ is some nonnegative real number, then X is called a Poisson random variable with parameter λ.

E13 Exercise. Verify that if X is a Poisson random variable of parameter λ,

$$\begin{aligned}E[X] = \lambda, \qquad E[(X - E[X])^2] = \lambda,\\ \phi_X(u) = E[\exp(iuX)] = \exp\{(e^{iu} - 1)\lambda\}, \qquad u \in R.\end{aligned} \tag{4.22}$$

Gaussian vectors. Let $X = (X_1, \ldots, X_n)$ be an R^n-valued random vector on $(\Omega, \mathscr{F}, P)$ such that

$$\begin{aligned}\phi_X(u_1, \ldots, u_n) &= E[e^{iu_1X_1 + \cdots + iu_nX_n}]\\ &= \exp\left(i\sum_{j=1}^n u_j m_j - \frac{1}{2}\sum_{i,j=1}^n r_{ij}u_i u_j\right),\end{aligned} \tag{4.23}$$

where the m_i's are real numbers and the r_{ij}'s are nonnegative numbers such that the matrix $R = \{r_{ij}\}$ is symmetric and nonnegative. It can be verified that

$$m_i = E[X_i],$$
$$r_{ij} = E[(X_i - m_i)(X_j - m_j)^t].$$

In the case where R is strictly positive (in particular $\det R > 0$), the distribution of X is given by the density

$$f(x_1, \ldots, x_n) = \frac{1}{(2\pi)^{n/2}(\det R)^{1/2}}\exp\left\{-\frac{1}{2}\sum_{i,j}(x_i - m_i)r_{ij}^{(-1)}(x_j - m_j)\right\}, \tag{4.24}$$

where $\{r_{ij}^{(-1)}\}$ is the inverse of $\{r_{ij}\}$.

Classical Inequalities

Chebyshev's inequality: If X is a random variable such that $|X|^p$ is P-integrable for some $p > 0$, then for any real $c > 0$

$$P[|X| \geq c] \leq c^{-p} E[|X|^p]. \tag{4.25}$$

Jensen's inequality: Let X be a random variable and $\phi: R \to R$ be a convex function. Suppose that X and $\phi(X)$ are integrable. Then

$$\phi(E[X]) \leq E[\phi(X)]. \tag{4.26}$$

Holder's inequality: Let p and q be two reals such that $p > 1$ and $1/p + 1/q = 1$. Let X and Y be real r.v.'s such that $|X|^p$ and $|Y|^q$ are integrable. Then XY is P-integrable and

$$E[|XY|] \leq E[|X|^p]^{1/p} E[|Y|^q]^{1/q}. \tag{4.27}$$

When $p = 2$, then $q = 2$ and the above inequality reduces to
Schwarz's inequality:

$$E[|XY|]^2 \leq E[|X|^2] E[|Y|^2]. \tag{4.28}$$

Minkowski's inequality: Let X and Y be two real random variables such that $|X|^p$ and $|Y|^p$ are P-integrable for some $p \geq 1$. Then we have the *triangle inequality*

$$E[|X + Y|^p]^{1/p} \leq E[|X|^p]^{1/p} + E[|Y|^p]^{1/p}. \tag{4.29}$$

L^p Spaces

Let $(\Omega, \mathcal{F}, P)$ be a probability space. We will not distinguish two numerical random variables such that $X = Y$, P-a.s. In other words we consider only equivalence classes of numerical random variables for the equivalence relation of P-a.s. equality. Let $p \geq 1$, and consider the set L^p of the (equivalence classes of) r.v. X such that $E[|X|^p] < \infty$. Clearly L^p is a vector space (use Minkowski's inequality to show that addition of two elements of L^p yields another element of L^p), and it turns out that with the norm $\|\cdot\|_p$ defined by

$$\|X\|_p = E[|X|^p]^{1/p},$$

L^p is a Banach space, i.e. a complete normed linear space. Also, if $q \geq p$, L^p is a subspace of L^q. In the case $p = 2$, L^2 is a Hilbert space, that is to say, a Banach space with a norm associated to a scalar product, namely

$$\langle X, Y \rangle = E[XY].$$

Remark. All the above definitions and results hold when P is replaced by a measure μ. However, a notable difference between the case μ finite and μ

4. Expectations

not finite is the following: If $q \geq p$, then L^p is not necessarily a subspace of L^q (but this is true if μ is finite).

With regard to notation, we shall adopt various types such as L^p, $L^p(\Omega, \mathscr{F}, \mu)$, $L^p(\mathscr{F}, \mu)$, $L^p(\mathscr{F})$, $L^p(\mu)$, depending on which element(s) of the triple $(\Omega, \mathscr{F}, \mu)$ need not be specified, the context providing enough information.

Radon–Nikodym Theorem

The Radon–Nikodym theorem is one of the main tools of measure theory, and particularly of probability theory, where it provides the mathematical basis for the notion of conditional expectation. It concerns absolute continuity, which we now define:

D9 Definition. Let μ and ν be two measures on a measurable space $(\Omega, \mathscr{F})$ such that whenever $\mu(A) = 0$ for $A \in \mathscr{F}$, then $\nu(A) = 0$. The measure ν is said to be *absolutely continuous* with respect to the measure μ, and this is denoted by $\nu \ll \mu$ (on $\mathscr{F}$).

T10 Theorem (Radon–Nikodym). *Let μ and ν be two measures given on the measurable space $(\Omega, \mathscr{F})$. Suppose that μ is σ-finite, ν is finite, and $\nu \ll \mu$ (on $\mathscr{F}$). Then there exists a nonnegative finite random variable, denoted $d\nu/d\mu$ and called the Radon–Nikodym derivative of ν with respect to μ, such that*

$$\nu(A) = \int_A \frac{d\nu}{d\mu} \, d\mu \quad \text{for all } A \in \mathscr{F}. \tag{4.30}$$

The random variable $d\nu/d\mu$ is μ-essentially unique. Also, if μ is finite, $d\nu/d\mu$ is μ-integrable.

E14 Exercise. Show that if X is a numerical random variable such that either one of the quantities $\int X \, d\nu$ or $\int X(d\nu/d\mu) \, d\mu$ can be defined, the other is defined and

$$\int X \, d\nu = \int X \frac{d\nu}{d\mu} \, d\mu. \tag{4.31}$$

(Use T3 and the monotone convergence theorem).

An elementary example: Let Ω be a set and $(A_n, n \geq 1)$ be a countable partition of Ω. Let $\mathscr{F}$ be the σ-field generated by the collection $(A_n, n \geq 1)$ [it is immediate to check that $\mathscr{F}$ consists of all unions of sets in $(A_n, n \geq 1)$, plus the void set]. Let μ and ν be two finite measures on $(\Omega, \mathscr{F})$.

E15 Exercise. Show that $v \ll \mu \Leftrightarrow (\mu(A_n) = 0 \Rightarrow v(A_n) = 0$ for all $n \geq 1)$, and verify that in this case

$$\frac{dv}{d\mu}(\omega) = \sum_{n \geq 1} \frac{v(A_n)}{\mu(A_n)} 1_{A_n}(\omega),$$

where $v(A_n)/\mu(A_n)$ is taken to be 0 when $\mu(A_n) = v(A_n) = 0$.

Fubini's Theorem

Essential to the modeling of doubly stochastic phenomena are the concepts of product of measurable spaces and of transition probability:

D11 Definition. Let $(\Omega_i, \mathscr{F}_i, i \in I)$ be a family of measurable spaces. Let Ω be the product $\Omega = \prod_{i \in I} \Omega_i$, and let $\mathscr{F}$ be the σ-field on Ω generated by the sets $\prod_{i \in I} A_i$, where $A_i \in \mathscr{F}_i$ for all $i \in J$ and $A_i = \Omega_i$ for all $i \in I - J$, J being a finite subset of I. Then $\mathscr{F}$ is denoted $\otimes_{i \in I} \mathscr{F}_i$, and $(\prod_{i \in I} E_i, \otimes_{i \in I} \mathscr{F}_i)$ is called the *product* of $((\Omega_i, \mathscr{F}_i), i \in I)$.

Notation. In the case where $I = \{1, 2\}$, the product is denoted $(\Omega_1 \times \Omega_2, \mathscr{F}_1 \otimes \mathscr{F}_2)$.

E16 Exercise. Let $(\Omega_i, \mathscr{F}_i, i \in I)$ be a family of measurable spaces, and let $(\Omega, \mathscr{F})$ be its product. Define the coordinate mappings $(X_i, i \in I)$ by $X_i(\omega) = \omega_i$, for all $\omega = (\omega_i, i \in I) \in \Omega$ and for all $i \in I$. Show that $\otimes_{i \in I} \mathscr{F}_i$ is the smallest σ-field on Ω that makes all the mappings $X_i : (\Omega, \mathscr{F}) \to (\Omega_i, \mathscr{F}_i)$ measurable.

D12 Definition. Let $(E, \mathscr{E})$ and $(\Omega, \mathscr{F})$ be two measurable spaces, and let $(P_x, x \in E)$ be a family of probability measures on $(\Omega, \mathscr{F})$ such that for all $A \in \mathscr{F}$, $x \to P_x(A)$ is $\mathscr{E}$-measurable. $(P_x, x \in E)$ is called a *transition probability* from $(E, \mathscr{E})$ into $(\Omega, \mathscr{F})$.

The above concept extends in an obvious way to measures, yielding the notion of *transition measure* $(\mu_x, x \in E)$ from $(E, \mathscr{E})$ into $(\Omega, \mathscr{F})$.

We can now state the Fubini theorem. We do it for probability measures, but the results are valid for *finite* measures.

T13 Theorem (Fubini). Let f be a numerical random variable defined on $(E \times \Omega, \mathscr{E} \otimes \mathscr{F}) = (U, \mathscr{U})$. Then $x \to f(x, \omega)$ is $\mathscr{E}$-measurable for all $\omega \in \Omega$, and $\omega \to f(x, \omega)$ is $\mathscr{F}$-measurable for all $x \in E$. Also, Q being a probability on $(E, \mathscr{E})$, there exists one and only one probability S on $(E \times \Omega, \mathscr{E} \otimes \mathscr{F}) = (U, \mathscr{U})$ and $(P_x, x \in E)$ being a transition probability from $(E, \mathscr{E})$ into $(\Omega, \mathscr{F})$ such that

$$S(A \times B) = \int_A P_x(B) Q(dx) \quad \text{for all } A \in \mathscr{E}, B \in \mathscr{F}. \tag{4.32}$$

4. Expectations

If f is a nonnegative random variable on $(E \times \Omega, \mathscr{E} \otimes \mathscr{F})$, then the mapping $x \to \int_\Omega f(x, \omega) P_x(d\omega)$ is $\mathscr{E}$-measurable and

$$\int_U f(u) S(du) = \int_E \left[\int_\Omega f(x, \omega) P_x(d\omega) \right] Q(dx). \tag{4.33}$$

The above relation is true if f is S-integrable. But then one can only assert that $\omega \to f(x, \omega)$ is P_x-integrable for Q-almost all x.

We will use in the sequel notation of the type

$$S(dx \times d\omega) = Q(dx) P_x(d\omega) \quad \text{for } S(du).$$

When $P_x = P$ for all $x \in E$, $S(du) = S(dx \times d\omega)$ is called the product measure of Q by P and is denoted

$$S = Q \times P$$

[or, as above, $S(dx \times d\omega) = Q(dx) P(d\omega)$].

Remarks

(α) Doubly stochastic phenomena can be described in the following way involving transition probabilities. A transition probability $(P_x, x \in E)$ summarizes the probabilistic nature of random experiments with outcomes $\omega \in \Omega$. The elements $x \in E$ are parameters that are also random, this randomness being modeled by Q, a probability on $(E, \mathscr{E})$. Let X be a nonnegative random variable on $(\Omega, \mathscr{F})$. If one were to know the exact value of the parameter x, one could compute the corresponding expectation for X:

$$E_x[X] = \int_\Omega X(\omega) P_x(d\omega). \tag{4.34}$$

Although X only depends upon ω, its randomness depends on the randomness of the parameter x as well. Intuitively, the mean of X is obtained by integrating $E_x[X]$ over $(E, \mathscr{E}, Q)$. That is to say,

$$(\text{mean of } X) = \int_E E_x[X] Q(dx) = \int_E \int_\Omega X(\omega) P_x(d\omega) Q(dx).$$

Fubini's theorem gives a precise meaning of the latter integration, specifying the probability S with respect to which (mean of X) is defined. In fact (4.34) is a particular case of (4.33) in Theorem T13 with $f(x, \omega) = X(\omega)$.

(β) A typical situation in which Fubini's theorem is used is the following one: In the notation of T13, let C be a set of $\mathscr{E} \otimes \mathscr{F}$ of S-measure 0. For any $x \in E$, the *section $C(x)$ of C by x* is the subset of Ω defined by $C(x) = \{\omega \in \Omega | (x, \omega) \in C\}$. Taking $f(x, \omega) = 1_C(x, \omega)$ in T13, we see that for all $x \in E$, $C(x)$ is in $\mathscr{F}$ and $P_x[C(x)] = 0$.

Image Probability and Distribution

This section is crucial for computational purposes: we will show how to calculate expectations of the type $\int_\Omega g(X) \, dP$ where X is a random variable on $(\Omega, \mathscr{F})$, and g a borelian function such that $g(X)$ is P-integrable. In fact we will see that

$$\int_\Omega g(X) \, dP = \int_R g(x) \, dF_X(x) \quad \text{(Stieltjes integral)},$$

where $F_X(x)$ is the distribution function of X. The above is a well-known rule that holds in more general situations.

D14 Definition. Let $X : (\Omega, \mathscr{F}) \to (E, \mathscr{E})$ be a measurable mapping, and let P be a probability on $(\Omega, \mathscr{F})$. The set function $Q : \mathscr{E} \to [0, 1]$ defined by

$$Q(C) = P[X^{-1}(C)], \quad C \in \mathscr{E}, \tag{4.35}$$

is called the *image probability* of P by X.

Remarks

(α) It is easy to check that Q is indeed a probability in view of the properties of the inverse mapping X^{-1} (that preserves set operations) and the fact that P is a probability.

(β) Sometimes Q is called the *distribution* of X on $(E, \mathscr{E})$. However, this appellation is generally reserved for the case where $(E, \mathscr{E}) = (R^n, \mathscr{B}^n)$. We then denote Q by Q_X, where $X = (X_1, \ldots, X_n)$. Clearly if $C = (X_1 \leq x_1, \ldots, X_n \leq x_n)$, where $x_i \in R$, $1 \leq i \leq n$, then $Q_X(C) = F_X(x_1, \ldots, x_n) = P(X_1 \leq x_1, \ldots, X_n \leq X_n)$.

T15 Theorem. Let $X : (\Omega, \mathscr{F}) \to (E, \mathscr{E})$ be a measurable mapping, P a probability on $(\Omega, \mathscr{F})$, and Q the image probability of P by X. Then, for any measurable mapping $g : (E, \mathscr{E}) \to (R, \mathscr{B})$ if either one of $\int_\Omega g(X(\omega))P(d\omega)$ or $\int_E g(x)Q(dx)$ has a meaning then they are equal:

$$\int_\Omega g(X(\omega))P(d\omega) = \int_E g(x)Q(dx). \tag{4.36}$$

Remark. In the case where $(E, \mathscr{E}) = (R^n, \mathscr{B}^n)$, then

$$\int_{R^n} g(x)Q_X(dx) = \int_{R^n} g(x_1, \ldots, x_n) \, dF_X(x_1, \ldots, x_n)$$

by definition of the Stieltjes–Lebesgue integral of the right-hand side. Therefore, as announced at the beginning of this subsection,

$$\int_\Omega g(X_1(\omega), \ldots, X_n(\omega))P(d\omega) = \int_{R^n} g(x_1, \ldots, x_n) \, dF_X(x_1, \ldots, x_n). \tag{4.37}$$

5. Conditioning and Independence

Conditional Expectation

D16 Definition. Let $(\Omega, \mathcal{F}, P)$ be a probability space on which is given an integrable random variable X. Let $\mathcal{G}$ be a sub-σ-field of $\mathcal{F}$. If there exists a random variable Y such that:

$$Y \text{ is } \mathcal{G}\text{-measurable and } P\text{-integrable,} \tag{5.1}$$

$$\int_C X \, dP = \int_C Y \, dP \quad \text{for all } C \in \mathcal{G}, \tag{5.2}$$

then Y is called a *version* of the conditional expectation of X given $\mathcal{G}$.

Note that if Y' is another such version, then $\int_C Y \, dP = \int_C Y' \, dP$ for all $C \in \mathcal{G}$, and therefore $Y' = Y$, P-a.s. Also clearly, if Y is a version of the conditional expectation of X given $\mathcal{G}$ and if Y' is any $\mathcal{G}$-measurable random variable such that $Y = Y'$, P-a.s., then Y' is a version of the conditional expectation of X given $\mathcal{G}$.

We will denote by $E[X|\mathcal{G}]$ the set of all versions Y of conditional expectations of X given $\mathcal{G}$, and practice the innocuous abuse of notation

$$Y = E[X|\mathcal{G}].$$

$E[X|\mathcal{G}]$ is called the conditional expectation of X given $\mathcal{G}$. In the case where $X = 1_A$, $E[1_A|\mathcal{G}]$ is also denoted $P[A|\mathcal{G}]$ and called the *conditional probability of A given $\mathcal{G}$*.

T17 Theorem. *Let X be an integrable or nonnegative random variable defined on $(\Omega, \mathcal{F}, P)$, and $\mathcal{G}$ be a sub-σ-field of $\mathcal{F}$. Then the conditional expectation $E[X|\mathcal{G}]$ does exist.*

PROOF

Case where $X \geq 0$ (not necessarily integrable): Consider the set function $v : \mathcal{G} \to R_+$ defined by

$$v(C) = \int_C X \, dP \quad \text{for all } C \in \mathcal{G}.$$

As we have already seen in §4, v is a measure on $(\Omega, \mathcal{G})$. Moreover this measure is σ-finite (finite if X is integrable), and v is absolutely continuous with respect to the restriction $P^\mathcal{G}$ of P to $(\Omega, \mathcal{G})$. Therefore there exists, by the Radon–Nikodym theorem T10, a P-essentially unique $\mathcal{G}$-measurable nonnegative random variable Y (which is in fact P-integrable if X is P-integrable) such that

$$v(C) = \int_C X \, dP = \int_C Y \, dP^{\mathscr{G}} \quad \text{for all } C \in \mathscr{G}. \tag{*}$$

Case where X is integrable: $X = X^+ - X^-$, and X^+ and X^- are non-negative and integrable, so that one can associate two integrable random variables Y^+ and Y^- such that

$$\int_C X^{\pm} \, dP = \int_C Y^{\pm} \, dP \quad \text{for all } C \in \mathscr{G}.$$

Take $Y = Y^+ - Y^-$. Then Y is $\mathscr{G}$-measurable and P-integrable, and $(*)$ is verified. $\square$

Independence

D18 Definition. Let $(\Omega, \mathscr{F}, P)$ be a probability space.

(i) If $A, B \in \mathscr{F}$ are such that

$$P(A \cap B) = P(A)P(B), \tag{5.3}$$

A and B are said to be *P-independent*.
(ii) If $\mathscr{C}_1$ and $\mathscr{C}_2$ are two subfamilies of $\mathscr{F}$ such that (5.3) holds for all $A \in \mathscr{C}_1, B \in \mathscr{C}_2$, then $\mathscr{C}_1$ and $\mathscr{C}_2$ are said to be *P-independent*.
(iii) If X and Y are random variables on $(\Omega, \mathscr{F})$ (taking their values in arbitrary measurable spaces), and if $\sigma(X)$ and $\sigma(Y)$ are P-independent, X and Y are said to be *P-independent*.

Remark. The trivial σ-field $\{\Omega, \varnothing\}$ is obviously P-independent of any sub-σ-field of $\mathscr{F}$.

In Definition D18 independence was defined relative to pairs (A and B, $\mathscr{C}_1$ and $\mathscr{C}_2$, X and Y). Clearly these definitions extend to arbitrary families, for instance.

If $(\mathscr{C}_i, i \in I)$ is a family of collection of sets of $\mathscr{F}$ such that for all sequences $(i_n, n \geq 1)$ of I, and all sequences $(A_n, n \geq 1)$ of events $A_n \subset \mathscr{C}_{i_n}$, $P(\bigcap_{n \geq 1} A_n) = \prod_{n \geq 1} P(A_n)$, then the $\mathscr{C}_i$'s are said to be *mutually P-independent*.

T19 Theorem. *Let $(\Omega, \mathscr{F}, P)$ be a measurable space, $(\mathscr{F}_n, n \in J)$ a finite collection of mutually P-independent sub-σ-fields of $\mathscr{F}$, and $(X_n, n \in J)$ a collection of random variables such that X_n is $\mathscr{F}_n$-measurable, $n \in J$. If each X_n is P-integrable, then the product $\prod_{n \in J} X_n$ is P-integrable and*

$$E\left[\prod_{n \in J} X_n\right] = \prod_{n \in J} E[X_n]. \tag{5.4}$$

5. Conditioning and Independence   277

PROOF. We prove the theorem for $J = \{1, 2\}$; the general case is obtained in exactly the same way, with more cumbersome writing.

(a) First we prove that if X and Y are bounded, then

$$E[XY] = E[X]E[Y]. \qquad (*)$$

Indeed, $(*)$ is verified for $X = 1_A$, $Y = 1_B$, $A \in \mathscr{F}_1$, $B \in \mathscr{F}_2$ by the definition of the P-independence of $\mathscr{F}_1$ and $\mathscr{F}_2$. By the linearity of expectation, (5.4) is verified for any step random variables X and Y. If X and Y are bounded, they can be approximated by bounded step random variables $X^{(n)}$, $Y^{(n)}$, $n \geq 1$, where $X^{(n)}$ is $\mathscr{F}_1$-measurable and $Y^{(n)}$ is $\mathscr{F}_2$-measurable for all $n \geq 1$, and

$$E[X^{(n)}Y^{(n)}] = E[X^{(n)}]E[Y^{(n)}]$$

yields $(*)$ by passage to the limit (bounded-convergence theorem).

(b) Now let X and Y be P-integrable. Let $|X|^{(n)} = |X|1(|X| \leq n)$ and $|Y|^{(n)} = |Y|1(|Y| \leq n)$. Then by the above

$$E[|X|^{(n)}|Y|^{(n)}] = E[|X|^{(n)}]E[|Y|^{(n)}].$$

By passage to the limit (monotone convergence theorem),

$$E[|X||Y|] = E[|XY|] = E[|X|]E[|Y|] < \infty,$$

that is to say, XY is P-integrable. Also by the same argument $E[XY] = E[X]E[Y]$.
$\square$

T20 Theorem. *Let $(\Omega, \mathscr{F}, P)$ be some probability space on which are given two sub-σ-fields $\mathscr{F}_1$ and $\mathscr{F}_2$. Let $\mathscr{C}_1$ and $\mathscr{C}_2$ be two π-systems on Ω such that $\mathscr{F}_1 = \sigma(\mathscr{C}_1)$ and $\mathscr{F}_2 = \sigma(\mathscr{C}_2)$. If $\mathscr{C}_1$ and $\mathscr{C}_2$ are P-independent, then $\mathscr{F}_1$ and $\mathscr{F}_2$ are P-independent.*

PROOF. By hypothesis, for all $A \in \mathscr{C}_1$ and all $B \in \mathscr{C}_2$,

$$P(A \cap B) = P(A)P(B). \qquad (*)$$

Fix $A \in \mathscr{C}_1$. Define $\mathscr{S}_2$ to be the set of finite disjoint sums of elements of $\mathscr{C}_2$. Clearly, $\mathscr{S}_2$ is a π-system and $(*)$ holds for all $B \in \mathscr{S}_2$. Let $\mathscr{V}_2$ be the largest subclass of $\mathscr{F}_2$ such that $(*)$ holds for all $B \in \mathscr{V}_2$. This $\mathscr{V}_2$ is a d-system, and since $\mathscr{V}_2 \supset \mathscr{S}_2$, $\mathscr{V}_2 = d(\mathscr{V}_2) \supset d(\mathscr{S}_2)$. By the monotone class theorem $d(\mathscr{S}_2) = \sigma(\mathscr{S}_2)$. Now $\sigma(\mathscr{S}_2) = \sigma(\mathscr{C}_2) = \mathscr{F}_2$. Therefore $\mathscr{V}_2 \supset \mathscr{F}_2$, and $\mathscr{V}_2 = \mathscr{F}_2$, and we have proved that $(*)$ holds for all $B \in \mathscr{F}_2$. A being arbitrary in $\mathscr{C}_1$, $(*)$ holds for all $A \in \mathscr{C}_1$ and all $B \in \mathscr{F}_2$. We now repeat the above argument, fixing $B \in \mathscr{F}_2$, and we obtain the announced result: that $(*)$ holds for all $A \in \mathscr{F}_1$ and all $B \in \mathscr{F}_2$.
$\square$

C21 Corollary. *Let X and Y be two random vectors defined on $(\Omega, \mathscr{F}, P)$, of dimension l and m respectively. For X and Y to be P-independent, it is necessary and sufficient that*

$$E[e^{iu'X}e^{iv'Y}] = E[e^{iu'X}]E[e^{iv'Y}] \qquad (5.5)$$

for all $u \in R^l$, $v \in R^m$.

PROOF. The necessity follows from Theorem T19. To prove sufficiency, it is enough to show that for all $a \in R^l$, $b \in R^m$,

$$P[\{X \leq a\} \cap \{Y \leq b\}] = P[\{X \leq a\}]P[\{Y \leq b\}], \quad (*)$$

since ($\{X \leq a\}$, $a \in R^l$) and ($\{Y \leq b\}$, $b \in R^m$) generate $\sigma(X)$ and $\sigma(Y)$ respectively. Let

$$f(x) = 1(x \leq a), \quad g(x) = 1(x \leq b).$$

One can construct two sequences $f^{(n)}: R^l \to [0, 1]$, $n \geq 1$, and $g^{(n)}: R^m \to [0, 1]$, $n \geq 1$, such that $f^{(n)} \to f$ uniformly, $g^{(n)} \to g$ uniformly, and $f^{(n)}$ and $g^{(n)}$ are, for all $n \geq 1$, infinitely differentiable. In turn, for all $n \geq 1$, $f^{(n)}$ can be approximated uniformly by a sequence $f^{(n, p)}: R^l \to R$, $p \geq 1$, of finite weighted sums of exponentials of the type $e^{iu'x}$, $u \in R$. And the same for $g^{(n)}$, which can be approximated uniformly by $g^{(n, p)}: R^m \to R$, $p \geq 1$, a sequence of finite weighted sums of exponentials $e^{iv'y}$, $v \in R^m$. By (5.5)

$$E[f^{(n, p)}(X)g^{(n, p)}(Y)] = E[f^{(n, p)}(X)]E[g^{(n, p)}(Y)].$$

By Lebesgue dominated convergence, letting p, then n, go to $+\infty$, we obtain (*), thus proving the announced result.

C22 Corollary. *The numerical r.v.'s X and Y defined on $(\Omega, \mathscr{F}, P)$ are P-independent if and only if*

$$F_{X, Y}(x, y) = F_X(x)F_Y(y), \quad (5.6)$$

where F_X, F_Y, $F_{X, Y}$ are the distribution functions of X, Y, and (X, Y) respectively.

PROOF. *Necessity*: Apply T20 to the case $\mathscr{F}_1 = \sigma(X)$, $\mathscr{F}_2 = \sigma(Y)$, $X_1 = 1(X \leq x)$, $X_2 = 1(Y \leq y)$. *Sufficiency*: Same argument as in the proof of Corollary C21.

Gaussian Vectors and Independence

If X and Y are two gaussian vectors defined on $(\Omega, \mathscr{F}, P)$ and of dimensions l and m respectively, the vector $Z = (X, Y)$ need not be gaussian, as simple counterexamples show. However, if X and Y are independent, then certainly Z is gaussian vector. Indeed, for any $w \in R^{l+m}$ of the form $w' = (u, v)$, $u \in R^l$, $v \in R^m$,

$$\phi_Z(w) = E[e^{iw'Z}] = E[e^{iu'X}] \times E[e^{iv'Y}]$$
$$= E[e^{iu'X}]E[e^{iv'Y}] = \phi_X(u)\phi_Y(v).$$

Since

$$\phi_X(u) = \exp\{iu'm_X - \tfrac{1}{2}u'R_{XX}u\}$$

5. Conditioning and Independence

and
$$\phi_Y(v) = \exp\{iv'm_Y - \tfrac{1}{2}v'R_{YY}v\},$$
we have
$$\phi_Z(w) = \exp\{iw'm_Z - \tfrac{1}{2}w'R_{ZZ}w\},$$
where
$$m_Z = \begin{pmatrix} m_X \\ m_Y \end{pmatrix}, \quad R_{ZZ} = \begin{pmatrix} R_{XX} & 0 \\ 0 & R_{YY} \end{pmatrix}.$$

Clearly R_{ZZ} is nonnegative. Therefore, by definition, Z is gaussian. □

For gaussian random vectors, noncorrelation implies equivalence:

T23 Theorem. *Let X and Y be a gaussian n-vector and m-vector respectively. Let m_X and m_Y (R_{XX} and R_{YY}) be their respective means (covariances). Suppose that the $(n+m)$-vector $Z = (X, Y)$ is gaussian and that X and Y are uncorrelated, i.e.*

$$R_{ZZ} = E[(Z - m_z)(Z - m_z)'] = \begin{pmatrix} R_{XX} & 0 \\ 0 & R_{YY} \end{pmatrix}. \tag{5.7}$$

Then X and Y are independent.

PROOF. By definition
$$\phi_Z(w) = \exp\{iw'm_Z - \tfrac{1}{2}w'R_{ZZ}w\},$$
where $w' = (u, v)'$, $u \in R^n$, $v \in R^m$. Now since $m'_Z = (m_X, m_Y)'$, we get, using (5.7),
$$\phi_Z(w) = \phi_X(u)\phi_Y(v),$$
and this by T21 is enough to ensure the independence of X and Y. □

Main Properties of Conditional Expectation

All random variables in this subsection are defined on the same $(\Omega, \mathscr{F}, P)$. The properties will be proven only when the proof is not obvious.

Linearity. X, Y integrable; $a, b, c \in R$; $\mathscr{G}$ a sub-σ-field of $\mathscr{F}$. Then
$$E[aX + bY + c | \mathscr{G}] = aE[X | \mathscr{G}] + bE[Y | \mathscr{G}] + c \quad P\text{-a.s.} \tag{5.8}$$

Monotonicity. X, Y integrable; $X \leq Y$, P-a.s.; $\mathscr{G}$ a sub-σ-field of $\mathscr{F}$. Then
$$E[X | \mathscr{G}] \leq E[Y | \mathscr{G}] \quad P\text{-a.s.} \tag{5.9}$$

Monotone Convergence. X_n integrable, $n \geq 1$; $\lim \uparrow X_n = X$, P-a.s.; X integrable; $\mathscr{G}$ a sub-σ-field of $\mathscr{F}$. Then

$$E[X|\mathscr{G}] = \lim E[X_n|\mathscr{G}] \quad P\text{-a.s.} \tag{5.10}$$

PROOF. By monotonicity, $E[X_n|\mathscr{G}]$ increases with n, and for each $n \geq 1$, $E[X_n|\mathscr{G}] \leq E[X|\mathscr{G}]$; therefore $\lim E[X_n|\mathscr{G}] = Y \leq E[X|\mathscr{G}]$.

By the monotone convergence theorem, for all $C \in \mathscr{G}$

$$\int_C X_n \, dP = \int_C E[X_n|\mathscr{G}] \, dP \uparrow \int_C Y \, dP.$$

Also

$$\int_C X_n \, dP \uparrow \int_C X \, dP = \int_C E[X|\mathscr{G}] \, dP.$$

Hence, for all $C \in \mathscr{G}$,

$$\int_C Y \, dP = \int_C E[X|\mathscr{G}] \, dP,$$

and therefore $Y = E[X|\mathscr{G}]$. □

Jensen's Inequality. Let $\phi: R \to R$ be convex, and let X be an integrable random variable such that $\phi(X)$ is integrable; then, if $\mathscr{G}$ is a sub-σ-field of $\mathscr{F}$,

$$\phi(E[X|\mathscr{G}]) \leq E[\phi(X)|\mathscr{G}] \quad P\text{-a.s.} \tag{5.11}$$

PROOF. By a classical result of real-function theory, there exist sequences $(a_n, n \geq 1)$ and $(b_n, n \geq 1)$ of real numbers such that

$$\phi(x) = \sup_n (a_n x + b_n) = \sup \phi_n(x), \quad x \in R,$$

where $\phi_n(x) = a_n x + b_n$. By linearity and monotonicity,

$$\phi_n(E[X|\mathscr{G}]) = E[\phi_n(X)|\mathscr{G}] \leq E[\phi(X)|\mathscr{G}].$$

The rest follows by letting $n \uparrow \infty$.

Remark. If X takes its values on an interval I of R, the property (5.10) holds for any ϕ that is convex on I (same proof as above).

Successive Conditioning. X a P-integrable random variable, $\mathscr{G}_1 \subset \mathscr{G}_2$ two sub-σ-fields of $\mathscr{F}$. Then

$$E[E[X|\mathscr{G}_2]|\mathscr{G}_1] = E[X|\mathscr{G}_1] \quad P\text{-a.s.} \tag{5.12}$$

PROOF. For all $C \in \mathscr{G}_2$

$$\int_C X \, dP = \int_C E[X|\mathscr{G}_2] \, dP.$$

5. Conditioning and Independence   281

If in addition C belongs to $\mathscr{G}_1$

$$\int_C E[X|\mathscr{G}_2] dP = \int_C E[E[X|\mathscr{G}_2]|\mathscr{G}_1] dP,$$

so that for all $C \in \mathscr{G}_1$,

$$\int_C X \, dP = \int_C E[E[X|\mathscr{G}_2]|\mathscr{G}_1] \, dP.$$

Now $E[E[X|\mathscr{G}_2]|\mathscr{G}_1]$ is $\mathscr{G}_1$-measurable, and therefore, by the definition of conditional expectation, we have the announced equality. □

Factoring. *Let X be a P-integrable random variable, $\mathscr{G}$ is a sub-σ-field of $\mathscr{F}$, and Y a $\mathscr{G}$-measurable random variable such that XY is P-integrable. Then*

$$E[XY|\mathscr{G}] = YE[X|\mathscr{G}] \quad P\text{-a.s.} \tag{5.13}$$

PROOF. If $Y = 1_D, D \in \mathscr{G}$, then (5.13) follows from the definition of conditional expectation. Indeed, for all $C \in \mathscr{G}$,

$$\int_C E[XY|\mathscr{G}] \, dP = \int_C XY \, dP = \int_{C \cap D} X \, dP$$

$$= \int_{C \cap D} E[X|\mathscr{G}] \, dP = \int_C YE[X|\mathscr{G}] \, dP \qquad (*)$$

(since $C \cap D \in \mathscr{G}$); C being arbitrary in $\mathscr{G}$, from $(*)$ we obtain the equality (5.13) for this special case. By linearity, (5.13) remains valid for step random variables Y that are $\mathscr{G}$-measurable. The rest follows by monotone convergence. In particular, if Y is P-integrable and $\mathscr{G}$-measurable, then $E[Y|\mathscr{G}] = Y$, P-a.s. (take $X = 1$).

Redundant Conditioning. *Let X be a P-integrable random variable, and $\mathscr{G}_1, \mathscr{G}_2$ be two sub-σ-fields of $\mathscr{F}$. Suppose that $\sigma(X)$ and $\mathscr{G}_1$ are P-independent of $\mathscr{G}_2$. Then*

$$E[X|\mathscr{G}_1 \vee \mathscr{G}_2] = E[X|\mathscr{G}_1] \quad P\text{-a.s.} \tag{5.14}$$

PROOF. Let $A \in \mathscr{G}_1, B \in \mathscr{G}_2$. Then, since $A \cap B \in \mathscr{G}_1 \vee \mathscr{G}_2$,

$$\int_{A \cap B} E[X|\mathscr{G}_1 \vee \mathscr{G}_2] \, dP = \int_{A \cap B} X \, dP = \int 1_A X 1_B \, dP.$$

By the independence hypothesis,

$$\int 1_A X 1_B \, dP = \int 1_A X \, dP \int 1_B \, dP = \int_A X \, dP \int 1_B \, dP$$

$$= \int_A E[X|\mathscr{G}_1] \, dP \int 1_B \, dP.$$

Again by the independence hypothesis,

$$\int_A E[X|\mathscr{G}_1] \, dP \int 1_B \, dP = \int_{A \cap B} E[X|\mathscr{G}_1] \, dP.$$

Hence
$$\int_{A \cap B} E[X|\mathcal{G}_1 \vee \mathcal{G}_2]\, dP = \int_{A \cap B} E[X|\mathcal{G}_1]\, dP.$$

Since the family $\{A \cap B,\ A \in \mathcal{G}_1,\ B \in \mathcal{G}_2\}$ generates $\mathcal{G}_1 \vee \mathcal{G}_2$, it follows by monotone-class arguments that the latter equality extends to

$$\int_C E[X|\mathcal{G}_1 \vee \mathcal{G}_2]\, dP = \int_C E[X|\mathcal{G}_1]\, dP$$

for all $C \in \mathcal{G}_1 \vee \mathcal{G}_2$, and therefore (5.14) holds. □

Gross Conditioning. *X is a P-integrable random variable. $\mathcal{G}$ is the trivial σ-field $\{\Omega, \emptyset\}$. Then*

$$E[X|\mathcal{G}] = E[X] \quad P\text{-a.s.} \tag{5.15}$$

PROOF. Clearly $E[X]$ is $\{\Omega, \emptyset\}$-measurable, being a constant. Also

$$\int_C E[X]\, dP = \int_C X\, dP, \quad C = \Omega \text{ or } \emptyset. \quad \square \tag{5.16}$$

Remark. From the properties (5.14) and (5.15) it follows that if X is P-integrable and if $\mathcal{G}$ is a sub-σ-field of $\mathcal{F}$, P-independent of X, then

$$E[X|\mathcal{G}] = E[X] \quad P\text{-a.s.} \tag{5.17}$$

PROOF. Apply (5.14) to $\mathcal{G}_1 = \mathcal{G}, \mathcal{G}_2 = \{\Omega, \emptyset\}$. Then $\mathcal{G} = \mathcal{G}_1 \vee \mathcal{G}_2$. Also $\mathcal{G}_1$ is independent of $\mathcal{G}_2$ and X. Therefore $E[X|\mathcal{G}] = E[X|\mathcal{G}_1 \vee \mathcal{G}_2] = E[X|\mathcal{G}_2] = E[X]$, by (5.13). □

Computation of Conditional Expectations

Let $Y_i : (\Omega, \mathcal{F}) \to (E_i, \mathcal{E}_i)$, $1 \le i \le n$, be measurable mappings, and let X be a numerical random variable on $(\Omega, \mathcal{F})$. Define $\mathcal{G}$ to be the σ-field generated by $(Y_i, 1 \le i \le n)$; then by Doob's representation theorem, for any borelian h such that $h(X)$ is P-integrable, there exists a measurable mapping $g : (\prod_{i=1}^n E_i, \otimes_{i=1}^n \mathcal{E}_i) \to (R, \mathcal{B})$ such that

$$E[h(X)|\mathcal{G}] = g(Y_1, \ldots, Y_n). \tag{5.18}$$

Remark. The notation $E[X|Y_1, \ldots, Y_{n-1}, Y_n = y_n]$, where $y_n \in E_n$, stands for $g(Y_1, \ldots, Y_{n-1}, y_n)$. In the case where E_n is a countable set, then

$$g(Y_1, \ldots, Y_{n-1}, Y_n) = \sum_{k=1}^{\infty} g(Y_1, \ldots, Y_{n-1}, y_n) 1(Y_n = y_n),$$

or, in the notation introduced above,

$$E[X|\sigma(Y_1, \ldots, Y_n)] = \sum_{k=1}^{\infty} E[X|Y_1, \ldots, Y_{n-1}, Y_n = y_n] 1(Y_n = y_n).$$

5. Conditioning and Independence

It is of practical interest to be able to compute g explicitly in terms of the joint distribution of $X, Y_1, \ldots, Y_n$. We consider the case $E_i = R$, $1 \leq i \leq n$, for simplification, and we define $F_{XY}(x, y_1, \ldots, y_n) = P[X \leq x, Y_1 \leq y_1, \ldots, Y_n \leq y_n]$. We denote by $Q_{XY}(dx\, dy_1 \cdots dy_n)$ the distribution of $(X_1, Y_1, \ldots, Y_n)$. In the case where Q_{XY} admits a density, that is to say, $Q_{XY}(dx\, dy_1 \cdots dy_n) = f_{XY}(x, y_1, \ldots, y_n)\, dx\, dy_1 \cdots dy_n$, then

$$g(y_1, \ldots, y_n) = \begin{cases} \dfrac{\int_R h(x) f_{XY}(x, y_1, \ldots, y_n)\, dx}{\int_R f_{XY}(x, y_1, \ldots, y_n)\, dx} & \text{if the denominator is strictly positive,} \\ 0 & \text{otherwise.} \end{cases} \quad (5.19)$$

PROOF. First we note that g is borelian function (this follows by Fubini's theorem). Therefore $g(Y_1, \ldots, Y_n)$, where g is defined by (5.19), is $\sigma(Y_1, \ldots, Y_n)$-measurable. In order to prove that $g(Y_1, \ldots, Y_n) = E[h(X)|\sigma(Y_1, \ldots, Y_n)]$, it remains to show that for any $C \in \sigma(X_1, \ldots, X_n)$

$$\int_C g(Y_1, \ldots, Y_n)\, dP = \int_C h(X)\, dP.$$

Now any $C \in \sigma(Y_1, \ldots, Y_n)$ can be written

$$C = \{(Y_1, \ldots, Y_n) \in D\} = \{(X, Y_1, \ldots, Y_n) \in R \times D\}$$

for some $D \in \mathscr{B}^n$. Therefore by Theorem T15 on image probabilities,

$$\int_C g(Y_1, \ldots, Y_n)\, dP$$
$$= \int \cdots \int_{R \times D} g(y_1, \ldots, y_n) f_{XY}(x, y_1, \ldots, y_n)\, dx\, dy_1 \cdots dy_n.$$

But also

$$\int_C h(X)\, dP = \int \cdots \int_{R \times D} h(x) f_{XY}(x, y_1, \ldots, y_n)\, dx\, dy_1 \cdots dy_n$$

(image probability, T15)

$$= \int \cdots \int_D \left[\int_R h(x) f_{XY}(x, y_1, \ldots, y_n)\, dx \right] dy_1 \cdots dy_n$$

(Fubini)

$$= \int \cdots \int_D \left[g(y_1, \ldots, y_n) \int_R f_{XY}(x, y_1, \ldots, y_n)\, dx \right] dy_1 \cdots dy_n$$

(definition of g)

$$= \int \cdots \int_{R \times D} g(y_1, \ldots, y_n) f_{XY}(x, y_1, \ldots, y_n)\, dx\, dy_1 \cdots dy_n$$

(Fubini). □

Conditional Distributions

Let $(\Omega, \mathscr{F}, P)$ be some probability space, and let $\mathscr{G}$ be a sub-σ-field of $\mathscr{F}$. For each $A \in \mathscr{F}$, we have defined $P[A|\mathscr{G}] = E[1_A|\mathscr{G}]$, the conditional probability of A given $\mathscr{G}$; in fact $P[A|\mathscr{G}]$ is an equivalence class (for the P-equivalence) of $\mathscr{G}$-measurable random variables.

Clearly, by the monotonicity property for conditional expectations,

$$0 \leq P[A|\mathscr{G}] \leq 1 \quad P\text{-a.s.},$$

which means that there exists a representation Y^A of $P[A|\mathscr{G}]$ such that

$$0 \leq Y(A, \omega) \leq 1, \quad \omega \in \Omega. \tag{*}$$

Let $(A_n, n \geq 1)$ be a sequence of disjoint sets in $\mathscr{F}$, and Y^{A_n} the corresponding representations of the $P[A_n|\mathscr{G}]$ satisfying the above sure equality. By the σ-additivity property for conditional expectations,

$$\sum_{i=1}^{n} Y(A, \omega) = Y\left(\bigcup_{i=1}^{n} A_i, \omega\right), \quad \omega \notin \mathscr{N}(A_i, i \leq n), \tag{**}$$

where $\mathscr{N}(A_i, i \leq n)$ is a set of $\mathscr{G}$ of P-measure 0. It should be insisted that we cannot guarantee the existence of $\mathscr{N}$ such that $\mathscr{N} \supset \mathscr{N}(A_i, i \leq n)$ for all sequences of disjoint sets in $\mathscr{F}$, and such that $P(\mathscr{N}) = 0$. If this were the case, then by redefining $Y(A)$ to be 0 on $\mathscr{N}$ for all $A \in \mathscr{F}$, (*) and (**) would be verified, and (**) would be true for *all* ω. This would be a pleasant situation because, for each ω, $A \to Y(A, \omega)$ would be a probability measure on $(\Omega, \mathscr{F})$. We could then define for any nonnegative random variable X on $(\Omega, \mathscr{F})$ quantities such as

$$\int_\Omega X(\omega_1) Y(d\omega_1, \omega) = V(\omega),$$

and we could expect $V(\omega)$ to be a version of $E[X|\mathscr{G}]$. Alas, this is not the general case; but fortunately, there is still something positive that can be said: Suppose that X is a (not necessarily integrable) random variable taking its values in $(R^n, \mathscr{B}^n)$, and for all $C \in \mathscr{B}_n$ define $Q(C, \omega)$ to be a version of $E[1_C(X(\omega))|\mathscr{G}]$. It turns out [and this is due to the particular topological structure of $(R^n, \mathscr{B}^n)$] that Q can always be chosen such that, for some $\mathscr{N} \in \mathscr{F}$, $P(\mathscr{N}) = 0$,

(1) for all $C \in \mathscr{B}^n$, $\omega \to Q(C, \omega)$ is $\mathscr{G}$-measurable,
(2) for all $\omega \notin \mathscr{N}$, $Q(\cdot, \omega)$ is a probability on $(R^n, \mathscr{B}^n)$.

Such Q is then called a *regular conditional distribution of X given $\mathscr{G}$*.

For any borelian function $g : R^n \to R$ such that $g(X)$ is P-integrable,

$$\int_{R^n} g(x) Q(dx, \omega) \text{ is a version of } E[g(X)|\mathscr{G}]. \tag{5.20}$$

Note that we have explicitly calculated such regular conditional distributions in the case $\mathscr{G} = \sigma(Y_1, \ldots, Y_n)$ [Equation (5.19)].

6. Convergence

Types of Convergence

We are going to recall some of the main types of convergence of a sequence $(X_n, n \geq 1)$ of numerical random variables defined on the same probability space $(\Omega, \mathscr{F}, P)$.

One says that X_n converges to the numerical r.v. X:

almost surely if
$$\lim X_n(\omega) = X(\omega) \quad P\text{-a.s.}; \tag{6.1}$$

in probability if
$$\lim P[|X_n(\omega) - X(\omega)| > \varepsilon] = 0 \quad \text{for all } \varepsilon > 0; \tag{6.2}$$

strongly in L^p if
$$X_n, X \in L^p \quad \text{and} \quad \lim E[|X_n - X|^p] = 0; \tag{6.3}$$

weakly in L^1 if
$$X_n, X \in L^1 \tag{6.4a}$$
and
$$\lim E[X_n Y] = E[XY] \quad \text{for all bounded random variables } Y; \tag{6.4b}$$

weakly in L^2 if
$$X_n, X \in L^2 \tag{6.5a}$$
and
$$\lim E[X_n Y] = E[XY] \quad \text{for all random variables } Y \text{ in } L^2. \tag{6.5b}$$

We recall that either almost sure convergence or strong convergence in L^p implies convergence in probability. Also, if X_n converges in probability to X, one can extract from $(X_n, n \geq 1)$ a subsequence $(X_{k_n}, n \geq 1)$ that converges almost surely to X.

For convergence in probability we have the following *Cauchy criterion for convergence in probability*: if $(X_n, n \geq 1)$ is such that for all $\varepsilon > 0$, $P[|X_n - X_m| > \varepsilon] \to 0$ as $n, m \to \infty$, then X_n converges in probability to some random variable X. Of course, we have for convergence in L^p the usual Cauchy criteria (for instance, we have already mentioned that L^p is a complete space under the norm $\|X\|_p = E[|X|^p]^{1/p}$).

Uniform Integrability

The notion of uniform integrability is useful in asserting the L^1 strong convergence of a sequence X_n of random variables.

D24 Definition. Let $(\Omega, \mathscr{F}, P)$ be a probability space and $(X_i, i \in I)$ a collection of (equivalence classes of) random variables of $L^1(\Omega, \mathscr{F}, P)$. This collection is said to be *uniformly integrable* (u.i.) if

$$\lim_{c \uparrow \infty} \int_{\{|X_i| > c\}} |X_i| \, dP = 0 \quad \text{uniformly in } i \in I. \tag{6.6}$$

A criterion for uniform integrability is given in the following:

T25 Theorem. *The collection $(X_i, i \in I)$ of random variables of $L^1(\Omega, \mathscr{F}, P)$ is uniformly integrable iff*

$$\sup_{i \in I} E[|X_i|] < \infty \tag{6.7}$$

and for all $\varepsilon > 0$ there exists $\delta > 0$ such that

$$A \in \mathscr{F}, \quad P[A] \le \delta \quad \Rightarrow \quad \int_A |X_i| \, dP < \varepsilon \quad \text{for all } i \in I. \tag{6.8}$$

Also:

T26 Theorem (Hadamard–La Vallée–Poussin Criterion). *The collection $(X_i, i \in I)$ of random variables of $L^1(\Omega, \mathscr{F}, P)$ is uniformly integrable iff there exists an increasing convex function $G: R^+ \to R^+$ such that*

$$\lim_{t \to +\infty} \frac{G(t)}{t} = +\infty \tag{6.9}$$

and

$$\sup_{i \in I} E[G(|X_i|)] < \infty. \tag{6.10}$$

The important result concerning uniform integrability is

T27 Theorem. *Let $(X_n, n \ge 1)$ be a sequence of $L^1(\Omega, \mathscr{F}, P)$ that converges in probability to a r.v. X. In order that $X \in L^1(\Omega, \mathscr{F}, P)$ and that $X_n \xrightarrow{L^1} X$, it is necessary and sufficient that $(X_n, n \ge 1)$ be uniformly integrable.*

T28 Theorem. *Let $(\Omega, \mathscr{F}, P)$ be a probability space, X a P-integrable random variable, and $(\mathscr{F}_i, i \in I)$ a collection of sub-σ-fields of $\mathscr{F}$. Then $(E[X|\mathscr{F}_i], i \in I)$ is uniformly integrable.*

PROOF. For any $i \in I$, by Jensen's inequality,

$$|E[X|\mathscr{F}_i]| \leq E[|X||\mathscr{F}_i],$$

so that for all $i \in I$ and all $c \geq 0$, defining $A_i(c) = \{|E[X|\mathscr{F}_i]| \geq c\}$, we have

$$\int_{A_i(c)} |E[X|\mathscr{F}_i]|\, dP \leq \int_{A_i(c)} E[|X||\mathscr{F}_i]\, dP.$$

Since $A_i(c)$ belongs to $\mathscr{F}_i$, the right-hand side of the latter inequality is

$$\int_{A_i(c)} |X|\, dP.$$

For any $b \geq 0$, the above integral can be decomposed into

$$\int_{A_i(c) \cap \{|X| \geq b\}} + \int_{A_i(c) \cap \{|X| \leq b\}}.$$

Thus

$$\int_{A_i(c)} |E[X|\mathscr{F}_i]|\, dP \leq bP[A_i(c)] + E[|X|1(|X| \geq b)].$$

Now, by Tchebyshev's inequality,

$$bP[A_i(c)] \leq \frac{b}{c} E[E[|X||\mathscr{F}_i]] = \frac{b}{c} E[|X|].$$

Therefore

$$\int_{A_i(c)} |E[X|\mathscr{F}_i]|\, dP \leq \frac{b}{c} E[|X|] + E[|X|1(|X| \geq b)].$$

Taking $b = \sqrt{c}$ and letting c go to infinity, we see that the right-hand side of the latter inequality tends to 0 uniformly in $i \in I$. □

7. Stochastic Processes

Histories, Measurability

D29 Definition. Let $(\Omega, \mathscr{F})$ be a measurable space. A *history* $(\mathscr{F}_t, t \geq 0)$ on $(\Omega, \mathscr{F})$ is a family of sub-σ-fields of $\mathscr{F}$ such that for all $0 \leq s \leq t$

$$\mathscr{F}_s \subset \mathscr{F}_t. \qquad (7.1)$$

In other words, a history is an increasing family of sub-σ-fields of $\mathscr{F}$ indexed by the nonnegative real numbers. It is also called a *filtration*.

Notation: $\mathscr{F}_\infty = \bigvee_{t \geq 0} \mathscr{F}_t$. If for all $t \geq 0$

$$\mathscr{F}_{t+} = \bigcap_{h > 0} \mathscr{F}_{t+h} = \mathscr{F}_t,$$

the history $(\mathscr{F}_t, t \geq 0)$ is said to be *right-continuous*. We will denote the whole family $(\mathscr{F}_t, t \geq 0)$ by $\mathscr{F}_t$. Similarly a family $(X_t, t \geq 0)$ of $(E, \mathscr{E})$-valued random variables defined on $(\Omega, \mathscr{F})$ will be denoted by X_t. The context will always dissipate confusion. The family X_t [i.e. $(X_t, t \geq 0)$] is called a *E-valued process* defined on $(\Omega, \mathscr{F})$. The history $\mathscr{F}_t^X$ defined by

$$\mathscr{F}_t^X = \sigma(X_s, s \in [0, t]), \qquad t \geq 0, \tag{7.2}$$

is called the *internal history* of X_t. Any history $\mathscr{F}_t$ such that

$$\mathscr{F}_t \supset \mathscr{F}_t^X, \qquad t \geq 0,$$

is called a *history* of X_t. One also says: X_t is *adapted* to $\mathscr{F}_t$. If X_t is a process on $(\Omega, \mathscr{F})$, and ω an element of Ω, the mapping

$$t \to X_t(\omega) \tag{7.3}$$

is called a *trajectory* (the ω-trajectory) of X_t, or equivalently a *path* (the ω-path) of X_t. If P is a probability measure on $(\Omega, \mathscr{F})$, and E a topological space, the process X_t is said to be *continuous* (*right-continuous, left-continuous*) iff the trajectories $t \to X_t(\omega)$ are continuous (right-continuous, left-continuous), P-a.s.

D30 Definition. Two E-valued processes X_t and Y_t defined on $(\Omega, \mathscr{F}, P)$ are *modifications* or *versions* of one another iff $P[\{\omega | X_t(\omega) \neq Y_t(\omega)\}] = 0$ for all $t \geq 0$. They are said to be *P-indistinguishable* if $P[\{\omega | X_t(\omega) \neq Y_t(\omega)$ for all $t \geq 0\}] = 0$, i.e., if they have identical trajectories except on a P-null set.

D31 Definition. The E-valued process X_t is said to be *measurable* iff the mapping from $R_+ \times \Omega$ into E defined by $t, \omega \to X_t(\omega)$ is $\mathscr{E}/\mathscr{B}_+ \otimes \mathscr{F}$-measurable.

By Fubini's theorem, for a trajectory $\omega \in \Omega$ the mapping $t \to X_t(\omega)$ is $\mathscr{E}/\mathscr{B}$-measurable. In particular, if $E = R$ and if X_t is bounded, or nonnegative, one can define Lebesgue integrals $\int_I X_t(\omega) \, dt$, $I \in \mathscr{B}_+$, for each $\omega \in \Omega$.

D32 Definition. An E-valued process X_t is said to be $\mathscr{F}_t$-*progressive* (or $\mathscr{F}_t$-*progressively measurable*) iff for all $t \geq 0$ the mapping $(t, \omega) \to X_s(\omega)$ from $[0, t] \times \Omega$ into E is $\mathscr{E}/\mathscr{B}([0, t]) \otimes \mathscr{F}_t$-measurable.

Clearly then X_t is adapted to $\mathscr{F}_t$ and measurable.

T33 Theorem. *If E is a metrizable topological space and if X_t is an E-valued process adapted to $\mathscr{F}_t$ and right-continuous (or left continuous), then X_t is $\mathscr{F}_t$-progressive.*

7. Stochastic Processes

PROOF. Let t be a nonnegative real number, and for all $n \geq 0$ and all $s \in [0, t]$, define $X_s^{(n)}$ by

$$X_0^{(n)} = X_0,$$
$$X_s^{(n)} = X_{kt/2^{-n}} \quad \text{if } s \in ((k-1)2^{-n}t, k2^{-n}t], 1 \leq k \leq 2^n.$$

This defines an application $[0, t] \times \Omega \to E$ which is obviously $\mathscr{E}/\mathscr{B}([0, t]) \otimes \mathscr{F}_t$-measurable. If X_t is right-continuous, $X_s(\omega)$ is the limit of $X_s^{(n)}(\omega)$ for all $(s, \omega) \in [0, t] \times \Omega$, and therefore $(s, \omega) \to X_s(\omega)$ is also $\mathscr{E}/\mathscr{B}([0, t]) \otimes \mathscr{F}_t$-measurable as an application of $[0, t] \times \Omega$ into E. The case where X_t is left-continuous can be treated in a similar way. □

Stopping Times (see also Appendix A2)

A measurable space $(\Omega, \mathscr{F})$ is given once and for all. All the σ-fields to be introduced are sub-σ-fields of $\mathscr{F}$, and all the random variables are defined on $(\Omega, \mathscr{F})$. We recall the definition of a stopping time.

D34 Definition. Let $\mathscr{F}_t$ be a history and T be a $\bar{R}_+$-valued random variable. T is called an $\mathscr{F}_t$-stopping time iff

$$\{T \leq t\} \subset \mathscr{F}_t, \quad t \geq 0. \tag{7.4}$$

Almost all the stopping times that we will encounter are first-passage times of the type described in the following theorem.

T35 Theorem. *Let X be a right-continuous (or left-continuous) R-valued process adapted to $\mathscr{F}_t$, and c be a given real number. Define T by*

$$T = \begin{cases} \inf\{t \,|\, X_t \geq c\} & \text{if } \{\cdots\} \neq \varnothing, \\ +\infty & \text{otherwise}. \end{cases} \tag{7.5}$$

T is an $\mathscr{F}_t$-stopping time.

PROOF. $\{T > t\} = \{X_s < c \text{ for all } s \in [0, t]\}$, and since X_t is right continuous (or left continuous),

$$\{X_s < c \text{ for all } s \in [0, t]\}$$
$$= \bigcap_{n=1}^{\infty} \{X_{kt/2^n} < c, k = 0, 1, \ldots, 2^n\} \in \mathscr{F}_t.$$

T36 Theorem. *Let T be a $\mathscr{F}_t$-stopping time. The family*

$$\mathscr{F}_T = \{A \in \mathscr{F}_\infty \,|\, A \cap \{T \leq t\} \in \mathscr{F}_t, t \geq 0\} \tag{7.6}$$

is a σ-field, and T is $\mathscr{F}_T$-measurable. Let X_t be an E-valued $\mathscr{F}_t$-progressive process, and let T be a finite $\mathscr{F}_t$-stopping time. If we define X_T by $X_T = X_{T(\omega)}(\omega)$, then X_T is $\mathscr{F}_T$-measurable.

PROOF. The verification that $\mathscr{F}_T$ is a σ-field is straightforward. In order to show that T is $\mathscr{F}_T$-measurable, it is enough to show that for all $c \geq 0$, $\{T \leq c\} \in \mathscr{F}_T$, and this follows from the definition of $\mathscr{F}_T$. Take any $A \in \mathscr{E}$ and $t \geq 0$. The set $\{X_T \in A\} \cap \{T \leq t\}$ is identical to $\{X_{T \wedge t} \in A\} \cap \{T \leq t\}$ and is in $\mathscr{F}_t$, as we now show: If we define $S = T \wedge t$, then S is an $\mathscr{F}_t$-stopping time and it is also a $\mathscr{F}_t$-measurable random variable, as one can check. The $\mathscr{F}_t$-measurability of X_S follows from the fact that X_S is obtained by composition of $\omega \to (S(\omega), \omega)$ from $(\Omega, \mathscr{F}_t)$ into $([0, t] \times \Omega, \mathscr{B}([0, t]) \otimes \mathscr{F}_t)$, and $(s, \omega) \to X_s(\omega)$ from $([0, t] \times \Omega, \mathscr{B}([0, t]) \otimes \mathscr{F}_t)$ into $(E, \mathscr{E})$, which are measurable: the first one by the definition of $\mathscr{F}_t$-stopping times, the second one by the definition of $\mathscr{F}_t$-progressive process. □

For more details on stopping times we refer to Appendix A2.

8. Markov Processes

Generalities

A Markov process is, roughly speaking, a process X_t such that the increments $X_{t+h} - X_t$ depend upon the past $\mathscr{F}_t^X$ only through X_t. More precisely, and also more generally:

D37 Definition. Let X_t be an E-valued process defined on $(\Omega, \mathscr{F}, P)$ and adapted to some history $\mathscr{F}_t$. It is called a $(P, \mathscr{F}_t)$-*Markov process* iff for all $t \geq 0$

$$\sigma(X_s, s \geq t) \text{ and } \mathscr{F}_t \text{ are } P\text{-independent given } X_t. \tag{8.1}$$

As usual, when there is no doubt about the probability P, we say "X_t is an $\mathscr{F}_t$-Markov process," and when we use a phrase such as "X_t is a Markov process," we mean that X_t is an $\mathscr{F}_t^X$-Markov process. If X_t is a $(P, \mathscr{F}_t)$-Markov process, then in particular

$$E[f(X_t)|\mathscr{F}_s] = E[f(X_t)|\sigma(X_s)] \tag{8.2}$$

for all $0 \leq s \leq t$ and all bounded measurable functions $f: E \to R$. All Markov processes to be encountered in this book admit a transition function, which we now define:

D38 Definition. *Let $(E, \mathscr{E})$ be some measurable space, and for each $0 \leq s \leq t$ let $P_{s,t}(x, A)$, $x \in E$, $A \in \mathscr{E}$ be a function from $E \times \mathscr{E}$ into R_+ such that:*

(a) *$A \to P_{s,t}(x, A)$ is a probability on $(E, \mathscr{E})$ for all $x \in E$,*
(b) *$x \to P_{s,t}(x, A)$ is in $\mathscr{B}_+/\mathscr{E}$ for all $A \in \mathscr{E}$,*

8. Markov Processes

(c) for all $0 \leq t \leq u \leq s$, all $x \in E$, and all $A \in \mathscr{E}$

$$P_{s,t}(x, A) = \int P_{s,u}(x, dy) P_{u,t}(y, A). \tag{8.3}$$

The function $P_{s,t}(x, A)$ is called a *Markov transition function* on $(E, \mathscr{E})$. If $P_{s,t}(x, A) = P_{t-s}(x, A)$, the Markov transition function is said to be *homogeneous*.

D39 Definition. Let X_t be an E-valued $(P, \mathscr{F}_t)$-Markov process, and $P_{t,s}(x, A)$ be a Markov transition function on $(E, \mathscr{E})$. If

$$E[f(X_t)|\mathscr{F}_s] = \int P_{s,t}(X_s, dy) f(y) \tag{8.4}$$

for all $0 \leq s \leq t$ and all bounded measurable $f: E \to R$, then one says that the $(P, \mathscr{F}_t)$-Markov process X_t *admits the Markov transition function* $P_{s,t}(x, A)$. If $P_{s,t}(x, A)$ is homogeneous, one says that X_t is a *homogeneous* $(P, \mathscr{F}_t)$-*Markov process* admitting the Markov transition function $P_{t-s}(x, A)$.

The meaning of $P_{s,t}(x, A)$ in D38 is clear; indeed, if we take $f = 1_A$, $A \in \mathscr{E}$, then (8.4) reads

$$P_{s,t}(X_t, A) = P[X_t \in A | \mathscr{F}_s]. \tag{8.4'}$$

Therefore $P_{s,t}(x, A)$ is the probability that starting from x at time s the process X_t ends up at time t in the set A, conditionally to $\mathscr{F}_s$ and $X_s = x$. Equation (8.3), known as the *Chapman–Kolmogorov equation*, expresses the intuitive fact that a transition $X_s = x \to X_t \in A$ can be decomposed into infinitesimal disjoint transitions $X_s = x \to X_u \in dy \to X_t \in A$, and that the corresponding probability is obtained by summation of the infinitesimal probabilities of the latter transitions.

EXAMPLE. Let X_t be a R-valued process defined on $(\Omega, \mathscr{F}, P)$ and with P-*independent increments*, i.e., for all sequences

$$0 \leq t_1 \leq t_2 \leq \cdots \leq t_n$$

the random variables

$$X_{t_2} - X_{t_1}, \ldots, X_{t_n} - X_{t_{n-1}}$$

are P-independent. Then X_t is a $(P, \mathscr{F}_t^X)$-Markov process, and this can be proven using part (β) of the following:

T40 Theorem. (α) *Let X_t be an E-valued stochastic process defined on $(\Omega, \mathscr{F}, P)$ and adapted to some history $\mathscr{F}_t$. The following statements are equivalent*:

(i) X_t is a $(P, \mathscr{F}_t)$-Markov process.
(ii) For each $t \geq 0$ and each bounded $\sigma(X_s, s \geq t)$-measurable random variable Y,

$$E[Y|\mathscr{F}_t] = E[Y|\sigma(X_t)]. \tag{8.5}$$

(iii) *For all $0 \leq s \leq t$ and all bounded measurable mappings $f: E \to R$,*

$$E[f(X_t)|\mathscr{F}_s] = E[f(X_t)|\sigma(X_s)]. \tag{8.6}$$

(β) X_t *is a $(P, \mathscr{F}_t)$-Markov process iff for each $0 \leq t_1 \leq \cdots \leq t_n \leq t$ and each bounded measurable mapping $f: E \to R$,*

$$E[f(X_t)|\sigma(X_{t_1}, \ldots, X_{t_n})] = E[f(X_t)|\sigma(X_{t_n})]. \tag{8.7}$$

Consider a homogeneous Markov process X_t admitting a (homogeneous) transition function $P_{t-s}(x, A)$. Define for each $t \geq 0$ the operator P_t, mapping onto itself the set of bounded measurable functions $f: E \to R$, by

$$P_t f(X) = \int_E P_t(x, dy) f(y). \tag{8.8}$$

Clearly, by the Chapman–Kolmogorov equations (8.3),

$$P_t P_s = P_{t+s}. \tag{8.9}$$

The family $(P_t, t \geq 0)$ is called the *transition semigroup* associated to X_t.

Markov Chains

Here $E = N_+$, and for each $t \geq 0$, P_t is described by an infinite matrix:

$$\{P_{ij}(t), i, j \in N_+\},$$

with the obvious interpretation that

$$P_{ij}(t) = P[X_{t+s} = j | X_s = i]. \tag{8.10}$$

The process X_t is then called a *homogeneous Markov chain*.

Infinitesimal Characteristics. By imposing a continuity condition:

$$\lim_{t \downarrow 0} P_{ij}(t) = \begin{cases} 0 & \text{if } i \neq j, \\ 1 & \text{if } i = j, \end{cases} \tag{8.11}$$

it can be shown that there exist limits for $[1 - P_{ii}(t)]/t$ and $P_{ij}(t)/t$ as t goes to 0:

$$\lim_{t \downarrow 0} \frac{1 - P_{ii}(t)}{t} = q_i \leq \infty,$$

$$\lim_{t \downarrow 0} \frac{P_{ij}(t)}{t} = q_{ij} < \infty. \tag{8.12}$$

If for each $i \in N_+$

$$q_i < \infty, \tag{8.13}$$

8. Markov Processes

the corresponding homogeneous Markov process is said to be *stable*. If moreover

$$q_i = \sum_{j \neq i} q_{ij}, \tag{8.14}$$

it is said to be *conservative*. The interpretation of the q_i's and q_{ij}'s is the following, in the case where X_t is right-continuous: if we denote by T_n the nth jump time of X_t with the convention $T_0 = 0$, then

$$P[T_{n+1} - T_n \geq t | \mathscr{F}^X_{T_n}] = 1(t \geq 0) \exp\{-tq_{X_{T_n}}\},$$

$$P[X_{T_{n+1}} = i, T_{n+1} - T_n \geq t | \mathscr{F}^X_{T_n}]$$

$$= 1(t \geq 0) \exp\{-tq_{X_{T_n}}\} \left(\frac{q_{X_{T_n} i}}{q_{X_{T_n}}}\right). \tag{8.15}$$

The set of parameters $Q = \{q_{ij}, i \in N_+, j \in N_+\}$ with the convention $q_{ii} = -q_i$ is the *Q-matrix* of the semigroup $(P_t, t \geq 0)$.

The q_{ij}'s are also called the infinitesimal characteristics of X_t.

Kolmogorov Equations. In the case where E is finite, the conditions (8.13) and (8.14) are always verified. For a countable-state-space homogeneous Markov process X_t with transition semigroup satisfying (8.11)–(8.14), the $P_{ij}(t)$'s satisfy the *forward Kolmogorov equations*:

$$P'_{ik}(t) = -P_{ik}(t)q_k + \sum_{j \neq k} P_{ij}(t)q_{jk}, \quad i \geq 0, \quad k \geq 0. \tag{8.16}$$

The *backward Kolmogorov equations*

$$P'_{ik}(t) = -q_i P_{ik}(t) + \sum_{j \neq i} q_{ij} P_{jk}(t), \quad i \geq 0, \quad k \geq 0, \tag{8.17}$$

are not always verified, even under the conditions (8.11)–(8.14). However, in the case of a finite state space, both the backward and the forward Kolmogorov equations are satisfied.

The distribution of X_t at time t, $P_k(t) = P[X_t = k]$, satisfies the equation

$$P'_k(t) = -P_k(t)q_k + \sum_{j \neq k} P_j(t)q_{jk}, \quad k \geq 0, \tag{8.18}$$

as can be seen by summing (8.16) over i.

If the initial distribution $P_0(k) = \Pi_k$ satisfies

$$\Pi_k q_k = \sum_{j \neq k} \Pi_j q_{jk}, \quad k \geq 0, \tag{8.19}$$

then clearly $P_k(t) = \Pi_k$ for all $t \geq 0$ and all k. The distribution Π_k is called a stationary distribution, or equilibrium distribution, of X_t. It is not always unique.

EXAMPLE (Birth-and-Death Process). A birth-and-death process is a Markov chain X_t with a Q-matrix having for nonzero elements

$$q_{n,n+1} = \lambda_n, \qquad n \geq 0, \tag{8.20}$$
$$q_{n+1,n} = \mu_{n+1}, \qquad n \geq 0.$$

If $\lambda_n \mu_{n+1} > 0$ for all $n > 0$, then the unique equilibrium distribution, when it exists, is given by

$$\Pi_n = C r_n, \tag{8.21}$$

where

$$r_0 = 1; \qquad r_n = \frac{\lambda_0 \cdots \lambda_n}{\mu_1 \cdots \mu_{n+1}}, \qquad n \geq 1, \tag{8.22}$$

and

$$C = \frac{1}{\sum_{j=0}^{\infty} r_j}. \tag{8.23}$$

A necessary and sufficient condition for this unique equilibrium distribution to exist is

$$\sum_{j=0}^{\infty} r_j < \infty. \tag{8.24}$$

Lévy's Formula. In the case where

$$q_i \leq K < \infty, \qquad i \geq 0,$$

there is no explosion, i.e., $T_n \uparrow \infty$, P-a.s. In that case, it can be shown that if X_t is a right-continuous $\mathscr{F}_t$-Markov process corresponding to the semigroup $(P_t, t \geq 0)$, then for all bounded functions $f: N_+^2 \to R$

$$E\left[\sum_{0 < s \leq t} f(X_{s-}, X_{s+})\right] = E\left[\int_0^t \sum_{i \neq X_s} q_{X_s i} f(X_s, i)\, ds\right], \tag{8.25}$$

or, taking $f(i, j) = g(i) - g(j)$ for some bounded $g: R_+^2 \to R$,

$$E[g(X_t) - g(X_0)] = E\left[\int_0^t \left(\sum_i q_{X_s i} g(i)\right) ds\right]. \tag{8.26}$$

The equality (8.25) is known as Lévy's formula, and (8.26) as Dynkin's formula.

Reversibility in Markov Chains

A stochastic process $(X_t, t \in R)$ is said to be *reversible* if for all $\tau \in R$ and all $t_1, t_2, \ldots, t_n \in R$, the distributions of $(X_{t_1}, \ldots, X_{t_n})$ and $(X_{\tau - t_1}, \ldots, X_{\tau - t_n})$ are the same. In particular the distribution of X_t is independent of t.

Let X_t be a stable and conservative Markov chain at equilibrium: $P[X_t = i] = \Pi(i)$. A necessary and sufficient condition for reversibility is

$$\Pi(i)q_{ij} = \Pi(j)q_{ji}, i, j \in N_+. \tag{8.27}$$

These equations are known as the detailed-balance equations. The above result is due to Kolmogorov {92}.

References

[1] Doob, J. L. (1953) *Stochastic Processes*, Wiley.
[2] Halmos, P. R. (1960) *Measure Theory*, Van Nostrand.
[3] Loève, M. (1963) *Probability Theory* (3rd ed.), Van Nostrand.
[4] Neveu, J. (1965) *Mathematical Foundations of the Calculus of Probability*, Holden Day.

APPENDIX A2

Stopping Times and Point-Process Histories

1. Stopping Times
2. Changes of Time and Meyer–Dellacherie's Integration Formula
3. Point-Process Histories
References

1. Stopping Times[1]

Definitions and Elementary Properties

In Appendix A1, D33, we have already given the definition of a stopping time, which we recall here for convenience:

D1 Definition. Let $\mathscr{F}_t$ be a history defined on $(\Omega, \mathscr{F})$. A $[0, \infty]$-valued random variable T defined on $(\Omega, \mathscr{F})$ is called an $\mathscr{F}_t$-stopping time iff

$$\{T \leq t\} \in \mathscr{F}_t \quad \text{for all } t \geq 0. \tag{1.1}$$

Let us examine a few immediate consequences of the definition. First, if T is an $\mathscr{F}_t$-stopping time, then necessarily

$$\{T < t\} \in \mathscr{F}_t \quad \text{for all } t \geq 0, \tag{1.2}$$

or equivalently, $\{T \geq t\} \in \mathscr{F}_t$ for all $t \geq 0$. Indeed, for each $n \geq 1$, $\{T \leq t - 1/n\} \in \mathscr{F}_{t-1/n} \subset \mathscr{F}_t$, and therefore $\{T < t\} = \bigcup_{n \geq 1} \{T \leq t - 1/n\} \in$

---
[1] In sections 1 and 2, we have borrowed intensively from Dellacherie's monograph [2].

1. Stopping Times

$\mathscr{F}_t$. It should be noted however that (1.2) does not imply (1.1) in general. All that can be said is that (1.2) implies

$$\{T \leq t\} \in \mathscr{F}_{t+} \quad \text{for all } t \geq 0.$$

Indeed, $\{T \leq t\} = \bigcap_{n \geq 1} \{T < t + 1/n\} \subset \bigcap_{n \geq 1} \mathscr{F}_{t+1/n} = \mathscr{F}_{t+}$. However, from the above, one sees that *if $\mathscr{F}_t$ is right-continuous, then (1.1) and (1.2) are equivalent.*

We now list a few obvious properties of stopping times:

(1) Any number a in $[0, \infty]$ is a stopping time (for any history $\mathscr{F}_t$).
(2) If T is an $\mathscr{F}_t$-stopping time, and $a \in [0, \infty]$, then $T + a$ is an $\mathscr{F}_t$-stopping time; indeed, $\{T + a \leq t\} = \phi$: if $a > t$, $= \{T \leq t - a\} \in \mathscr{F}_t$ if $a \leq t$.
(3) If T and S are $\mathscr{F}_t$-stopping times, then $S \wedge T = \inf(S, T)$ and $S \vee T = \sup(S, T)$ are $\mathscr{F}_t$-stopping times, since $\{S \wedge T \leq t\} = \{S \leq t\} \cup \{T \leq t\}$ and $\{S \vee T \leq t\} = \{S \leq t\} \cap \{T \leq t\}$.
(4) If T is an $\mathscr{F}_t$-stopping time, then for any $a \in [0, \infty)$, $T \wedge a$ is $\mathscr{F}_a$-measurable; indeed, if $a \geq t$, $\{T \wedge a \leq t\} = \{T \leq t\} \in \mathscr{F}_t \subset \mathscr{F}_a$, and if $a \leq t$, $\{T \wedge a \leq t\} = \Omega \in \mathscr{F}_a$.

T2 Theorem. *Let $(T_n, n \geq 1)$ be a family of $\mathscr{F}_t$-stopping times. Then $\sup_{n \geq 1} T_n$ is an $\mathscr{F}_t$-stopping time and $\inf_{n \geq 1} T_n$ is an $\mathscr{F}_{t+}$-stopping time.*

Remark. In particular, if $\mathscr{F}_t$ is right continuous, then $\sup T_n$, $\inf T_n$, $\limsup T_n$, $\liminf T_n$, and $\lim T_n$ (whenever it exists) are $\mathscr{F}_t$-stopping times.

PROOF OF T2. The proof follows from

$$\left\{\sup_{n \geq 1} T_n \leq t\right\} = \bigcap_{n \geq 1} \{T_n \leq t\} \in \mathscr{F}_t$$

$$\left\{\inf_{n \geq 1} T_n \leq t\right\} = \bigcap_{m \geq 1} \bigcup_{n \geq 1} \left\{T_n < t + \frac{1}{m}\right\} \in \bigcap_{m \geq 1} \mathscr{F}_{t+1/m} = \mathscr{F}_{t+}. \quad \square$$

We now recall the definitions of the σ-fields which are related respectively to the *past* and the *strict past at time T*, where T is a stopping time:

D3 Definition. Let T be an $\mathscr{F}_t$-stopping time. The *past at time T* (relative to $\mathscr{F}_t$) is the σ-field $\mathscr{F}_T$ defined by

$$\mathscr{F}_T = \{A \in \mathscr{F}_\infty | A \cap \{T \leq t\} \in \mathscr{F}_t \text{ for all } t \geq 0\}. \tag{1.3}$$

The *strict past at time T* (relative to $\mathscr{F}_t$) is the σ-field $\mathscr{F}_{T-}$ defined in terms of generators by

$$\mathscr{F}_{T-} = \sigma\{A_0 \in \mathscr{F}_0; A_s \cap \{s < T\}, s \geq 0, A_s \in \mathscr{F}_s\}. \tag{1.4}$$

In the above definition, we have claimed that $\mathscr{F}_T$ defined by (1.3) is a σ-field. This can be checked: the stability under countable union or countable intersection is obvious, and if $A \in \mathscr{F}_T$, then $\bar{A} \in \mathscr{F}_T$, since $\bar{A} \cap \{T \leq t\} = \{T \leq t\} - A \cap \{T \leq t\}$ is in $\mathscr{F}_t$ for all $t \geq 0$.

Inclusion Properties

T4 Theorem. *If T is an $\mathscr{F}_t$-stopping time, then $\mathscr{F}_{T-} \subset \mathscr{F}_T$ and T is $\mathscr{F}_{T-}$-measurable (therefore $\mathscr{F}_T$-measurable).*

Proof. In order to prove that $\mathscr{F}_{T-} \subset \mathscr{F}_T$, it suffices to show that the generators of $\mathscr{F}_{T-}$ are in $\mathscr{F}_T$; this is true because $\mathscr{F}_0 \subset \mathscr{F}_T$ and $A_s \cap \{s < T\} \cap \{T \leq t\} = A_s \cap \{s < T \leq t\}$ is in $\mathscr{F}_t$ for all $s, t \in [0, \infty)$ whenever $A_s \in \mathscr{F}_s$.

The sets $\{T = 0\}$ and $\{T > a\}$, $a \in [0, \infty)$, are generators of $\mathscr{F}_{T-}$, and therefore T is $\mathscr{F}_{T-}$-measurable. □

T5 Theorem. *Let T be an $\mathscr{F}_t$-stopping time, and S be a $[0, \infty]$-valued $\mathscr{F}_T$-measurable random variable such that $S \geq T$; then S is an $\mathscr{F}_t$-stopping time.*

Proof. $\{S \leq t\} = \{S \leq t\} \cap \{T \leq t\} + \{S \leq t\} \cap \{T > t\}$. Since $S \geq T$, $\{S \leq t\} \cap \{T > t\} = \emptyset$. Since S is $\mathscr{F}_T$-measurable, we have $\{S \leq t\} \in \mathscr{F}_T$, and therefore, by the definition of $\mathscr{F}_T$, $\{S \leq t\} \cap \{T \leq t\} \in \mathscr{F}_t$. □

An immediate consequence of the above result is:

C6 Corollary. *Any $\mathscr{F}_t$-stopping time can be approximated by a decreasing sequence $(S_n, n \geq 1)$ of $\mathscr{F}_t$-stopping times taking a countable number of values.*

Proof. Take

$$S_n = +\infty \cdot 1(T = +\infty) + \sum_{k \geq 1} \frac{k}{2^n} 1(k - 1 \leq 2^n T < k).$$
□

T7 Theorem. *Let S and T be two $\mathscr{F}_t$-stopping times. For all $A \in \mathscr{F}_S$, we have $A \cap \{S \leq T\} \in \mathscr{F}_T$ and $A \cap \{S < T\} \in \mathscr{F}_{T-}$.*

Proof. For all $t \geq 0$,

$$A \cap \{S \leq T\} \cap \{T \leq t\} = A \cap \{S \leq t\} \cap \{T \leq t\} \cap \{S \wedge t \leq T \wedge t\}.$$

Also, $\{S \wedge t \leq T \wedge t\} \in \mathscr{F}_t$, since $S \wedge t$ and $T \wedge t$ are $\mathscr{F}_t$-measurable (see remarks after Definition D1), $A \cap \{S \leq t\} \in \mathscr{F}_t$ (because $A \in \mathscr{F}_S$), and $\{T \leq t\} \in \mathscr{F}_t$. Thus $A \cap \{S \leq T\} \cap \{T \leq t\} \in \mathscr{F}_t$ for all $t \geq 0$, and therefore $A \cap \{S \leq T\} \in \mathscr{F}_T$ by the definition of $\mathscr{F}_T$ (D3).

$A \cap \{S < T\} = \bigcup_{r \in Q} A \cap \{S \leq r\} \cap \{r < T\} \in \mathscr{F}_{T-}$, since for each $r \in Q$ we have $A \cap \{S \leq r\} \in \mathscr{F}_r$, and therefore $A \cap \{S \leq r\} \cap \{r < T\}$ is a generator of $\mathscr{F}_{T-}$, by the definition of $\mathscr{F}_{T-}$ (D3). □

Remarks. In particular, $\{S \leq T\}$ and $\{S < T\}$ are in $\mathscr{F}_T$; therefore also $\{S = T\}$, $\{S \geq T\}$, and $\{S > T\}$. By symmetry, these events are also in $\mathscr{F}_S$. The event $\{S < T\}$ is in $\mathscr{F}_{T-}$ (and also in $\mathscr{F}_T$ and $\mathscr{F}_S$); however, it does not necessarily belong to $\mathscr{F}_{S-}$.

T8 Theorem. *Let S and T be two $\mathscr{F}_t$-stopping times such that $S \leq T$. Then $\mathscr{F}_S \subset \mathscr{F}_T$ and $\mathscr{F}_{S-} \subset \mathscr{F}_{T-}$.*

PROOF. The inclusion $\mathscr{F}_S \subset \mathscr{F}_T$ follows from the fact that $A \cap \{S \leq T\} \in \mathscr{F}_S$ for all $A \in \mathscr{F}_S$ and from $\{S \leq T\} = \Omega$. As to the inclusion $\mathscr{F}_{S-} \subset \mathscr{F}_{T-}$, it is a consequence of the fact that the generators of $\mathscr{F}_{S-}$ are also generators of $\mathscr{F}_{T-}$; indeed, if $A \in \mathscr{F}_t$, then $A \cap \{t < S\} = A \cap \{t < S\} \cap \{t < T\} = B \cap \{t < T\}$, where $B \in \mathscr{F}_t$. □

Continuity Properties

T9 Theorem. *Let $(T_n, n \geq 1)$ be a monotone sequence of $\mathscr{F}_t$-stopping times, where $\mathscr{F}_t$ is a right-continuous history. Then:*

(a) *if $(T_n, n \geq 1)$ is decreasing, $\mathscr{F}_T = \bigcap_{n \geq 1} \mathscr{F}_{T_n}$;*
(b) *if $(T_n, n \geq 1)$ is increasing, $\mathscr{F}_{T-} = \bigvee_{n \geq 1} \mathscr{F}_{T_n-}$.*

Remark. $T = \lim T_n$ is an $\mathscr{F}_t$-stopping time, by the remark following T2.

PROOF

(a): By T8, $\mathscr{F}_{T_n} \supset \mathscr{F}_T$ and therefore $\bigcap_{n \geq 1} \mathscr{F}_{T_n} \supset \mathscr{F}_T$. Conversely, let $A \in \mathscr{F}_{T_n}$ for all $n \geq 1$. Hence for all $n \geq 1$ and all t, we have $A \cap \{T_n < t\} \in \mathscr{F}_t$; therefore $A \cap \{T < t\} \in \mathscr{F}_t$ for all $t \geq 0$. Since $\mathscr{F}_t$ is right-continuous, this implies $A \cap \{T \leq t\} \in \mathscr{F}_t$ for all $t \geq 0$; therefore $A \in \mathscr{F}_T$ by D3.

(b) By T8, $\mathscr{F}_{T_n-} \subset \mathscr{F}_{T-}$ and therefore $\bigvee_{n \geq 1} \mathscr{F}_{T_n-} \subset \mathscr{F}_{T-}$. Conversely, let $A_s \cap \{s < T\}$ be a generator of $\mathscr{F}_{T-}$, where $A_s \in \mathscr{F}_s$. This generator is contained in $\bigvee_{n \geq 1} \mathscr{F}_{T_n-}$ since $A_s \cap \{s < T\} = \lim_n A_s \cap \{s < T_n\}$. □

T10 Theorem. *Same hypotheses as in T9 (in particular, $\mathscr{F}_t$ is a right-continuous history).*

(a) *If $(T_n, n \geq 1)$ is decreasing and if for all $n \geq 1$, $T < T_n$ on $\{0 < T_n < \infty\}$, then $\mathscr{F}_T = \bigcap_{n \geq 1} \mathscr{F}_{T_n-}$.*
(b) *If $(T_n, n \geq 1)$ is increasing and if for all $n \geq 1$, $T_n < T$ on $\{0 < T < \infty\}$, then $\mathscr{F}_{T-} = \bigvee_{n \geq 1} \mathscr{F}_{T_n}$.*

PROOF. T10 is a consequence of T9 and of the next two results. □

T11 Theorem. *If T is an $\mathscr{F}_t$-stopping time and A belongs to $\mathscr{F}_\infty = \bigvee_{t \geq 0} \mathscr{F}_t$, the event $A \cap \{T = \infty\}$ is in $\mathscr{F}_{T-}$.*

PROOF. The family $\mathcal{G} = \{B \in \mathcal{F}_\infty | B \cap \{T = \infty\} \in \mathcal{F}_{T-}\}$ is a σ-field. If we can show that for all $n \geq 1$, $\mathcal{F}_n \subset \mathcal{G}$, then $\mathcal{F}_\infty \subset \mathcal{G}$, q.e.d. Let then A be in $\mathcal{F}_n$. We have $A \cap \{T = \infty\} = \bigcap_{m \geq n} (A \cap \{m < T\}) \in \mathcal{F}_{T-}$, since $A \cap \{m < T\}$ is a generator of $\mathcal{F}_{T-}$ whenever $m \geq n$. This terminates the proof. □

T12 Theorem. *Let S and T be two $\mathcal{F}_t$-stopping times such that $S \leq T$. If moreover $S < T$ on $\{0 < T < \infty\}$, then $\mathcal{F}_S \subset \mathcal{F}_{T-}$.*

PROOF. For all $A \in \mathcal{F}_S$, A can be written $(A \cap \{S = 0\}) \cup \{S < T\} \cup (A \cap \{T = \infty\})$. But $A \cap \{S = 0\}$ is in $\mathcal{F}_0$ by the definition of $\mathcal{F}_S$, $\{S < T\}$ is in $\mathcal{F}_{T-}$ by T7, and $A \cap \{T = \infty\}$ is in $\mathcal{F}_{T-}$ by T11. □

2. Changes of Time and Meyer–Dellacherie's Integration Formula

Changes of Time

D13 Definition. Let $\mathcal{F}_t$ be a given history. A right-continuous increasing $[0, \infty]$-valued process T_t such that for each $s \geq 0$, T_s is a $\mathcal{F}_t$-stopping time is called an $\mathcal{F}_t$-*change of time*.

T14 Theorem. *Let T_t be an $\mathcal{F}_t$-change of time, where $\mathcal{F}_t$ is a right-continuous history. The family $\mathcal{F}_{T_t}$ is a right-continuous history. If X_t is an $\mathcal{F}_t$-progressive process, then the process $Y_t = X_{T_t} 1(T_t < \infty)$ is adapted to $\mathcal{F}_{T_t}$.*

PROOF. The right continuity of $\mathcal{F}_{T_t}$ follows from the right continuity of $\mathcal{F}_t$, the right continuity and monotony of the process T_t, and T9, part (a). The adaptation of X_{T_t} follows from what has been recalled in Appendix A1, §7, i.e., $X_T 1(T < \infty)$ is $\mathcal{F}_T$-measurable whenever X_t is $\mathcal{F}_t$-progressive and T is an $\mathcal{F}_t$-stopping time. □

Remark. If X_t is right-continuous, X_{T_t} is right-continuous (recall that T_t is assumed right-continuous).

From now on we assume that $\mathcal{F}_t$ is right-continuous.
We now define increasing processes and their associated changes of times.

D15 Definition. A process A_t defined on $(\Omega, \mathcal{F}, P)$ is called an *increasing process* if its trajectories are P-a.s. right-continuous, increasing, and such that $A_0 \equiv 0$.

2. Changes of Time and Meyer–Dellacherie's Integration Formula

With an increasing process A_t one can associate the "inverse" process C_t defined by

$$C_t(\omega) = \inf\{s \mid A_s(\omega) > t\}. \tag{2.1}$$

If A_t is adapted to $\mathscr{F}_t$, then C_t is an $\mathscr{F}_t$-change of time. Indeed, for all $t \geq 0$

$$\{C_t < s\} = \bigcup_{n \geq 1} \{A_{s-1/n} > t\} \in \bigvee_{n \geq 1} \mathscr{F}_{s-1/n} = \mathscr{F}_{s-} \subset \mathscr{F}_s.$$

The two next results concern changes of times in Stieltjes integrals and are preliminaries for the integration-by-parts formula of next subsection.

T16 Theorem. *Let $t \to a(t)$ be a nonnegative right-continuous increasing function defined on $[0, \infty)$ and $[0, \infty]$-valued. For all $t \in [0, \infty)$ define*

$$c(t) = \inf\{s \mid a(s) > t\}. \tag{2.2}$$

The function $t \to c(t)$ is nonnegative, right-continuous, and increasing, and $c(t) < \infty$ holds if and only if $a(\infty) > t$. Also, for all $s \in [0, \infty)$,

$$a(s) = \inf\{t \mid c(t) > s\}. \tag{2.3}$$

Suppose moreover that a is $[0, \infty)$-valued and that $a(0) = 0$. Then for all nonnegative borelian functions f defined on $[0, \infty)$,

$$\int_0^\infty f(t)\, da(t) = \int_0^{a(\infty)} f(c(t))\, dt = \int_0^\infty 1(c(t) < \infty) f(c(t))\, dt. \tag{2.4}$$

PROOF. See Dellacherie [2]. □

C17 Corollary. *Let $t \to a(t)$ be a nonnegative continuous increasing function from $[0, \infty)$ into $[0, \infty)$ and such that $a(0) = 0$. For all nonnegative borelian functions g defined on $[0, \infty)$,*

$$\int_0^\infty g(a(t))\, da(t) = \int_0^{a(\infty)} g(t)\, dt. \tag{2.5}$$

PROOF. Remark that since $t \to a(t)$ is continuous, $a(c(t)) = t$ if $t \in [0, a(\infty))$, and apply the equality (2.4) to $f = g \circ a$. □

Integration-by-Parts Formula for Martingales and Increasing Processes

T18 Theorem. *Let X_t and Y_t be two nonnegative measurable processes such that for any $\mathscr{F}_t$-stopping time S,*

$$E[X_S 1(S < \infty)] = E[Y_S 1(S < \infty)]. \tag{2.6}$$

Then for any increasing process A_t adapted to $\mathscr{F}_t$ and any $\mathscr{F}_t$-stopping time T,

$$E\left[\int_0^T X_s \, dA_s\right] = E\left[\int_0^T Y_s \, dA_s\right]. \tag{2.7}$$

PROOF

First case: $T \equiv \infty$. By T16 [Equation (2.4)] and Fubini's theorem,

$$E\left[\int_0^\infty X_s \, dA_s\right] = E\left[\int_0^\infty X_{C_s} 1(C_s < \infty) \, ds\right] = \int_0^\infty E[X_{C_s} 1(C_s < \infty) \, ds],$$

where C_t is defined by (2.1). The process Y_t verifies a similar equation, and this suffices to prove the theorem in this particular case.

General case. Replace A_t by B_t, where

$$B_t(\omega) = \begin{cases} A_t(\omega) & \text{if } t < T(\omega), \\ A_T(\omega) & \text{if } t \geq T(\omega). \end{cases} \qquad \square$$

The main result of this section is:

T19 Theorem [3]. *Let A_t be an increasing process adapted to $\mathscr{F}_t$, and let M_t be a nonnegative $(P, \mathscr{F}_t)$-martingale, right-continuous and uniformly integrable. Then for any $\mathscr{F}_t$-stopping time T*

$$E\left[\int_0^T M_t \, dA_t\right] = E[M_T A_T]. \tag{2.8}$$

Before proving T19, we recall a few facts from continuous-time martingale theory; for proofs see Meyer [6].

T20 Theorem. *If $(\Omega, P, \mathscr{F}_t)$ satisfies the usual conditions [i.e., $(\Omega, \mathscr{F}, P)$ is complete, $\mathscr{F}_t$ is right-continuous, and $\mathscr{F}_0$ contains the P-null sets of $\mathscr{F}_\infty$] and if M_t is a $(P, \mathscr{F}_t)$-martingale, there always exists a right-continuous modification of M_t; that is to say, there exists a right-continuous process $\tilde{M}_t$ such that $M_t(\omega) = \tilde{M}_t(\omega)$, P-a.s., for all $t \in [0, \infty)$, and $\tilde{M}_t$ is also a $(P, \mathscr{F}_t)$-martingale.*

T21 Theorem. *If M_t is a right-continuous $(P, \mathscr{F}_t)$-martingale which is uniformly integrable [i.e., the family $(M_t, t \geq 0)$ is u.i.], then there exists a P-integrable random variable M_∞ such that $M_\infty = \lim_{t \uparrow \infty} M_t$, P-a.s., and $M_t = E[M_\infty | \mathscr{F}_t]$.*

The latter result gives a meaning to M_T when T is an $\mathscr{F}_t$-stopping time (which, we recall, can take the value $+\infty$).

We now proceed to the

PROOF OF T19. Define X_t and Y_t by $X_t = M_t 1(t \leq T)$ and $Y_t = M_T 1(t \leq T)$. By Doob's optional-sampling theorem (I, T2)

$$E[X_S 1(S < \infty)] = E[Y_S 1(S < \infty)]$$

for any $\mathscr{F}_t$-stopping time S. Theorem T18 therefore applies, yielding the announced result. □

3. Point-Process Histories

Recall the definition

D22 Definition. Let $(\Omega, \mathscr{F})$ be a measurable space, and $(T_n, n \geq 1)$ and $(Z_n, n \geq 1)$ be two sequences of random variables on $(\Omega, \mathscr{F})$. The T_n's are real-valued, and the Z_n's take their values in some measurable space $(E, \mathscr{E})$. Suppose moreover that whenever $T_n < \infty$, then $0 < T_n \leq T_{n+1}$. The sequence $(T_n, Z_n, n \geq 1)$ is called an *E-marked point process*.

Also recall the following conventions, notation, and definitions:

$$T_\infty = \lim \uparrow T_n \quad \text{[conventions:} \quad T_0 \equiv 0, \quad Z_0 \equiv \delta \text{ (some point)]},$$

$$N_t(A) = \sum_{n \geq 1} 1(T_n \leq t) 1(Z_n \in A), \quad A \in \mathscr{E},$$

$$N_t(E) = N_t,$$

$$p((0, t] \times A) = N_t(A), \quad (3.1)$$

$$\mathscr{F}_t^p = \sigma(N_s(A); 0 \leq s \leq t, A \in \mathscr{E}),$$

$$S_{n+1} = \begin{cases} T_{n+1} - T_n & \text{if } T_n < \infty, \\ \infty & \text{if } T_n = \infty. \end{cases}$$

A few obvious facts are gathered in the following statement:

T23 Theorem. *Let $(T_n, Z_n, n \geq 1)$ be an E-marked point process defined on $(\Omega, \mathscr{F})$. With the notation given in (3.1):*

(i) *for all $n \geq 0$, T_n is an $\mathscr{F}_t^p$-stopping time,*
(ii) *for all $n \geq 1$, $\sigma(T_i, Z_i; 1 \leq i \leq n) \subset \mathscr{F}_{T_n}^p$,*
(iii) *for all $t \geq 0$, $\mathscr{F}_t^p = \sigma(1(Z_n \in A) 1(T_n \leq s); n \geq 1, 0 \leq s \leq t, A \in \mathscr{E})$.*

PROOF. (i) Observe that $\{T_n \leq t\} = \{N_t \geq n\}$.

(ii) By T4, T_i is $\mathscr{F}_{T_i}^p$-measurable for all $1 \leq i$; if moreover $i \leq n$, then $\mathscr{F}_{T_i}^p \subset \mathscr{F}_{T_n}^p$, by T8; hence for all $1 \leq i \leq n$, T_i is $\mathscr{F}_{T_n}^p$-measurable. Now, for any $1 \leq i$, $A \in \mathscr{E}$, we have $1(Z_i \in A) = N_{T_i}(A) - N_{T_{i-1}}(A)$; also, $N_t(A)$ is a right-continuous process adapted to $\mathscr{F}_t^p$ and hence $\mathscr{F}_t^p$-progressive (Appendix A1, T32), and A1, T35, $N_{T_i}(A)$ is $\mathscr{F}_{T_i}^p$-measurable for all $i \geq 0$. Therefore $1(Z_i \in A)$ is $\mathscr{F}_{T_n}^p$-measurable for all $1 \leq i \leq n$.

(iii) The inclusion $\supset$ is immediate. The inclusion $\subset$ follows from the observation that for all $0 \leq s \leq t$, $A \in \mathscr{E}$, we have

$$\{T_n \leq s\} \cap \{Z_n \in A\} = \{T_n \wedge t \leq s\} \cap \{T_n \leq t\} \cap \{Z_n \in A\}. \quad \square$$

In order to understand the next statement, one needs to know the definition of predictability; see I, §3.

T24 Theorem. Let $(T_n, Z_n, n \geq 1)$ be an E-marked point process defined on $(\Omega, \mathscr{F})$. With the notation given in (3.1), the following holds:
(i) The process $X_t = 1(T_\infty \leq t)$ is $\mathscr{F}_t^p$-predictable.
(ii) For each $n \geq 0$, let $f^{(n)}$ be a mapping from $\Omega \times [0, \infty)$ into R_+ that is $\mathscr{F}_{T_n}^p \otimes \mathscr{B}_+$-measurable; then the process X_t defined by

$$X_t(\omega) = \sum_{n \geq 1} f^{(n)}(t, \omega) 1(T_n(\omega) < t \leq T_{n+1}(\omega)) \tag{3.2}$$

is $\mathscr{F}_t^p$-predictable.

PROOF. (i) Observe that

$$X_t(\omega) = \lim \downarrow 1(T_n(\omega) < t),$$

and that for each $n \geq 1$, the process $1(T_n < t)$ is adapted to $\mathscr{F}_t^p$ and left-continuous, and hence $\mathscr{F}_t^p$-predictable, by I, T5.

(ii) If suffices to show that for each $n \geq 0$, $f^{(n)}(t) 1(T_n < t \leq T_{n+1})$ is $\mathscr{F}_t^p$-predictable; by the usual arguments, it is enough to show this for $f^{(n)}$ of the form

$$f^{(n)}(t, \omega) = 1_A(\omega) 1_B(t), \qquad A \in \mathscr{F}_{T_n}^p, \quad B \in \mathscr{B}_+.$$

Then, defining $T_n^A(\omega) = +\infty \cdot 1_{A^c}(\omega) + T_n \cdot 1_A(\omega)$,

$$f^{(n)}(t, \omega) 1(T_n < t \leq T_{n+1}) = 1_B(t) 1(T_n^A < t \leq T_{n+1}^A).$$

But the deterministic process $1_B(t)$ is $\mathscr{F}_t^p$-predictable. As for $1(T_n^A < t \leq T_{n+1}^A)$, it is a left-continuous process; the theorem would be proven if one could argue that T_n^A and T_{n+1}^A are $\mathscr{F}_t^p$-stopping times. Since $A \in \mathscr{F}_{T_n}^p \subset \mathscr{F}_{T_{n+1}}^p$ for all $n \geq 0$, it suffices to show that for all $n \geq 0$ and all $A \in \mathscr{F}_{T_n}^p$, T_n^A is an $\mathscr{F}_t^p$-stopping time. This verification is easy, since

$$\{T_n^A \leq t\} = A \cap \{T_n \leq t\} \in \mathscr{F}_t^p,$$

by the definition of $\mathscr{F}_{T_n}^p$. □

T25 Theorem. Let $(T_n, Z_n, n \geq 1)$ be a E-marked point process defined on $(\Omega, \mathscr{F})$. The internal history as defined in (3.1) is right-continuous, i.e. $\mathscr{F}_t^p = \bigcap_{h > 0} \mathscr{F}_{t+h}^p$.

This theorem is a special case of a more general result:

T26 Theorem. Let Y_t be an $(E, \mathscr{E})$-valued process defined on $(\Omega, \mathscr{F})$, and suppose that for all $t \geq 0$ and all $\omega \in \Omega$, there exists a strictly positive real number $\varepsilon(t, \omega)$ such that

$$Y_{t+s}(\omega) = Y_t(\omega) \quad \text{on } [t, t + \varepsilon(t, \omega)). \tag{3.3}$$

Then the history $\mathscr{F}_t^Y$ is right-continuous.

3. Point-Process Histories   305

PROOF. To be proven: $\mathscr{F}_t^Y = \bigcap_{h>0} \mathscr{F}_{t+h}^Y$, or equivalently, since $\mathscr{F}_t^Y$ increases with t, $\mathscr{F}_t^Y = \bigcap_{n\geq 1} \mathscr{F}_{t+1/n}^Y$. The inclusion $\subset$ is obvious. It therefore remains to show that whenever Z is a nonnegative bounded random variable that is $\mathscr{F}_{t+1/n}^Y$-measurable for all $n \geq 1$, then Z is $\mathscr{F}_t^Y$-measurable.

Using A1, T6 and A1, T7, we are granted the existence of a denumerable subset U of $[0, \infty)$ such that

$$Z = \phi_n(Y_s, s \in U) \tag{3.4}$$

for some measurable mapping $\phi_n : E^{|U|} \to R_+$ with the following property: whenever $(y_s, s \in U)$ and $(\tilde{y}_s, s \in U)$ are such that

$$y_s = \tilde{y}_s \quad \text{when} \quad s \in U \cap \left[0, t + \frac{1}{n}\right],$$

then $\phi_n(y_s, s \in U) = \phi_n(\tilde{y}_s, s \in U)$. This will be represented in symbols by

$$Z = \phi_n(Y_s, s \in U \cap [0, t + 1/n]).$$

Now for each $n \geq 1$, let A_n be defined by

$$A_n = \{\omega \mid Y_{t+h}(\omega) = Y_t(\omega), h \in [0, 1/n]\}. \tag{3.5}$$

By the hypothesis of "strong right continuity" on Y_t [i.e. (3.3)],

$$A_n \uparrow \Omega. \tag{3.6}$$

Also

$$Z 1_{A_n} = \psi_n(Y_s, s \in U \cap [0, t]) 1_{A_n}, \tag{3.7}$$

where, for instance,

$$\psi_n(y_s, s \in U \cap [0, t]) = \phi_n(\tilde{y}_s, s \in U \cap [0, t + 1/n]) \tag{3.8}$$

where $(\tilde{y}_s, s \in U \cap [0, t + 1/n]) = (y_s, s \in U \cap [0, t]; \bar{y}_s, s \in U \cap (t, t + 1/n])$ and $[\bar{y}_s, s \in U \cap (t, t + 1/n])$ is arbitrarily fixed. Define ψ by

$$\psi = \liminf \psi_n. \tag{3.9}$$

Then, by (3.7) and (3.6),

$$Z = \psi(Y_s, s \in U \cap [0, t]), \tag{3.10}$$

that is to say, Z is $\mathscr{F}_t^Y$-measurable. $\square$

PROOF OF T25. In T26 take

$$Y_t = \begin{cases} \delta & \text{if } t \in [0, T_1] \\ Z_n & \text{if } t \in [T_n, T_{n+1}), \\ \delta & \text{if } t \geq T_\infty, \end{cases} \tag{3.11}$$

where δ is some arbitrary point of E, and observe that $\mathscr{F}_t^p \equiv \mathscr{F}_t^Y$. $\square$

T27 Theorem. *Let $(T_n, Z_n, n \geq 1)$ be an E-marked point process defined on $(\Omega, \mathscr{F})$. With the notation of (3.1), for all $\mathscr{F}_t^p$-stopping times T,*

$$\mathscr{F}_T^p = \sigma(N_{s \wedge T}^A; s \geq 0, A \in \mathscr{E}). \tag{3.12}$$

This is a special case of :

T28 Theorem. *Under the conditions of T26, for any $\mathscr{F}_t^Y$-stopping time T:*

$$\mathscr{F}_T^Y = \sigma(Y_{s \wedge T}, s \geq 0). \tag{3.13}$$

PROOF. The inclusion $\supset$ is obvious. We prove the inclusion $\subset$ in two steps.

First case: T has only a denumerable set of values $(a_n, n \geq 0)$ where $0 \leq a_0 < a_1 < \cdots \leq \infty$. Any event $A \in \mathscr{F}_T^Y$ can be decomposed into

$$A = \sum_{p=0}^{\infty} A_p,$$

where $A_p = A \cap \{T = a_p\} \subset \mathscr{F}_{a_p}^Y$. From Appendix A1, T6 and A1, T7, there exists a denumerable set $V \subset [0, \infty]$ and a sequence $(\phi_p, p \geq 0)$ of appropriately measurable mappings such that

$$1_{A_p} = \phi_p(Y_t, t \in V \cap [0, a_p]).$$

Now if $\omega \in A_p$, then $\omega \in \{T = a_p\}$; hence if $t \in [0, a_p]$, then $t \wedge T(\omega) = t$. Therefore for all $p \geq 0$

$$1_{A_p} = \phi_p(Y_{t \wedge T}, t \in V \cap [0, a_p]);$$

in particular, $A_p \in \sigma(Y_{s \wedge T}, s \geq 0)$.

General case: T being an $\mathscr{F}_t^Y$-stopping time, define for all $k \geq 1$ another $\mathscr{F}_t^Y$-stopping time T_k by

$$T_k = \sum_{p \geq 1} \frac{p}{2^{-k}} 1\left(\frac{p-1}{2^{-k}} \leq T < \frac{p}{2^{-k}}\right) + \infty \cdot 1(T = \infty).$$

By the first part of the proof,

$$\mathscr{F}_{T_k}^Y = \sigma(Y_{s \wedge T_k}, s \geq 0).$$

Also, for all $k \geq 1$, we have $T_k \geq T$; therefore $\mathscr{F}_T^Y \subset \mathscr{F}_{T_k}^Y$ and

$$\mathscr{F}_T^Y \subset \sigma(Y_{s \wedge T_k}, s \geq 0) \quad \text{for all } k \geq 1.$$

Define A_k by

$$A_k = \{\omega \mid Y_t = Y_T, t \in [T, T + 2^{-k}]\}.$$

By the "strong right continuity" assumption [(3.3) of T25],

$$A_k \uparrow \Omega$$

on A_k, $Y_{t \wedge T_k} = Y_{t \wedge T}$, and therefore

$$\mathscr{F}_T^Y \cap A_k \subset \sigma(Y_{s \wedge T_k}, s \geq 0) \cap A_k = \sigma(Y_{s \wedge T}, s \geq 0) \cap A_k$$

(where $\mathscr{C} \cap A = \{C \cap A, C \in \mathscr{C}\}$). The rest of the proof follows from

L29 Lemma. *If $\mathscr{F}_1$ and $\mathscr{F}_2$ are two σ-fields on Ω, and $(A_n, n \geq 1)$ is a sequence of subsets of Ω increasing to Ω, and if $\mathscr{F}_1 \cap A_n \subset \mathscr{F}_2 \cap A_n$ for all $n \geq 1$, then $\mathscr{F}_1 \subset \mathscr{F}_2$.*

3. Point-Process Histories

PROOF. Let $C \in \mathscr{F}_1$. By hypothesis, for all $n \geq 1$ there exists $C_n \in \mathscr{F}_2$ such that
$$C \cap A_n = C_n \cap A_n.$$
Writing the above equality with $n + k$ replacing n, and then intersecting with A_n, we obtain
$$C \cap A_n = C_{n+k} \cap A_n, \quad n \geq 1, \quad k \geq 0.$$
Let $D = \varliminf C_n$; then $D \in \mathscr{F}_2$, and by the latter equality
$$C \cap A_n = D \cap A_n.$$
By letting n go to ∞, we obtain $C = D$; hence $C \in \mathscr{F}_2$. □

T30 Theorem. *Let $(T_n, Z_n, n \geq 1)$ be an E-marked point process defined on $(\Omega, \mathscr{F})$. With the notation of (3.1), one has*

(a) $\mathscr{F}^p_{T_n} = \sigma(T_i, Z_i, 0 \leq i \leq n)$,
(b) $\mathscr{F}^p_{T_n-} = \sigma(T_i, Z_i, 0 \leq i \leq n-1; T_n)$,
(c) $\mathscr{F}^p_{T_\infty} = \mathscr{F}^p_{T_\infty-} = \mathscr{F}^p_\infty = \sigma(T_i, Z_i, i \geq 0)$.

PROOF. (a) We have already obtained $\supset$ in T23(i). By T27, it suffices in order to prove the inclusion $\subset$ to show that for all $t \geq 0$ and all $C \in \mathscr{E}$, $N_{t \wedge T_n}(C)$ is $\sigma(T_i, Z_i, 0 \leq i \leq n)$-measurable, and this is obvious.

(b) The inclusion $\supset$ is not difficult. As for $\subset$, we will show that each generator of $\mathscr{F}^p_{T_n-}$, $A \cap \{T_n > t\}$ where $A \in \mathscr{F}^p_t$ $(t \geq 0)$, is $\sigma(T_i, Z_i, 0 \leq i \leq n-1; T_n)$-measurable. It is enough to show this for A of the form $\{N_s(C) = k\}$ where $s \leq t$, $C \in \mathscr{E}$, $k \geq 0$. But then
$$\{N_s(C) = k\} \cap \{T_n > t\} = \{T_n > t\} \cap \left\{ \sum_{i=1}^{n-1} 1(T_i \leq s) 1(Z_i \in C) = k \right\},$$
and it is in $\sigma(T_i, Z_i, 0 \leq i \leq n-1; T_n)$.

(c) By T10 (b)
$$\mathscr{F}^p_{T_\infty-} = \bigvee \mathscr{F}^p_{T_n}.$$
Therefore, using (a) for the first equality,
$$\mathscr{F}^p_{T_\infty-} = \sigma(T_n, Z_n, n \geq 0) \subset \mathscr{F}^p_{T_\infty} \subset \mathscr{F}^p_\infty.$$
And the result is proven since $\mathscr{F}^p_\infty = \sigma(T_n, Z_n, n \geq 0)$. □

T31 Theorem. *If for some $n \geq 1$, $A \in \mathscr{F}_{T_n-}$, there exists a $\mathscr{F}^p$-predictable process X_t such that $X_{T_n} = 1_A$ if $T_n < \infty$ and X_t is null outside $(T_{n-1}, T_n]$.*

PROOF. By (b) of T30 and the representation theorem A1, T6,
$$1_A(\omega) = f(T_1(\omega), Z_1(\omega), \ldots, T_{n-1}(\omega), Z_{n-1}(\omega), T_n(\omega))$$
for some measurable function f. Define
$$X_t = 1(T_{n-1} < t \leq T_n) f(T_1, Z_1, \ldots, T_{n-1}, Z_{n-1}, t).$$

By (ii) of T24, X_t is indeed $\mathscr{F}_t^p$-predictable; also, by the definition of X_t, $X_{T_n} = 1_A$. $\square$

We introduce the convention $T_{\infty+1} = \infty$.

T32 Theorem. *Let $(T_n, Z_n, n \geq 1)$ be an E-marked point process, and let S be a finite $\mathscr{F}_t^p$-stopping time. Then for all $n \in \overline{N}_+$*

$$\mathscr{F}_S^p \cap \{T_n \leq S < T_{n+1}\} = \mathscr{F}_{T_n}^p \cap \{T_n \leq S < T_{n+1}\}. \qquad (3.14)$$

PROOF. It suffices to show that for any generator U of $\mathscr{F}_S^p$ there exists a generator V of $\mathscr{F}_{T_n}^p$ such that $U \cap \{T_n \leq S < T_{n+1}\} = V \cap \{T_n \leq S < T_{n+1}\}$, and vice versa. The observation

$$\{N_{t \wedge S}(C) = k\} \cap \{T_n \leq S < T_{n+1}\} = \{N_{t \wedge T_n}(C) = k\} \cap \{T_n \leq S < T_{n+1}\}$$

therefore constitutes the proof of T32. $\square$

T33 Theorem. *Let S be an $\mathscr{F}_t^p$-stopping time. There exists a sequence $(R_n, n \in \overline{N}_+)$ of R_+-valued $\mathscr{F}_{T_n}^p$-measurable random variables such that on $S \geq T_n$*

$$S \wedge T_{n+1} = (T_n + R_n) \wedge T_{n+1}. \qquad (3.15)$$

PROOF. S is $\mathscr{F}_S^p$-measurable, and by T27, $\mathscr{F}_S^p = \sigma(N_{t \wedge S}(A), t \geq 0, A \in \mathscr{E})$. Therefore, by the countable-dependence theorem A1, T7, there exist a countable set I, times $t_i \geq 0$, sets $A_i \in \mathscr{E}$, and a borelian function $\Phi : R^{|I|} \to [0, \infty]$ such that

$$S = \Phi(N_{t_i \wedge S}(A_i), i \in I). \qquad (*)$$

Now

$$S = \sum_{n=0}^{\infty} S \cdot 1(T_n \leq S < T_{n+1}) + S \cdot 1(T_\infty \leq S).$$

By $(*)$ it is clear that there exist measurable mappings ψ_n ($n \geq 1$), and ψ_∞ such that

$$S \cdot 1(T_n \leq S < T_{n+1}) = \psi_n(T_0, T_1, Z_1, \ldots, T_n, Z_n) 1(T_n \leq S < T_{n+1})$$

and

$$S \cdot 1(T_\infty \leq S) = \psi_\infty(T_0, T_1, Z_1, \ldots) 1(T_\infty \leq S).$$

The announced result follows by taking

$$R_n = (\psi_n(T_0, T_1, Z_1, \ldots, T_n, Z_n) - T_n)^+$$

and

$$R_\infty = (\psi_\infty(T_0, T_1, Z_1, \ldots) - T_\infty)^+. \qquad \square$$

3. Point-Process Histories

T34 Theorem. *In order for the process X_t to be $\mathscr{F}_t^P$-predictable it is necessary and sufficient that it admits the representation*

$$X_t(\omega) = \sum_{n \geq 1} f^{(n)}(t, \omega) 1(T_n(\omega) < t \leq T_{n+1}(\omega))$$

$$+ f^{(\infty)}(t, \omega) 1(T_\infty(\omega) < t < \infty) \quad (3.16)$$

where for each $n \in N_+$ the mapping $(t, \omega) \to f^{(n)}(t, \omega)$ is $\mathscr{F}_{T_n}^P \otimes B_+$-measurable.

PROOF. The sufficiency was proven in T24. By I, E9, it suffices to prove the necessity of representation (3.16) for X_t of the form $X_t(\omega) = 1(t \leq S(\omega))$, where S is a finite $\mathscr{F}_t^P$-stopping time. This is done by using the representation (3.15) of T33.

Completion of the Histories

Let $(\Omega, \mathscr{F}, P)$ be a complete probability space on which is defined a history $\mathscr{F}_t$. The completed history $\bar{\mathscr{F}}_t$ is defined by

$$\bar{\mathscr{F}}_t = \mathscr{F}_t \vee \sigma(\mathscr{N}) = \{A_t \cup N | A_t \in \mathscr{F}_t, N \in \mathscr{N}\}, \quad (3.17)$$

where $\mathscr{N}$ is the set of P-negligible sets of $\mathscr{F}_\infty$.

We want to convince the reader that one can replace $\mathscr{F}_t$ by $\bar{\mathscr{F}}_t$ without changing things "too much."

T35 Theorem. *If $\mathscr{F}_t$ is right-continuous, then $\bar{\mathscr{F}}_t$ is right-continuous.*

PROOF. Let $\bar{A} \in \bar{\mathscr{F}}_{t+}$. In particular $\bar{A} \in \bar{\mathscr{F}}_{t+1/n}$ for all $n \geq 1$. Therefore, for all $n \geq 1$, $\bar{A} = A_{t+1/n} \cup N_n$, where $A_{t+1/n} \in \mathscr{F}_{t+1/n}$, $N_n \in \mathscr{N}$. In other words, $\bar{A} = A_{t+1/n}$, P-a.s., or $1_{\bar{A}} = 1_{A_{t+1/n}}$, P-a.s. Hence $\bar{A} = \varliminf_{n \uparrow \infty} A_{t+1/n}$, P-a.s. (i.e. $1_{\bar{A}} = \varliminf_{n \uparrow \infty} 1_{A_{t+1/n}}$, P-a.s.). But $\varliminf_{n \uparrow \infty} A_{t+1/n} \in \mathscr{F}_{t+} = \mathscr{F}_t$, and therefore $\bar{A} \in \bar{\mathscr{F}}_t$ (being of the form $\varliminf_{n \uparrow \infty} A_{t+1/n} \cup N$). □

T36 Theorem. *Let $\mathscr{F}_t$ be a right-continuous history, and let $\mathscr{N}$ be the family of P-negligible events of $\mathscr{F}_\infty$. Then $\bar{\mathscr{F}}_T = \mathscr{F}_T \vee \sigma(\mathscr{N})$ for any $\mathscr{F}_t$-stopping time T.*

PROOF. The nonobvious part is $\bar{\mathscr{F}}_T \subset \mathscr{F}_T \vee \sigma(\mathscr{N})$. To prove this, we will show that for any $A \in \bar{\mathscr{F}}_T$, A is P-a.s. equal to some $A^{(n)}$ in $\mathscr{F}_{T^{(n)}}$, where $T^{(n)}$ is a sequence of $\mathscr{F}_t$-stopping times decreasing towards T. The result then follows, since by the right continuity of $\mathscr{F}_t$, $\bigcap_{n \geq 1} \mathscr{F}_{T^{(n)}} = \mathscr{F}_T$.

The set $A^{(n)}$ is constructed as follows: Write

$$A = \sum_{k=0}^{\infty} A \cap \left\{ \frac{k}{2^n} \leq T < \frac{k+1}{2^n} \right\} + A \cap \{T = \infty\}.$$

Therefore

$$A = \sum_{k=0}^{\infty} A_{(k+1)/2^n} \cap \left\{\frac{k}{2^n} \leq T < \frac{k+1}{2^n}\right\} + A_\infty \cap \{T = \infty\} \quad P\text{-a.s.,}$$

where $A_{(k+1)/2^n} \in \mathscr{F}_{(k+1)/2^n}$, $A_\infty \in \mathscr{F}_\infty$. Hence, defining

$$T^{(n)} = \begin{cases} \dfrac{k+1}{2^n} & \text{if } \dfrac{k}{2^n} \leq T < \dfrac{k+1}{2^n}, k \geq 0, \\ +\infty & \text{if } T = \infty, \end{cases}$$

we see that

$$A = \sum_{k=1}^{\infty} A_{(k+1)/2^n} \cap \left\{T^{(n)} = \frac{k+1}{2^n}\right\} + A_\infty \cap \{T^{(n)} = \infty\} \quad P\text{-a.s.}$$

Clearly the set $A^{(n)}$ on the right-hand side of the above equality is in $\mathscr{F}_{T^{(n)}}$, and $T^{(n)} \downarrow T$. □

T37 Theorem. *Let $\bar{T}$ be an $\bar{\mathscr{F}}_t$-stopping time. There exists an $\mathscr{F}_t$-stopping time T such that $T = \bar{T}'$, P-a.s.*

PROOF. Let $\bar{T}_m$ be the following approximation of $\bar{T}$:

$$\bar{T}_m = \sum_{k=0}^{\infty} \frac{k+1}{2^m} 1\left(\frac{k}{2^m} \leq \bar{T} < \frac{k+1}{2^m}\right) + \infty \cdot 1(\bar{T} = \infty).$$

The set $\{k/2^m \leq \bar{T} < k+1/2^m\}$ is in $\bar{\mathscr{F}}_{(k+1)/2^m}$, and therefore of the form $A_{(k+1)/2^m} \cup N_{k,m}$, where $A_{(k+1)/2^m} \in \mathscr{F}_{(k+1)/2^m}, N_{k,m} \in \mathcal{N}$. Similarly $\{\bar{T} = \infty\}$ has the form $A_\infty \cup N_\infty$, where $A_\infty \in \mathscr{F}_\infty$ and $N_\infty \in \mathcal{N}$. Define T_m by

$$T_m = \frac{k+1}{2^m} \quad \text{on} \quad A_{(k+1)/2^m} \cap \left[\bigcup_{l \leq k} A_{l/2^m}\right]$$

$$T_m = \infty \quad \text{on} \quad A_\infty \cap \left[\bigcup_{l \geq 1} A_{l/2^m}\right].$$

Thus $T_m = \bar{T}_m$, P-a.s. Then, clearly, if we define T by $T = \underline{\lim}_{m \uparrow \infty} T_m$, we have $T = \bar{T}'$, P-a.s.

T38 Theorem. *Let T and T' be two $\bar{\mathscr{F}}_t$-stopping times, and suppose that $\mathscr{F}_t$ is right-continuous. If $\bar{T} = T'$, P-a.s., then $\bar{\mathscr{F}}_T = \bar{\mathscr{F}}_{T'}$.*

PROOF. Take $A \in \mathscr{F}_T$ and decompose A into

$$A = \sum_{k=0}^{\infty} A \cap \left\{\frac{k}{2^m} \leq T < \frac{k+1}{2^m}\right\} + A \cap \{T = \infty\}.$$

Now

$$A \cap \left\{\frac{k}{2^m} \leq T < \frac{k+1}{2^m}\right\} = A_{(k+1)/2^m} \cap \left\{\frac{k}{2^m} \leq T < \frac{k+1}{2^m}\right\},$$

where $A_{(k+1)/2^m} \in \overline{\mathscr{F}}_{(k+1)/2^m}$. Similarly $A \cap \{T = \infty\} = A_\infty \cap \{T = \infty\}$, where $A_\infty \in \overline{\mathscr{F}}_\infty$. Therefore

$$A = \sum_{k=0}^{\infty} A_{(k+1)/2^m} \cap \left\{\frac{k}{2^m} \leq T' < \frac{k+1}{2^m}\right\} + A_\infty \cap \{T' = \infty\} \quad P\text{-a.s.}$$

But $A_{(k+1)/2^m} \cap \{k/2^m \leq T' < (k+1)/2^m\}$ and $A_\infty \cap \{T' = \infty\}$ are in $\overline{\mathscr{F}}_{T'+1/2^m}$; therefore $A \in \overline{\mathscr{F}}_{T'+1/2^m}$ for all $m \geq 1$. Since $\overline{\mathscr{F}}_t$ is right continuous, $A \in \overline{\mathscr{F}}_{T'}$. Thus $\overline{\mathscr{F}}_T \subset \overline{\mathscr{F}}_{T'}$. Symmetrically $\overline{\mathscr{F}}_{T'} \subset \overline{\mathscr{F}}_T$, and therefore

$$\overline{\mathscr{F}}_T = \overline{\mathscr{F}}_{T'}.$$
□

The last four theorems show that all the representation results of the present section concerning the history $\mathscr{F}_t^p$ also hold when $\mathscr{F}_t^p$ is replaced by its completion $\overline{\mathscr{F}}_t^p$.

References

[1] Courrège, Ph. and Priouret, P. (1965) Temps d'arrêt d'une fonction aléatoire, Publ. Inst. Stat. Univ. Paris, pp. 245–274.
[2] Dellacherie, C. (1972) *Capacités et Processus Stochastiques*, Springer, Berlin.
[3] Dellacherie, C. and Meyer, P. A. (1976) *Probabilités et Potentiel* (version refondue), Hermann, Paris.
[4] Jacod, J. (1975) Multivariate point processes: predictable projections, Radon–Nikodym derivatives, representation of martingales, Z. für W. **31**, pp. 235–253.
[5] Lazaro, J. (1974) Sur les hélices du flot spécial sous une fonction, Z. für W. **30**, pp. 279–302.
[6] Meyer, P. A. (1966) *Probability and Potential*, Blaisdell, San Francisco.
[7] Itmi, M. (1980) Histoire interne des processus ponctuels, Thèse 3ème cycle, U. de Rouven.

APPENDIX A3
Wiener-Driven Dynamical Systems

1. Ito's Stochastic Integral
2. Square-Integrable Brownian Martingales
3. Girsanov's Theorem
References

1. Ito's Stochastic Integral

Wiener Process and White Noise

Although point-process systems and Wiener-driven systems are qualitatively different, they present striking similarities when the martingale point of view is adopted. The theory of filtering and control for stochastic systems described by Ito differential equations was developed in the '50s and '60s, long before it was realized that martingale theory provided a unifying approach to point-process systems. In fact, when it was discovered that the tools of stochastic Stieltjes integration played in point-process theory a role similar to Ito stochastic integration in Wiener-type system theory, almost all the results of the latter immediately found counterparts in the former. It is therefore interesting that the theory of stochastic systems described by Ito equations should be sketched at this point, even if it is not the central topic of this book. But this historical reason is not the only one for this insertion. Many applications involve systems of the mixed type, where point processes and Wiener (or Ito) processes naturally intermingle, as is the case in optical communication (see Chapter IV, §3). We will now summarize Ito's stochastic

1. Ito's Stochastic Integral

integration theory, referring to various textbooks for the unproven results. Our aim in this Appendix is to make the reader aware of the parallelism between point-process systems and systems driven by Ito differential equations when the martingale approach is used.

We first recall the definition of a Wiener process.

D1 Definition. Let $(\Omega, \mathscr{F}, P)$ be a probability space endowed with a history $\mathscr{F}_t$. A $(P, \mathscr{F}_t)$-*Wiener process* is a continuous process W_t adapted to $\mathscr{F}_t$ and such that for all $0 \leq s \leq t$

$$W_t - W_s \text{ is } P\text{-independent of } \mathscr{F}_s, \qquad (1.1)$$

$$W_t - W_s \text{ is gaussian, mean 0, variance } t - s. \qquad (1.2)$$

An *m-vector* $(P, \mathscr{F}_t)$-*Wiener process* is an *m*-vector process W_t with coordinates that are *P*-independent $(P, \mathscr{F}_t)$-*Wiener processes*.

If W_t is a $(P, \mathscr{F}_t)$ Wiener process, it is easy to verify that W_t and $W_t^2 - t$ are $(P, \mathscr{F}_t)$-martingales (see I, E1).

It turns out that these martingale properties entirely characterize the Wiener processes. More precisely:

T2 Theorem (Lévy's Characterization of Wiener Process). *Let W_t be a continuous process defined on $(\Omega, \mathscr{F}, P)$, adapted to a history $\mathscr{F}_t$, and such that*

$$W_t \text{ is a } (P, \mathscr{F}_t)\text{-local martingale}, \quad W_0 = 0 \quad P\text{-a.s.}, \qquad (1.3)$$

$$(W_t^2 - t) \text{ is a } (P, \mathscr{F}_t)\text{-local martingale}. \qquad (1.4)$$

Then W_t is a $(P, \mathscr{F}_t)$-Wiener process.

PROOF. Postponed to §2. ☐

Remark. Theorem T2 is the analogue of Watanabe's characterization theorem for Poisson processes (II, T5).

Even though they have continuous trajectories, Wiener processes behave extremely badly. In particular:

T3 Theorem. *Let W_t be a $(P, \mathscr{F}_t)$-Wiener process. Then there exists a P-null set N such that*:

(1) *For all $\omega \in \Omega - N$, $t \to W_t(\omega)$ is nowhere differentiable.*
(2) *For all $\omega \in \Omega - N$, $t \to W_t(\omega)$ is of unbounded variation on any bounded interval.*

PROOF. See [1, Chapter 12, §8] for instance. ☐

The above result has very negative consequences from the applied point of view of engineering literature. First, one cannot define a "white noise" process ξ_t by

$$W_t = \int_0^t \xi_s \, ds. \tag{*}$$

[If (*) defined the "white noise" ξ_t, one could give a mathematical meaning to the phrase "the white noise is the derivative of the Wiener process."]

Secondly, one cannot define integrals of the type

$$\int_0^t \phi_s \, dW_s$$

as Stieltjes–Lebesgue integrals [as we can do for $\int_0^t \phi_s \, dM_s$, when $M_t = N_t - \int_0^t \lambda(s) \, ds$, and N_t is Poisson process with intensity $\lambda(t)$]. From an engineer's model-building point of view, one wishes to consider extensions of ordinary differential equations

$$\dot{X}_t = f(t, X_t)$$

to include in the second member a disturbance of a stochastic nature, as follows:

$$\dot{X}_t = f(t, X_t) + \sigma(t, X_t)\xi_t,$$

where ξ_t is some "noise." Under certain regularity properties for the trajectories of ξ_t and for the coefficients $f(t, x)$ and $\sigma(t, x)$, the differential equations

$$\dot{X}_t(\omega) = f(t, X_t(\omega)) + \sigma(t, X_t(\omega))\xi_t(\omega), \tag{1.5}$$

parametrized by $\omega \in \Omega$, might have solutions. But what really interests a certain category of modelers is to have for ξ_t a "white noise." The problem is then: how to give a meaning to (1.5)?

Proceeding formally, and writing that ξ_t is the derivative of W_t ($dW_t = \xi_t \, dt$), we can write (1.5) as

$$dX_t = f(t, X_t) \, dt + \sigma(t, X_t) \, dW_t, \tag{1.5'}$$

or

$$X_t = X_0 + \int_0^t f(s, X_s) \, ds + \int_0^t \sigma(s, X_s) \, dW_s. \tag{1.6}$$

But then, we are faced with the impossibility of using Stieltjes–Lebesgue theory in order to give a meaning to the integral $\int_0^t \sigma(s, X_s) \, dW_s$. Such an integral has to be defined in a new way, and this is the object of the integration theory developed by Ito, who defined, for any measurable process ϕ_t adapted to $\mathscr{F}_t$ and any $0 \le a \le b$ such that

$$\int_a^b |\phi_s|^2 \, ds < \infty \quad P\text{-a.s.,}$$

a quantity denoted

$$\int_a^b \phi_s \, dW_s$$

which enjoys very pleasant properties from a mathematical point of view. For instance, if ϕ_t is such that $E[\int_0^t |\phi_s|^2 \, ds] < \infty$ for all $t \geq 0$, then the process $\int_0^t \phi_s \, dW_s$ admits continuous modifications and is a $(P, \mathscr{F}_t)$-martingale. However, $\int_0^t \phi_s \, dW_s$ still possesses the bad properties mentioned in Theorem T3: its paths are almost surely nowhere differentiable and of unbounded variation on any finite interval. It therefore has no physical meaning and must be considered, like W_t itself, as the limit (in some sense) of a sequence of stochastic processes with "physical" trajectories (of bounded variation on finite intervals for instance). Moreover, the definition of $\int$ is such that the rules of "ordinary calculus" do not apply. For instance, anticipating §2, we have

$$W_t^2 = 2 \int_0^t W_s \, dW_s + t \tag{1.7}$$

instead of the expected

$$W_t^2 = 2 \int_0^t W_s \, dW_s \quad (wrong).$$

In spite of all these apparent drawbacks, Ito's stochastic integral has become a popular tool for modeling stochastic systems, mainly because of its nice mathematical property of yielding martingales and also because of its associated stochastic calculus.

Ito's Integral: The Isometric Construction

We proceed along a path parallel to that followed in the definition of Stieltjes integrals with respect to Poisson processes in III, §4.

Elementary Predictable Integrands

L4 Lemma. *Let $W_t(1)$ and $W_t(2)$ be two independent $(P, \mathscr{F}_t)$-Wiener processes, and let $C_t(1)$ and $C_t(2)$ be two elementary $\mathscr{F}_t$-predictable processes of the form*

$$C_t(i, \omega) = H_{a_i}(\omega) 1_{(a_i, b_i]}(t), \tag{1.8}$$

$0 \leq a_i \leq b_i$, H_{a_i} *bounded and $\mathscr{F}_{a_i}$-measurable, $i = 1, 2$. Define, for $i = 1, 2$,*

$$\int_0^t C_s(i) \, dW_s(i) = \begin{cases} 0 & \text{if } t \leq a_i, \\ H_{a_i}[W_t - W_{a_i}] & \text{if } a_i \leq t \leq b_i, \\ H_{a_i}[W_{b_i} - W_{a_i}] & \text{if } b_i \leq t, \end{cases} \tag{1.9}$$

and extend the definition of $\int_0^t$ to all elementary $\mathcal{F}_t$-predictable processes by linearity. Then for all elementary $\mathcal{F}_t$-predictable processes $C_t(1)$ and $C_t(2)$,

$$\int_0^t C_s(i)\, dW_s(i) \text{ is a } (P, \mathcal{F}_t)\text{-martingale}, \qquad i = 1, 2, \tag{1.10}$$

and

$$E\left[\int_0^t C_s(i)\, dW_s(i) \int_0^t C_s(j)\, dW_s(j)\right] = \delta_{ij} E\left[\int_0^t C_s(i) C_s(j)\, ds\right], \tag{1.11}$$

$i, j = 1, 2$, where δ_{ij} is the Kronecker symbol ($\delta_{ij} = 1$ if $i = j$, $\delta_{ij} = 0$ if $i \neq j$).

E1 Exercise. Prove L4. (Hint: just compute.)

The analogy between Wiener and Poisson processes seems to dry out at this point: indeed, for Poisson processes, an integral such as $\int_0^t C_s\, dM_s$ was defined ω-pointwise an an ordinary Stieltjes–Lebesgue integral. For Wiener processes, such a pathwise definition is no longer possible, since by T3, the paths of W_t are almost surely of unbounded variation on finite intervals. However, the Ito integral $\int_0^t \phi_s\, dW_s$ can still be defined in a way very similar to what was done in III, §4, by means of an isometry between two Hilbert spaces.

The Fundamental Isometry. Let $\mathcal{M}^2[0, T]$ be the space of equivalence classes of $(P, \mathcal{F}_t)$-square integrable martingales over the interval $[0, T]$, for the equivalence $\sim$ defined by

$$m_t \sim m'_t \quad \Leftrightarrow \quad m_t(\omega) = m'_t(\omega) \quad P\text{-a.s.}, \qquad t \in [0, T].$$

As usual, we do not distinguish between an equivalence class and a specific member of this class. We have already seen that $\mathcal{M}^2[0, T]$ endowed with the scalar product

$$\langle m_t(1), m_t(2) \rangle_{\mathcal{M}^2[0, T]} = E[m_T(1) m_T(2)] \tag{1.12}$$

is a Hilbert space (I, E13).

Now let $\mathcal{G}[0, T]$ be the space of equivalence classes of $\mathcal{F}_t$-measurable processes ϕ_t such that

$$E\left[\int_0^T \phi_t^2\, dt\right] < \infty$$

for the equivalence $\mathcal{R}$ defined by

$$\phi_t \mathcal{R} \phi'_t \quad \Leftrightarrow \quad \phi_t(\omega) = \phi'_t(\omega) \quad P(d\omega)\, dt\text{-a.e.}$$

With the scalar product

$$\langle \phi_t(1), \phi_t(2) \rangle_{\mathcal{G}[0, T]} = E\left[\int_0^T \phi_t(1) \phi_t(2)\, dt\right], \tag{1.13}$$

1. Ito's Stochastic Integral

$\mathscr{G}[0, T]$ is a Hilbert space. We denote by $\|\cdot\|_{\mathscr{M}^2[0,T]}$ and $\|\cdot\|_{\mathscr{G}[0,T]}$ the norms induced by the scalar products in $\mathscr{M}^2[0,T]$ and $\mathscr{G}[0, T]$ respectively.

Let $\mathscr{G}'[0, T]$ be the subset of $\mathscr{G}[0, T]$ consisting of elementary $\mathscr{F}_t$-predictable processes. Then

L5 Lemma (Approximation Lemma). $\mathscr{G}'[0, T]$ is dense in $\mathscr{G}[0, T]$.

PROOF. Let ϕ_t be in $\mathscr{G}[0, T]$, i.e., ϕ_t is adapted to $\mathscr{F}_t$, measurable, and such that $E[\int_0^T \phi_s^2 \, ds] < \infty$. Such ϕ_t can be approximated in $\mathscr{G}[0, T]$ by a sequence of bounded processes of $\mathscr{G}[0, T]$: take $\phi_t^{(n)} = \phi_t 1(|\phi_t| \le n)$. One can therefore suppose without loss of generality that ϕ_t is bounded, say by a constant c. Define for each $n \ge 1$

$$\phi_t^{(n)} = n \int_0^t e^{n(u-t)} \phi_u \, du.$$

Clearly, $\phi_t^{(n)}$ is, for each $n \ge 1$, a continuous bounded process. □

E2 Exercise. Proof that $\lim \phi_t^{(n)}(\omega) = \phi_t(\omega)$, $P(d\omega) \, dt$-a.e.

By the dominated-convergence theorem,

$$\lim E\left[\int_0^T |\phi_t^{(n)} - \phi_t|^2 \, dt\right] = 0.$$

We can therefore suppose that ϕ_t is continuous and bounded. But clearly, for such ϕ_t, the sequence $(\phi_t^{(n)}, n \ge 1)$ defined by

$$\phi_t^{(n)} = \phi_{k/n} \quad \text{if } \frac{kT}{n} \le t < \frac{(k+1)T}{n}, \quad 0 \le k \le n-1,$$

approximates ϕ_t in $\mathscr{G}[0, T]$ and is in $\mathscr{G}'[0, T]$. □

T6 Theorem (Ito's Integration Theorem [9]). *Let $W_t(1)$ and $W_t(2)$ be two independent $(P, \mathscr{F}_t)$ Wiener processes, and let ϕ_t and ψ_t be two measurable processes adapted to $\mathscr{F}_t$ and such that*

$$E\left[\int_0^T |\phi_s|^2 \, ds\right] < \infty \quad \text{and} \quad E\left[\int_0^T |\psi_s|^2 \, ds\right] < \infty. \tag{1.14}$$

Let $(\phi_t^{(n)}, n \ge 0)$ and $(\psi_t^{(n)}, n \ge 0)$ approximate ϕ_t and ψ_t respectively in $\mathscr{G}'[0, T]$. Then $\int_0^T \phi_s^{(n)} \, dW_s(1)$ and $\int_0^T \psi_s^{(n)} \, dW_s(2)$ have limits in quadratic mean, which we denote by $\int_0^T \phi_s \, dW_s(1)$ and $\int_0^T \psi_s \, dW_s(2)$. These limits do not depend upon the approximating sequences $(\phi_t^{(n)}, n \ge 0)$ and $(\psi_t^{(n)}, n \ge 0)$. Also, the mappings $\mathscr{G}[0, T] \to \mathscr{M}^2[0, T]$ defined by $\phi_t \to \int_0^T \phi_s \, dW_s(1)$ and $\psi_t \to \int_0^T \psi_s \, dW_s(2)$ are isometries. Moreover,

$$E\left[\int_0^T \phi_s \, dW_s(1) \int_0^T \psi_s \, dW_s(2)\right] = 0. \tag{1.15}$$

PROOF. The proof follows immediately from L4, L5, and the Banach extension theorem for isometries between Hilbert spaces.

It is easy to show that if W_t is a $(P, \mathscr{F}_t)$-Wiener process and if ϕ_t is in $\mathscr{G}[0, T]$, then for any $0 \le t_1 \le t_2 \le T$

$$\oint_0^{t_1} \phi_s\, dW_s = \oint_0^{t_2} \phi_s 1(s \le t_1)\, dW_s. \tag{1.16}$$

(In other words, the definition of $\oint$ is consistent with localization on deterministic intervals.) We have thus defined a stochastic process on $[0, T]$:

$$X_t = \oint_0^t \phi_s\, dW_s, \tag{1.17}$$

which is adapted to $\mathscr{F}_t$ and square-integrable.

Properties of Ito Integral Processes

The most interesting property is

T7 Theorem (Martingale Property of Ito's Integral). *For any $(P, \mathscr{F}_t)$-Wiener process W_t and for any measurable process ϕ_t adapted to $\mathscr{F}_t$ and such that $E[\int_0^T |\phi_s|^2\, ds] < \infty$, the process $X_t = \oint_0^t \phi_s\, dW_s$ defined on $[0, T]$ is a $(P, \mathscr{F}_t)$-square-integrable martingale.*

PROOF. Let $(\phi_t^{(n)}, n \ge 0)$ be a sequence of $\mathscr{G}'[0, T]$ approximating $\mathscr{G}[0, T]$. By L5, $\int_0^t \phi_s^{(n)}\, dW_s$ is a $(P, \mathscr{F}_t)$-square-integrable martingale on $[0, T]$, for all $n \ge 0$. Therefore for any $0 \le s \le t \le T$ and any $A \in \mathscr{F}_s$,

$$E\left[1_A \oint_0^t \phi_u^{(n)}\, dW_u\right] = E\left[1_A \oint_0^s \phi_u^{(n)}\, dW_u\right], \tag{$*$}$$

and since $\oint_0^x \phi_u^{(n)}\, dW_u$ tends to $\oint_0^x \phi_u\, dW_u$ in quadratic mean for all $x \in [0, T]$, the equality $(*)$ holds in the limit, i.e.,

$$E\left[1_A \oint_0^t \phi_u\, dW_u\right] = E\left[1_A \oint_0^s \phi_u\, dW_u\right]. \qquad \square$$

We will not prove the next theorem. See [6], [10], [12], [13], [14], [15].

T8 Theorem (Continuity Property of Ito's Integral). *Let X_t be defined as in Theorem T7. There exists a version which is continuous.*

1. Ito's Stochastic Integral

Ito's integrals are defined *globally* over $(\Omega, \mathcal{F}, P)$, as limits in quadratic mean, and *not trajectorywise* (ω by ω). If they were defined trajectorywise (say as Stieltjes integrals as in the Poisson case), then for any random time τ we would have

$$\int_0^{t \wedge \tau(\omega)} \phi_s(\omega) \, dW_s(\omega) = \int_0^t 1(s \leq \tau(\omega))\phi_s(\omega) \, dW_s(\omega), \quad \omega \in \Omega;$$

that is to say, for each $\omega \in \Omega$, the value at $t \wedge \tau(\omega)$ of the process $X_t = \int_0^t \phi_s \, dW_s$ is equal to the value at t of the process $Y_t = \int_0^t 1(s \leq \tau)\phi_s \, dW_s$. In fact this is true, although not obvious, when τ is a stopping time.

More precisely:

L9 Lemma (Localization Lemma). *Let W_t be a $(P, \mathcal{F}_t)$-Wiener process, and ϕ_t be a measurable process adapted to $\mathcal{F}_t$ and such that $E[\int_0^t \phi_s^2 \, ds] < \infty$, $t \geq 0$. Let τ be an $\mathcal{F}_t$-stopping time, and denote $\int_0^t \phi_s \, dW_s$ by X_t. Then*

$$X_{t \wedge \tau} = \int_0^t 1(s \leq \tau)\phi_s \, dW_s. \tag{1.18}$$

PROOF. See [13].

Ito's Integrals Defined as Limits in Probability

It is possible to define the Ito integral $\int_0^t \phi_s \, dW_s$ when W_t is a $(P, \mathcal{F}_t)$-Wiener process and ϕ_t is a measurable process adapted to $\mathcal{F}_t$ and such that

$$\int_0^t |\phi_s|^2 \, ds < \infty \quad P\text{-a.s.}, \quad t \geq 0 \tag{1.19}$$

(a weaker condition than $E[\int_0^t |\phi_s|^2 \, ds] < \infty$, $t \geq 0$). In order to define $\int_0^t \phi_s \, dW_s$ for such integrands, for each $n \geq 1$ let T_n be the $\mathcal{F}_t$-stopping time

$$T_n = \begin{cases} \inf\left\{t \, \Big| \, \int_0^t |\phi_s|^2 \, ds \geq n\right\} & \text{if } \{\cdots\} \neq \emptyset \\ +\infty & \text{otherwise.} \end{cases} \tag{1.20}$$

Clearly $T_n \uparrow \infty$, P-a.s. Also, if we define

$$\phi_t^{(n)} = \phi_t 1(t \leq T_n),$$

then $E[\int_0^t |\phi_s^{(n)}|^2 \, ds] < \infty$, $t \geq 0$, and $\int_0^t \phi_s^{(n)} \, dW_s$ is thus well defined for each $t \geq 0$.

Now, by the localization lemma L9,

$$\int_0^{t \wedge T_n} \phi_s^{(n)} \, dW_s = \int_0^t \phi_s^{(n)} \, dW_s$$

whenever $n \leq m$. Therefore for any n, m, and for any $\varepsilon > 0$,

$$P\left[\left|\oint_0^t \phi_s^{(n)} \, dW_s - \oint_0^t \phi_s^{(m)} \, dW_s\right| \geq \varepsilon\right] \leq P\left[\int_0^t |\phi_s|^2 \, ds > \min(n, m)\right] \to 0$$

$$\text{as } n, m \uparrow \infty. \quad (1.21)$$

By Cauchy's criterion for convergence in probability, for each $t \geq 0$, $\oint_0^t \phi_s^{(n)} \, dW_s$ converges in probability to some random variable, which we denote by $\oint_0^t \phi_s \, dW_s$, thus extending the meaning of the symbol $\oint$. For this extension of the Ito integral we have the analogues of T7, T8, and L9, which we state without proof:

T10 Theorem. *For any $(P, \mathscr{F}_t)$-Wiener process W_t and for any measurable process ϕ_t adapted to $\mathscr{F}_t$ and such that*

$$\int_0^t |\phi_s|^2 \, ds < \infty \quad P\text{-a.s.}, \quad t \geq 0,$$

the process $X_t = \int_0^t \phi_s \, dW_s$ is a $(P, \mathscr{F}_t)$-local martingale which admits a continuous version. Moreover, for any $\mathscr{F}_t$-stopping time τ

$$X_{t \wedge \tau} = \oint_0^t \phi_s 1(s \leq \tau) \, dW_s. \quad (1.22)$$

PROOF. See Neveu [13], Priouret [14]. □

Remark. Before we close this section, we want to make a comparison between Ito integrals and Stieltjes integrals with respect to a Poisson process. First recall that all that was required in the definition of $\oint_0^t \phi_s \, dW_s$ as a $(P, \mathscr{F}_t)$-square-integrable martingale was that W_t and $W_t^2 - t$ be $(P, \mathscr{F}_t)$-martingales and that ϕ_t be some measurable process adapted to $\mathscr{F}_t$ and such that $E[\int_0^t |\phi_s|^2 \, ds] < \infty$, $t \geq 0$. Therefore, in Ito's integration theory we could very well replace W_t by $M_t = N_t - t$, where N_t is a $(P, \mathscr{F}_t)$-Poisson process of intensity 1 [indeed, M_t and $M_t^2 - t$ are $(P, \mathscr{F}_t)$-martingales]. We could then define, for all ϕ_t as above, an integral $\oint_0^t \phi_s \, dM_s$ that is a $(P, \mathscr{F}_t)$-martingale. Note that we are still using the special symbol $\oint$ for this integral, even though M_t has bounded variation and $\int_0^t \phi_s \, dM_s$ has therefore a meaning as a Stieltjes integral: indeed, Ito's theory developed for $M_t = N_t - t$ as it was for W_t does not say that $\oint_0^t \phi_s \, dM_s = \int_0^t \phi_s \, dM_s$. In fact that is not true in general. In the integration theory for point processes of III, §4 we have required that ϕ_t be in addition $\mathscr{F}_t$-predictable, and for such a class of integrands, we have been able to show that $\oint_0^t \phi_s \, dM_s$ (defined by the Hahn–Banach extension theorem) is indeed equal to $\int_0^t \phi_s \, dM_s$.

2. Square-Integrable Brownian Martingales

Ito's Differentiation Rule

Now that Ito's integral has been defined, we will present the stochastic calculus associated with it. Ito calculus, as it is called, is based upon a differentiation rule which does not look like the familiar differentiation rules (Lebesgue, Stieltjes). Very roughly, one writes this rule as if the integrals $\oint \phi_s \, dW_s$ were integrals with respect to a continuous function of bounded variation (recall that in reality, the trajectories of W_t are *not* of bounded variation), and one adds a term which compensates for the unbounded variation of W_t. More precisely:

T11 Theorem (Ito's Differentiation Rule [9]). *Let W_t be a $(P, \mathscr{F}_t)$-Wiener process, let ϕ_t be a measurable process adapted to $\mathscr{F}_t$ and such that*

$$\int_0^t |\phi_s|^2 \, ds < \infty \quad P\text{-a.s.}, \quad t \geq 0,$$

and let f_t be a measurable process adapted to $\mathscr{F}_t$ satisfying

$$\int_0^t |f_s| \, ds < \infty \quad P\text{-a.s.}, \quad t \geq 0.$$

Define

$$X_t = X_0 + \int_0^t f_s \, ds + \oint_0^t \phi_s \, dW_s, \tag{2.1}$$

where X_0 is some random variable, summable and $\mathscr{F}_0$-measurable. Let $F: R^2 \to C$ be twice continuously differentiable in the first variable x, and once continuously differentiable in the second variable t. Then

$$F(X_t, t) = F(X_0, 0) + \int_0^t \frac{\partial F}{\partial t}(X_s, s) \, ds$$
$$+ \oint_0^t \frac{\partial F}{\partial x}(X_s, s) \, dX_s + \frac{1}{2} \int_0^t \frac{\partial^2 F}{\partial x^2}(X_s, s) \phi_s^2 \, ds, \tag{2.2}$$

where, by definition,

$$\oint_0^t \frac{\partial F}{\partial x}(X_s, s) \, dX_s = \oint_0^t \frac{\partial F}{\partial x}(X_s, s) \phi_s \, dW_s + \int_0^t \frac{\partial F}{\partial x}(X_s, s) f_s \, ds.$$

PROOF. See [6], [9], [10], [12], [13], [14], [15]. □

Remarks

(1) One checks that all quantities on the right-hand side of the equality (2.2) are well defined: for instance the integrals

$$\int_0^t \frac{\partial F}{\partial x}(X_s, s) f_s \, ds \quad \text{and} \quad \int_0^t \frac{\partial^2 F}{\partial x^2}(X_s, s) \phi_s^2 \, ds$$

are defined trajectorywise because $t \to X_t(\omega)$ is continuous (and therefore

$$t \to \frac{\partial F}{\partial x}(X_t(\omega), t) \quad \text{and} \quad t \to \frac{\partial^2 F}{\partial x^2}(X_t(\omega), t)$$

are continuous, F being twice continuously differentiable in x); also

$$\oint_0^t \frac{\partial F}{\partial x}(X_s, x) \phi_s \, dW_s$$

is well defined because

$$\int_0^t \left|\frac{\partial F}{\partial x}(X_s, s)\right|^2 |\phi_s|^2 \, ds < \infty \quad P\text{-a.s.}, \quad t \geq 0,$$

this inequality following from the trajectorywise continuity of

$$t \to \frac{\partial F}{\partial x}(X_t(\omega), t)$$

and the hypothesis $\int_0^t |\phi_s|^2 \, ds < \infty$, P-a.s., $t \geq 0$.

(2) The equality (2.2) reads as if it were an ordinary differentiation rule written for X_t continuous and of bounded variation, *except for the additional term*

$$\frac{1}{2} \int_0^t \frac{\partial^2 F}{\partial x^2}(X_s, s) \, ds.$$

A few examples will familiarize the reader with such "pathology."

EXAMPLE 1. Let $F(x, t) = x^2$ and $X_t = W_t$ in T11. Then (2.2) reads

$$W_t^2 = 2 \oint_0^t W_s \, dW_s + t.$$

EXAMPLE 2. Let $F(x, t) = e^x$ and $X_t = \int_0^t \phi_s \, dW_s - \frac{1}{2}\int_0^t \phi_s^2 \, ds$, where ϕ_s satisfies the condition $\int_0^t \phi_s^2 \, ds < \infty$, P-a.s., $t \geq 0$. Applying the rule (2.2), we obtain

$$L_t = \exp\left(\oint_0^t \phi_s \, dW_s - \frac{1}{2}\int_0^t \phi_s^2 \, ds\right) = 1 + \oint_0^t L_s \phi_s \, dW_s. \quad (2.3)$$

The above formula is a very important one in the theory of Wiener-type systems: L_t gives the general form of likelihood ratios with respect to brownian measure (see §3).

2. Square-Integrable Brownian Martingales

Remark. Going back to Example 1 above, we see that

$$W_t^2 - t = 2 \oint_0^t W_s \, dW_s \quad [\text{a } (P, \mathscr{F}_t)\text{-martingale}], \tag{\dag}$$

which we can compare (formally) to

$$M_t^2 - t = 2 \int_0^t M_{s-} \, dM_s \quad [\text{a } (P, \mathscr{F}_t)\text{-martingale}], \tag{\dag\dag}$$

where $M_t = N_t - t$ and N_t is a $(P, \mathscr{F}_t)$ standard Poisson process. This is worth noting, since the pathology of Ito calculus is always discussed with respect to Lebesgue calculus (integration with respect to measure dt); but then Stieltjes calculus is equally pathological, since we do *not* have in (††) that $M_t^2 = 2 \int_0^t M_s \, dM_s$. However, the reason for the additional terms in Stieltjes calculus is obvious, and it has nothing to do with probability theory, whereas in Ito calculus things are entirely different. Since we have not given the derivation of Ito's differentiation rule, the technical reason for the additional term is not so clear; we wish only to mention that in the case of the Ito calculus it is the quadratic-variation property of the Wiener process:

$$E[(W_t - W_s)^2] = t - s,$$

which also has a trajectorywise counterpart due to Lévy (see [13]):

$$\lim_{|\mathscr{D}^{(k)}| \downarrow 0} \sum_{\mathscr{D}^{(k)}} (W_{t_{i+1}^{(k)}} - W_{t_i^{(k)}})^2 = t - s, \quad P\text{-a.s.},$$

where $\mathscr{D}^{(k)} = \{t_0^{(k)} = s < t_1^{(k)} < \cdots < t_n^{(k)} = t\}$, and where $|\mathscr{D}^{(k)}| \downarrow 0$ means that subdivision $\mathscr{D}^{(k+1)}$ contains at least all the points of subdivision $\mathscr{D}^{(k)}$ for each $k \geq 0$, and that

$$\lim_{k \uparrow \infty} \max |t_{i+1}^{(k)} - t_i^{(k)}| = 0.$$

The latter equality can be written formally:

$$\int_0^t (dW_t)^2 = t - s.$$

From this, the additional term is to be expected.

A spectacular application of Ito's differentiation rule is the following proof of Doob's characterization theorem due to Kunita and Watanabe [11].

Proof of Lévy's Characterization Theorem

Before starting the proof, we mention that all the results previously stated in §1 and §2 depend only upon the fact that a $(P, \mathscr{F}_t)$-Wiener process is a continuous process satisfying (1.3) and (1.4) in the statement of Theorem T2.

The reader can easily check this in the proofs when these proofs are given in the present text, and will have to assume it for the other results not proven in the text (we refer to Neveu [13] and Priouret [14] for details). In particular, when W_t is a continuous process such that (1.3) and (1.4) of T2 are satisfied, Ito's differentiation rule is applicable. Therefore, taking $F(x) = e^{iux}$, $u \in R$,

$$e^{iuW_t} - e^{iuW_s} = iu \oint_s^t e^{iuW_z} \, dW_z - \tfrac{1}{2}u^2 \int_s^t e^{iuW_z} \, dz, \qquad 0 \le s \le t. \quad (2.4)$$

Now, since $E[\int_0^t |e^{iuW_s}|^2 \, ds] = t < \infty$, it follows that $\int_0^t e^{iuW_s} \, dW_s$ is a $(P, \mathscr{F}_t)$-martingale, and therefore, for all $A \in \mathscr{F}_s$,

$$\int_A (e^{iuW_t} - e^{iuW_s}) \, dP = -\tfrac{1}{2}u^2 \int_A \int_s^t e^{iuW_z} \, dz \, dP.$$

Dividing both sides of the above equation by e^{iuW_s} and applying Fubini's theorem on the right-hand side, we obtain

$$\int_A e^{iu(W_t - W_s)} \, dP = P(A) - \tfrac{1}{2}u^2 \int_s^t \int_A e^{iu(W_z - W_s)} \, dP \, ds,$$

the solution of which is

$$\int_A e^{iu(W_t - W_s)} \, dP = P(A)e^{-(1/2)/u^2/(t-s)}. \quad (2.5)$$

This equality is valid for all $0 \le s \le t$, all $u \in R$, and all $A \in \mathscr{F}_s$. Taking $A = \Omega$, we obtain

$$E[e^{iu(W_t - W_s)}] = e^{-(1/2)/u^2/(t-s)},$$

that is to say, $W_t - W_s$ is gaussian, mean 0, variance 1. Equation (2.5) then reads

$$E[1_A e^{iu(W_t - W_s)}] = P(A)E[e^{iu(W_t - W_s)}], \qquad A \in \mathscr{F}_s,$$

from which follows that $W_t - W_s$ is P-independent of $\mathscr{F}_s$.[1]

Representation of Brownian Functionals as Ito Integrals

A result similar to the representation of square-integrable Poissonian martingales of III, §4 is available: every square-integrable martingale with respect to the internal history of a Wiener process is, up to an additive constant, a stochastic integral of the Ito type.

---

[1] Here we have presented the proof in the case where W_t and $W_t^2 - t$ are $(P, \mathscr{F}_t)$-martingales. The general case (when W_t and $W_t^2 - t$ are *local* martingales) is proven in exactly the same way, at the expense of replacing t by $t \wedge T_n$ where appropriate (T_n is a localizing sequence).

2. Square-Integrable Brownian Martingales

T12 Theorem. *Let W_t be a $(P, \mathscr{F}_t^W)$-Wiener process, and let m_t be a $(P, \mathscr{F}_t^W)$-square-integrable martingale on $[0, 1]$. Then there exists a measurable process ϕ_t adapted to $\mathscr{F}_t^W$ such that*

$$E\left[\int_0^1 |\phi_s|^2 \, ds\right] < \infty \tag{2.6}$$

and

$$m_t = m_0 + \int_0^t \phi_s \, dW_s, \qquad t \in [0, 1]. \tag{2.7}$$

The method of proof is exactly the same as in III, T12. We start with the lemma

L13 Lemma. *Let W_t be a $(P, \mathscr{F}_t^W)$-Wiener process, and let*

$$\mathscr{K} = \left\{\overline{M}_1^f = \exp\left\{\int_0^1 f(s) \, dW_s\right\} \middle| f \text{ nonrandom, measurable, bounded}\right\}.$$

$\mathscr{K}$ is total in the Hilbert space $L^2(\mathscr{F}_1^W, P)$.

PROOF. We have to show that if Y is an element of $L^2(\mathscr{F}_1^W, P)$ such that

$$E[Y\overline{M}_1^f] = 0, \qquad \overline{M}_1^f \in \mathscr{K}, \tag{$*$}$$

then necessarily

$$Y = 0 \quad P\text{-a.s.}$$

Now, when f has the form

$$f(t) = u_1 1_{[0, t_1]}(t) + \cdots + u_k 1_{(t_{k-1}, t_k]}(t),$$

where the u_i's are real numbers, then

$$\overline{M}_1^f = \prod_{j=1}^k \exp\{u_j(W_{t_j} - W_{t_{j-1}})\}.$$

Since the collection of the above $\overline{M}_1^f$ when n, $(u_i, 1 \leq i \leq n)$, and $(t_i, 1 \leq i \leq n)$ vary is complete in $L^2(\mathscr{F}_1^W, P)$, the result follows as in III, T14. □

Before proving T12, we need a few preliminaries. Let $f: R \to R$ be bounded and measurable, and define

$$X_t = \exp\left\{\int_0^t f(s) \, dW_s - \frac{1}{2} \int_0^t f(s)^2 \, ds\right\}.$$

By Ito's differentiation rule (see Example 2 after T11),

$$X_t = 1 + \int_0^t X_s f(s) \, dW_s.$$

For each $n \geq 1$, let T_n be the $\mathscr{F}_t$-stopping time defined by

$$T_n = \begin{cases} \inf\{t \mid X_s \geq n\} & \text{if } \{\cdots\} \neq \varnothing, \\ +\infty & \text{otherwise.} \end{cases}$$

By the localization lemma L9,

$$X_{t \wedge T_n} = 1 + \int_0^{t \wedge T_n} X_s f(s)\, dW_s = 1 + \int_0^t X_s f(s) 1(s \leq T_n)\, dW_s.$$

By the boundedness of f,

$$E\left[\int_0^t |X_s f(s) 1(s \leq T_n)|^2\, ds\right] < \infty,$$

and therefore, by the isometry of T6,

$$E[|X_{t \wedge T_n} - 1|^2] = E\left[\int_0^t |X_s f(s) 1(s \leq T_n)|^2\, ds\right] < \infty. \qquad (*)$$

In particular

$$E[|X_{t \wedge T_n}|^2] \leq 1 + \int_0^t E[|X_{s \wedge T_n}|^2] f(s)^2\, ds.$$

Therefore, by Gronwall's lemma,

$$E[|X_{t \wedge T_n}|^2] \leq \exp\left\{\int_0^t f(s)^2\, ds\right\}.$$

A fortiori,

$$E[|X_t 1(t \leq T_n)|^2] \leq \exp\left\{\int_0^t f(s)^2\, ds\right\},$$

and therefore, by monotone convergence,

$$E[|X_t|^2] \leq \exp\left\{\int_0^t f(s)^2\, ds\right\} < \infty.$$

Hence, from $(*)$

$$E\left[\int_0^t |X_s f(s)|^2\, ds\right] < \infty, \qquad t \geq 0,$$

which implies that X_t is a $(P, \mathscr{F}_t)$-square-integrable martingale.

Define $L_0^2(\mathscr{F}_1^W, P)$ to be the set of square-integrable random variables of $(\Omega, \mathscr{F}_1^W, P)$ that have mean 0. We are now ready for the

PROOF OF THEOREM T12. Let m_t be a $(P, \mathscr{F}_t^W)$-square-integrable martingale of mean 0. The random variable m_1 is in $L_0^2(\mathscr{F}_1^W, P)$, and by the totality of $\mathscr{K}$, it is a limit in quadratic mean of finite linear combinations of random variables of the form

$$M_1^f - 1 = \exp\left\{\oint_0^1 f(s)\, d\bar{W}_s - \frac{1}{2}\int_0^1 f(s)^2\, ds\right\} - 1,$$

where f is deterministic bounded and measurable. Indeed M_1^f is obtained from $\bar{M}_1^f$ by multiplication by a deterministic number, and $\bar{M}_1^f$ is in $\mathcal{K}$ by definition.

Now, by Ito's differentiation rule

$$M_1^f - 1 = \exp\left\{\oint_0^t f(s)\, dW_s - \frac{1}{2}\int_0^t (f(s))^2\, ds\right\} = \int_0^t M_s^f f(s)\, dW_s.$$

Therefore m_1 is the limit in quadratic mean of elements of the form

$$\oint_0^1 \phi_s^{(n)}\, dW_s,$$

where $\phi_t^{(n)}$ is a measurable process adapted to $\mathscr{F}_t^W$ and such that

$$E\left[\int_0^1 |\phi_s^{(n)}|^2\, ds\right] < \infty$$

(see discussion before beginning of the proof). By the fundamental isometry, M_1 is of the form

$$m_1 = \oint_0^1 \phi_s\, dW_s, \tag{*}$$

where ϕ_t is a measurable process adapted to $\mathscr{F}_t^W$ and such that

$$E\left[\int_0^1 |\phi_s|^2\, ds\right] < \infty.$$

Now, $m_t = E[m_1 | \mathscr{F}_t^W]$ and also, by T6, $m_t' = \int_0^t \phi_s\, dW_s$ is a $(P, \mathscr{F}_t^W)$-square-integrable martingale. By (*) $m_1 = m_1'$, P-a.s., and therefore $m_t = m_t'$, P-a.s., $t \geq 0$. □

3. Girsanov's Theorem

Section 1 of the present appendix is to be compared with Chapter II, in that it presents a stochastic calculus. Section 2 is concerned with representations of brownian martingales as Ito stochastic integrals, whereas Chapter III deals with the representations of point-process martingales.

In this section, we will continue with this analogy, and present the Girsanov–Cameron–Martin theorem of absolutely continuous change of probability measures, which is akin to the Radon–Nikodym-derivative theorems of Chapter VI for point processes.

Historically, Girsanov's result was anterior to the analogous result concerning point processes (VI, T3). It can be considered as a construction of the abstract model of signal corrupted by white noise, as will be discussed after the statement of Theorem T14.

T14 Theorem (Girsanov [8]). *Let X_t be a $(P, \mathscr{F}_t)$-Wiener process, and ϕ_t be a measurable process adapted to $\mathscr{F}_t$ and such that*

$$\int_0^1 |\phi_s|^2 \, ds < \infty \quad \text{P-a.s.} \tag{3.1}$$

Define, for all $t \in [0, 1]$,

$$L_t = \exp\left\{\int_0^t \phi_s \, dX_s - \frac{1}{2}\int_0^t \phi_s^2 \, ds\right\}. \tag{3.2}$$

Then L_t is a $(P, \mathscr{F}_t)$-local martingale and also a $(P, \mathscr{F}_t)$-supermartingale, and

$$L_t = 1 + \int_0^t L_s \phi_s \, dX_s, \qquad t \in [0, 1]. \tag{3.3}$$

If

$$E[L_1] = 1, \tag{3.4}$$

then L_t is a $(P, \mathscr{F}_t)$-martingale over $[0, 1]$. Moreover, under (3.4), one can define a probability $\tilde{P}$ on $(\Omega, \mathscr{F})$ by

$$\frac{d\tilde{P}}{dP} = L_1,$$

and then

$$X_t - \int_0^t \phi_s \, ds \text{ is a } (\tilde{P}, \mathscr{F}_t)\text{-Wiener process over } [0, 1]. \tag{3.5}$$

Comment (White noise and point process noise). Equation (3.5) can be read

$$X_t = \int_0^t \phi_s \, ds + \tilde{W}_t, \tag{A}$$

$\tilde{W}_t$ is a $(\tilde{P}, \mathscr{F}_t)$-Wiener process.

This model is a popular one for signal corrupted additively by white noise; ϕ_t is then called the *signal process*, and the condition

$$\int_0^t \phi_s^2 \, ds < \infty \quad \tilde{P}\text{-a.s.} \tag{3.6}$$

(which is a consequence of $\int_0^t \phi_s^2 \, ds < \infty$, P-a.s., since $\tilde{P} \ll P$) is a natural assumption of finite energy. $\tilde{W}_t$ is the *integrated white noise*, and X_t is the *observation process*.

Let us now consider the following model in a point-process setting:

$$N_t = \int_0^t \lambda_s \, ds + \tilde{W}_t, \tag{B}$$

$\tilde{W}_t$ is a $(\tilde{P}, \mathscr{F}_t)$-martingale,

where N_t is a counting process adapted to $\mathscr{F}_t$ and λ_t is a nonnegative $\mathscr{F}_t$-progressive process satisfying

$$\int_0^t \lambda_s \, ds < \infty \quad \tilde{P}\text{-a.s.} \tag{3.7}$$

By definition, ϕ_t is the $(\tilde{P}, \mathscr{F}_t)$-intensity of N_t. Analogously to the model (A), we call N_t and $\phi_t = \sqrt{\lambda_t}$ in (B) the observation process and the signal process respectively.

Note that ϕ_t satisfies the energy condition (3.6). One can think of ϕ_t as the amplitude of a light ray falling onto a photodetector (see discussion of VI, §1). If the photodetector is a photoelectronic converter, then the light ray triggers the emission of electrons and N_t counts the number of electrons absorbed by a cathodic plate during the interval $[0, T]$. Under various physical hypotheses which will not be discussed here, the statistics of the emission are such that (A) holds: in other words, the intensity of N_t is proportional to the intensity ϕ_t^2 of the signal.

The analogy between (A) and (B) is remarkable and pervades the whole theory of Ito-type systems (A) and point-process systems (B).

It may be worthwhile to insist on the fact that the process $\tilde{W}_t$ in the model (B) is *not* a "white noise," not having independent increments in general. However, its increments are *uncorrelated*, since for any $(P, \mathscr{F}_t)$-martingale M_t, $E[(M_{t_4} - M_{t_3})(M_{t_2} - M_{t_1})] = 0$ for all $0 \leq t_1 \leq t_2 \leq t_3 \leq t_4$. Although the increments of $\tilde{W}_t$ in the model (B) are not independent, we will say that N_t is a noisy representation of ϕ_t, and that the noise is a "point-process noise." Nevertheless one should be very careful in using such expressions, because there may be a conflict with existing terminology (concerning shot noise for instance).

In order to prove T14 we need a technical lemma which is a particular form of Lévy's characterization theorem T2.

L15 Lemma. *Let X_t be a continuous process defined on $(\Omega, \mathscr{F}, P)$, adapted to a history $\mathscr{F}_t$, and such that for all $u \in R$, the process $Z_t(u)$ defined by*

$$Z_t(u) = \exp\{uX_t - \tfrac{1}{2}u^2 t\} \tag{3.8}$$

is a $(P, \mathscr{F}_t)$-local martingale. Then X_t is a $(P, \mathscr{F}_t)$-Wiener process.

PROOF. By definition, there exists a sequence $(T_n, n \geq 0)$ of $\mathcal{F}_t$-stopping times such that $T_n \uparrow \infty$, P-a.s., and for all $0 \leq s \leq t$ and $A \in \mathcal{F}_s$

$$\int_A Z_{s \wedge T_n}(u) \, dP = \int_A Z_{t \wedge T_n}(u) \, dP. \tag{3.9}$$

Differentiate once with respect to u:

$$\int_A X_{t \wedge T_n} - u(t \wedge T_n))Z_{t \wedge T_n}(u) \, dP = \int_A (X_{s \wedge T_n} - u(s \wedge T_n))Z_{s \wedge T_n}(u) \, dP, \tag{3.10}$$

and a second time:

$$\int_A ((X_{t \wedge T_n} - u(t \wedge T_n))^2 - t \wedge T_n)Z_{t \wedge T_n}(u) \, dP$$
$$= \int_A ((X_{s \wedge T_n} - u(s \wedge T_n))^2 - s \wedge T_n)Z_{s \wedge T_n}(u) \, dP. \tag{3.11}$$

Letting $u = 0$ in (3.10) and (3.11), we see that X_t and $X_t^2 - t$ are $(P, \mathcal{F}_t)$-local martingales; the lemma then follows by Lévy's characterization theorem. □

PROOF OF T14. Ito's differentiation rule (T11) applied to $F(x) = e^x$ yields (3.3).

Now for each $n \geq 1$, let S_n be the $\mathcal{F}_t$-stopping time defined by

$$S_n = \begin{cases} \inf\left\{ t \mid \int_0^t \phi_s^2 \, ds + L_t \geq n \right\} & \text{if } \{\cdots\} \neq \emptyset, \\ +\infty & \text{otherwise.} \end{cases} \tag{3.12}$$

Since L_t and $\int_0^t \phi_s^2 \, ds$ are finite continuous processes, we have $S_n \uparrow \infty$, P-a.s. Also

$$\int_0^t |L_s \phi_s 1(s \leq S_n)|^2 \, ds \leq n^2,$$

and therefore (T6) $L_{t \wedge S_n}$ is a $(P, \mathcal{F}_t)$-square-integrable martingale, and therefore L_t is a $(P, \mathcal{F}_t)$-local martingale. Being nonnegative, it is also a supermartingale by I, E8, and I, E6 ensures that L_t is a $(P, \mathcal{F}_t)$-martingale over $[0, 1]$ when (3.4) is satisfied. For the second part of the proof, it suffices to show, by L15, that for all real numbers u,

$$Z_t^u = \exp\left(u\left(X_t - \int_0^t \phi_s \, ds \right) - \frac{u^2}{2} t \right) \text{ is a } (\tilde{P}, \mathcal{F}_t)\text{-local martingale over } [0, 1]. \tag{3.13}$$

3. Girsanov's Theorem

Since

$$\left.\frac{d\tilde{P}}{dP}\right|_{\mathscr{F}_t} = E\left[\frac{d\tilde{P}}{dP}\Big|\mathscr{F}_t\right] = E[L_1|\mathscr{F}_t] = L_t,$$

it suffices to show that

$$Z_t^u L_t \text{ is a } (P, \mathscr{F}_t)\text{-local martingale over } [0, 1]. \tag{3.14}$$

Now

$$Z_t^u L_t = \exp\left\{\oint_0^t (\phi_s + u) \, dX_s - \frac{1}{2}\int_0^t (\phi_s + u)^2 \, ds\right\}, \tag{3.15}$$

and the $(P, \mathscr{F}_t)$-local-martingale property follows the first part of the theorem. □

The following result is a sufficient condition for the hypothesis (3.4) to hold.

T16 Theorem. *If in T14 ϕ_t is bounded, then (3.4) holds. Moreover L_t thereof is a $(P, \mathscr{F}_t)$-square-integrable martingale over $[0, 1]$.*

PROOF. The argument that follows has already been given, before the proof of T11. Briefly: S_n being defined by (3.12), we see in view of (3.3) that $L_{t \wedge S_n}$ is a square-integrable martingale, and that by the fundamental isometry of T6

$$E[|L_{t \wedge S_n} - 1|^2] = E\left[\int_0^t |L_s \phi_s|^2 1(s \leq S_n) \, ds\right].$$

In particular,

$$E[|L_{t \wedge S_n}|^2] \leq 1 + \int_0^t E[|L_s \phi_s|^2 1(s \leq S_n)] \, ds$$

$$\leq 1 + K \int_0^t E[|L_{s \wedge S_n}|^2] \, ds,$$

where K is an upper bound of ϕ_t. The rest follows by Bellman and Gronwall's lemma.

The next result is an obvious extension of T14.

E4 Exercise. Prove the following:

T17 Theorem. *Let X_t be a process with the representation*

$$X_t = \int_0^t \psi_s \, ds + W_t, \tag{3.16}$$

where ψ_t is some measurable process adapted to a history $\mathcal{F}_t$, and W_t is a $(P, \mathcal{F}_t)$-Wiener process. Let ϕ_t be a measurable process adapted to $\mathcal{F}_t$ and satisfying

$$\int_0^t (\phi_s - \psi_s)^2 \, ds < \infty \quad P\text{-a.s.}, \quad t \geq 0. \tag{3.17}$$

Define L_t by

$$L_t = \exp\left\{\int_0^t (\phi_s - \psi_s) \, dX_s - \frac{1}{2} \int_0^t (\phi_s^2 - \psi_s^2) \, ds\right\}. \tag{3.18}$$

Then L_t is a $(P, \mathcal{F}_t)$-local martingale and a $(P, \mathcal{F}_t)$-supermartingale; also, if

$$E[L_1] = 1, \tag{3.19}$$

L_t is a $(P, \mathcal{F}_t)$-martingale over $[0, 1]$. Moreover, if we then define probability $\tilde{P}$ by

$$\frac{d\tilde{P}}{dP} = L_1, \tag{3.20}$$

the process X_t has the representation

$$X_t = \int_0^t \phi_s \, ds + \tilde{W}_t, \tag{3.21}$$

where $\tilde{W}_t$ is a $(\tilde{P}, \mathcal{F}_t)$-Wiener process.

It is time to conclude this very brief and partial presentation of stochastic systems driven by white noise; but it is worth mentioning that for such systems a filtering theory can be developed which is parallel to that of point-process systems (innovations method: see Fujisaki, Kallianpur, and Kunita [7]; reference method: see Duncan [5] Zakaï [16]). Also a control theory along the lines of Chapter VII is available (see in particular Davis and Varaiya [3]). The historical facts are the following: the filtering and control theory for Wiener-driven stochastic systems was developed, *and then* the analogous theory for point processes followed after it was discovered that the martingale point of view was the natural one for the latter type of dynamical systems.

References

[1] Breiman, L. (1968) *Probability*, Addison-Wesley.
[2] Cameron, H. and Martin, J. (1944) Transformation of Wiener integrals under translations, Ann. Math. **45**, pp. 386–396.
[3] Davis, M. H. A. and Varaiya, P. (1973) Dynamic programming conditions for partially observable stochastic systems, SIAM J. Control **11**, pp. 226–261.

References

[4] Doob, J. L. (1953) *Stochastic Processes*, Wiley.
[5] Duncan, T. (1968) Evaluation of likelihood functions, Inf. and Control **13**, pp. 62–74.
[6] Friedman, A. (1975) *Stochastic Differential Equations and Applications*, Vols. I and II, Academic Press, New York.
[7] Fujisaki, M., Kallianpur, G., and Kunita, H. (1972) Stochastic differential equations for non-linear filtering problem, Osaka J. Math. **9**, pp. 19–40.
[8] Girsanov, I. (1960) On transforming a certain class of stochastic processes by absolutely continuous substitution of probability measure, Theor. Probab. Appl. **5**, pp. 285–301.
[9] Ito, K. (1944) Stochastic integral, Proc. Imp. Acad. Tokyo **20**, pp. 519–524.
[10] Ito, K. and McKean, H. P. (1965) *Diffusion Processes and Their Sample Paths*, Springer.
[11] Kunita, H. and Watanabe, S. (1967) On square integrable martingales, Nagoya Math. J. **30**, pp. 209–245.
[12] Lipčer, R. and Shyriayev, A. (1978) *Statistics of Random Processes*, Vols. I and II, Springer, Berlin.
[13] Neveu, J. (1972) Intégrales stochastiques, Cours de 3ème cycle, Univ. de Paris VI.
[14] Priouret, P. (1974) Equations différentielles stochastiques, *Ecole d'Eté de Probabilité de Saint Flour III*, Lect. Notes in Math., Springer.
[15] Wong, E. (1971) *Stochastic Processes in Information and Dynamical Systems*, McGraw-Hill.
[16] Zakai, M. (1969) On the optimal filtering of diffusion processes, Z. für W. **11**, pp. 230–243.

APPENDIX A4

Stieltjes–Lebesgue Calculus

1. The Stieltjes–Lebesgue Integral
2. The Product and Exponential Formulas
References

1. The Stieltjes–Lebesgue Integral

All functions considered in this Appendix are from $[0, \infty)$ into R.

D1 Definition. Let $f(t)$ be a function such that for all $t \geq 0$

$$V_f(t) = \sup_{\mathscr{D}} \sum_{i=1}^{N} |f(t_i) - f(t_{i-1})| < \infty, \qquad (1.1)$$

where $\mathscr{D}$ ranges over all the subdivisions of $[0, t]$:

$$0 = t_0 < t_1 < \cdots < t_N = t. \qquad (1.2)$$

The function $f(t)$ is said to be *of bounded variation over finite intervals*; $V_f(t)$ is called the *variation* of $f(t)$ over $(0, t]$.

For instance, $f(t) = \int_0^t g(s)\, ds$, where $g(t)$ is a locally integrable function from $[0, \infty)$ into R, defines a function $f(t)$ of bounded variation over finite intervals. Also $f(t) = a(t)$, where $a(t)$ is increasing, is of bounded variation over finite intervals, and in that case $V_f(t) = a(t) - a(0)$.

A classical result of calculus (see [3] for instance) tells us that any right-continuous function $f: [0, \infty) \to R$ of bounded variation over finite intervals

1. The Stieltjes–Lebesgue Integral

can be expressed as the difference of two right-continuous increasing functions:

$$f(t) = f(0) + a(t) - b(t). \tag{1.3}$$

For instance,

$$a(t) = V_f(t), \qquad b(t) = -f(t) + f(0) + V_f(t). \tag{1.4}$$

We call the decomposition (1.4) the *canonical decomposition* of $f(t)$. For instance, if $f(t) = \int_0^t g(s)\,ds$, then $f(t) = \int_0^t |g(s)|\,ds - \int_0^t 2g^-(s)\,ds$, where $g^-(s)$ is the nonpositive part of the locally integrable function $g(t)$.

Unless explicit mention of the contrary is made, we will consider only functions of bounded variation over finite intervals which are right-continuous with left-hand limits and null at the origin; we will call them b.v. functions. Also all increasing functions considered will be standard, that is to say right-continuous and null at the origin.

Let $f(t)$ be a b.v. function with the canonical decomposition $f(t) = f(0) + a(t) - b(t)$. To $a(t)$ and $b(t)$ correspond two σ-finite measures on $((0, \infty), \mathcal{B}((0, \infty)))$, μ_a and μ_b respectively, defined by

$$\mu_a((0, t]) = a(t) = V_f(t), \quad \mu_b((0, t]) = b(t). \tag{1.5}$$

Note that $\mu_b \leq 2\mu_a$. We will use the notation $d\mu_a(t) = \mu_a(dt) = |df(t)|$.

Let $u(t)$ be a measurable function such that

$$\int_{(0, \infty)} |u(s)|\,d\mu_a(s) = \int_{(0, \infty)} |u(s)|\,|df(s)| < \infty \tag{1.6}$$

(and therefore $\int_{(0, \infty)} |u(s)|\,d\mu_b(s) < \infty$). Then $u(t)$ is said to be *Stieltjes–Lebesgue integrable* with respect to $f(t)$, and one defines the *Stieltjes–Lebesgue integral* $\int_{(0, \infty)} u(s)\,df(s)$ by

$$\int_{(0, \infty)} u(s)\,df(s) = \int_{(0, \infty)} u(s)\,d\mu_a(s) - \int_{(0, \infty)} u(s)\,d\mu_b(s). \tag{1.7}$$

Also with the notation just introduced we have $\int_{(0, \infty)} u(s)\,d\mu_a(s) = \int_{(0, \infty)} u(s)\,da(s)$ and $\int_{(0, \infty)} u(s)\,d\mu_b(s) = \int_{(0, \infty)} u(s)\,db(s)$; and therefore

$$\int_{(0, \infty)} u(s)\,df(s) = \int_{(0, \infty)} u(s)\,da(s) - \int_{(0, \infty)} u(s)\,db(s). \tag{1.8}$$

If A is in $\mathcal{B}(0, \infty)$ the notation $\int_A u(s)\,df(s)$ stands for $\int_{(0, \infty)} u(s)1_A(s)\,df(s)$. Also, the notation $\int_a^b u(s)\,df(s)$ stands for $\int_{(a, b]} u(s)\,df(s)$. Observe that $\int_{(0, t]} df(s) = f(t) - f(0)$ and $\int_{(0, t)} df(s) = f(t-) - f(0)$.

When $u(t)$ is locally Stieltjes–Lebesgue integrable with respect to the b.v. function $f(t)$, i.e.

$$\int_0^t |u(s)|\,|df(s)| < \infty, \qquad t \geq 0, \tag{1.9}$$

then the function $t \to \int_0^t u(s)\,df(s)$ is b.v. and its variation over $[0,t]$ is

$$V(t) = \int_0^t |u(s)|\,|df(s)|. \qquad (1.10)$$

Stieltjes–Lebesgue Integral in R^2

We will consider a very particular case, that of integrals of the form

$$\iint_{(0,\infty)\times(0,\infty)} u(x,y)\,df_1(x)\,df_2(y), \qquad (1.11)$$

where $f_1(t)$ and $f_2(t)$ are b.v. functions with the canonical decompositions $a_1(t) - b_1(t)$ and $a_2(t) - b_2(t)$ respectively. The definition of (1.11) is given by the decomposition

$$df_1(x)\,df_2(y) = d\mu_{a_1}(x)\,d\mu_{a_2}(y) + d\mu_{b_1}(x)\,d\mu_{b_2}(y)$$
$$- d\mu_{a_1}(x)\,d\mu_{b_2}(y) - d\mu_{b_1}(x)\,d\mu_{a_2}(y). \qquad (1.12)$$

In view of this decomposition, the meaning and conditions of applications of Fubini's theorem,

$$\int_{(0,\infty)\times(0,\infty)} u(x,y)\,df_1(x)\,df_2(y)$$
$$= \int_{(0,\infty)} \left(\int_{(0,\infty)} u(x,y)\,df_1(x) \right) df_2(y), \qquad (1.13)$$

are clear, and we can proceed to give the formulas of interest in this book.

2. The Product and Exponential Formulas

T2 Theorem (Product Formula). *Let $f(t)$ and $g(t)$ be two functions of bounded variation over finite intervals, right-continuous and with left-hand limits. Then*

$$f(t)g(t) = f(0)g(0) + \int_0^t f(s)\,dg(s) + \int_0^t g(s-)\,df(s). \qquad (2.1)$$

PROOF [2]. By Fubini's theorem,

$$(f(t) - f(0))(g(t) - g(0)) = \int_0^t df(x) \int_0^t dg(y)$$
$$= \iint_{(0,t]\times(0,t]} df(x)\,dg(y). \qquad (2.2)$$

2. The Product and Exponential Formulas

Let D be the set $(0, t] \times (0, t]$, and define D_1 and D_2 by
$$D_1 = \{(x, y) \in D \mid x \le y\},$$
$$D_2 = \{(x, y) \in D \mid x > y\}.$$
Then, since $D = D_1 + D_2$,
$$\iint_{(0,t]\times(0,t]} df(x)\, dg(y) = \iint_{D_1} df(x)\, dg(y) + \iint_{D_2} df(x)\, dg(y). \tag{2.3}$$
Making use of Fubini's theorem,
$$\iint_{D_1} df(x)\, dg(y) = \int_{(0,t]} \left(\int_{(0,y]} df(x) \right) dg(y) = \int_0^t (f(y) - f(0))\, dg(y)$$
$$= \int_0^t f(y)\, dg(y) - f(0)(g(t) - g(0)). \tag{2.4}$$
Similarly
$$\iint_{D_2} df(x)\, dg(y) = \int_{(0,t]} \left(\int_{(0,x)} dg(y) \right) df(x) = \int_0^t (g(x-) - g(0))\, df(x)$$
$$= \int_0^t g(x-)\, df(x) - g(0)(f(t) - f(0)). \tag{2.5}$$
Equation (2.1) then follows from (2.2), (2.3), (2.4), and (2.5). □

C3 Corollary. *Let $a(t)$ be a right-continuous increasing process. The following inequality holds for all $n \ge 1$:*
$$\int_0^t a^{n-1}(s-)\, da(s) \le \frac{a^n(s) - a^n(0)}{n} \le \int_0^t a^{n-1}(s)\, da(s). \tag{2.6}$$

PROOF. With obvious notation, Theorem T2 reads
$$d(fg) = f_- \, dg + g \, df.$$
Applying the above rule to $f = a^{n-1}, g = a$, we obtain successively
$$d(a^n) = a_-^{n-1}\, da + a\, d(a^{n-1}) = a_-^{n-1}\, da + aa_-^{n-2}\, da + a^2\, d(a^{n-2})$$
$$= (a_-^{n-1} + aa_-^{n-2} + \cdots + a^{n-1})\, da,$$
and (2.6) follows, since $a(t-) \le a(t)$ for all $t \ge 0$.

T4 Theorem (Exponential Formula). *Let $a(t)$ be a right-continuous increasing function with $a(0) = 0$, and let $u(t)$ be such that*
$$\int_0^t |u(s)|\, da(s) < \infty, \qquad t \ge 0. \tag{2.7}$$

Then the equation

$$x(t) = x(0) + \int_0^t x(s-)u(s)\,da(s) \quad (2.8)$$

admits a unique locally bounded ($\sup_{s\in[0,t]} |x(s)| < \infty$, $t \geq 0$) solution given by

$$x(t) = x(0) \prod_{0<s\leq t} (1 + u(s)\Delta a(s)) \exp\left(\int_0^t u(s)\,da^c(s)\right), \quad (2.9)$$

where $\Delta a(t) = a(t) - a(t-)$ and $a^c(t)$ is the continuous part of $a(t)$: $a^c(t) = a(t) - \sum_{s\leq t} \Delta a(s)$.

PROOF [1]. Define $f(t)$ and $g(t)$ by

$$f(t) = x(0) \prod_{0<s\leq t}(1 + u(s)\Delta a(s)),$$

$$g(t) = \exp\left(\int_0^t u(s)\,da^c(s)\right), \quad (2.10)$$

where $\prod_{0<s\leq t}(1 + u(s)\Delta a(s)) = 1$ if $\Delta a(s) = 0$ for all $s \in (0, t]$.

Let $x(t) = f(t)g(t)$. By the product formula,

$$x(t) = x(0) + \int_0^t f(s-)\,dg(s) + \int_0^t g(s)\,df(s)$$

$$= x(0) + \int_0^t f(s-)g(s)u(s)\,da^c(s) + \sum_{0<s\leq t} g(s)f(s-)u(s)\Delta a(s)$$

$$= x(0) + \int_0^t x(s-)u(s)\,da(s),$$

and therefore (2.9) is a solution of (2.8).

Let us now prove uniqueness. Let $x(t)$ and $x'(t)$ be two locally bounded solutions of (2.8), and define

$$\tilde{x}(t) = x(t) - x'(t),$$

$$M(t) = \sup_{s\in[0,t]} |\tilde{x}(s)|, \quad (2.11)$$

$$\alpha(t) = \int_0^t |u(s)|\,da(s).$$

Fix t. Then, for all $s \in [0, t]$

$$|\tilde{x}(s)| \leq \int_0^s |\tilde{x}(v-1)||u(v)|\,da(v) \leq M(t)\alpha(s),$$

and therefore, making use of C3,

$$|\tilde{x}(s)| \leq \int_0^s |\tilde{x}(v-)||u(v)|\,da(v) \leq M(t)\int_0^s \alpha(v-)\,d\alpha(v) \leq \frac{M(t)}{2}\alpha^2(s),$$

and reiterating the above process,

$$|\tilde{x}(s)| \leq \frac{M(t)}{2} \int_0^s \alpha^2(v-)\,d\alpha(v) \leq \frac{M(t)}{3!} \alpha^3(s),$$

and so on, the general inequality being

$$|\tilde{x}(s)| \leq \frac{M(t)}{n!} \alpha^n(s), \qquad (2.12)$$

from which it follows that $\tilde{x}(s)$ must be null for all $s \in [0, t]$. The uniqueness then follows, since t is arbitrary. □

References

[1] Lipster, R. and Shiryayev, A. (1978) *Statistics of Random Processes, II: Applications*, Springer, New York.
[2] Meyer, P. A. (1966) *Probability and Potential*, Blaisdell, San Francisco.
[3] Riesz, F. and Sz.-Nagy, B. (1955) *Functional Analysis*, Frederick Ungar, New York.

General Bibliography

[1] Barbour, A. (1976) Networks of queues and the method of stages, Adv. Appl. Probab. **8**, pp. 584–591.

[2] Baskett, F., Chandy, K. M., Muntz, R. R., and Palacios, F. G. (1975) Open, closed and mixed networks of queues with different classes of customers, J.A.C.M. **22**, pp. 248–260.

[3] Beneš, V. (1971) Existence of optimal stochastic control laws, SIAM J. Control **9**, pp. 446–475.

[4] Bensoussan, A. and Lions, J. L. (1975) Nouvelles méthodes en contrôle impulsionnel, J. Appl. Math. Optimization **1**, pp. 289–312.

[5] Beutler, F. J. and Dolivo, F. (1976) Recursive integral equations for the detection of counting processes, J. Appl. Math. Optimization **3**, pp. 65–72.

[6] Beutler, F., Melamed, B., and Zeigler, B. (1977) Equilibrium properties of arbitrarily interconnected queueing networks, *Multivariate Analysis, IV*, P. R. K. Krishnaiah, Ed., North-Holland, pp. 351–370.

[7] Bismut, J. M. (1975) Contrôle des processus de sauts, C.R. Acad. Sci. Paris **A281**, pp. 767–770.

[8] Boel, R. (1974) Control of jump processes, Ph. D. Dissertation, Univ. of Calif., Berkeley.

[9] Boel, R. and Varaiya, P. (1977) Optimal control of jump processes, SIAM J. Control **15**, pp. 92–119.

[10] Boel, R., Varaiya, P., and Wong, E. (1975) Martingales on jump processes; Part I: Representation results; Part II: Applications, SIAM J. Control **13**, pp. 999–1061.

[11] Breiman, L. (1968), *Probability*, Addison-Wesley.

[12] Brémaud, P. (1972) A martingale approach to point processes, Ph.D. Dissertation, Univ. of Calif., Berkeley.

[13] Brémaud, P. (1974) The martingale theory of point processes over the real half line, *Control Theory, Numerical Methods and Computer System Modelling*, Lecture Notes in Economics and Mathematical Systems **107**, Springer, Berlin, pp. 519–542.

[14] Brémaud, P. (1975) An extension of Watanabe's theorem of characterization of Poisson processes, J. Appl. Probab. **12**, pp. 396–399.

[15] Brémaud, P. (1975) Estimation de l'état d'une file d'attente et du temps de panne d'une machine par la méthode des semi-martingales, Adv. Appl. Probab. **7**, 845–863.

[16] Brémaud, P. (1975) La méthode des semi-martingales en filtrage lorsque l'observation est un processus ponctuel marqué, *Séminaire de Probabilités*, Lecture Notes in Mathematics **511**, Springer, Berlin, pp. 1–18.

[17] Brémaud, P. (1975) On the information carried by a stochastic point process, Cahiers du CETHEDEC **45**, pp. 43–70.

[18] Brémaud, P. (1976) Bang-bang controls of point processes. Adv. Appl. Probab. **8**, pp. 385–394.

[19] Brémaud, P. (1976) Prédiction, filtrage et détection pour une observation mixte: méthode de la probabilité de référence, Thèse Doctorat, Univ. Paris VI.

[20] Brémaud, P. (1978) On the output theorem of queueing theory, J. Appl. Probab. **15**, pp. 397–405.

[21] Brémaud, P. (1978) Streams of a $M/M/1$ feedback queue in equilibrium, Z. für W. **45**, pp. 21–33.

[22] Brémaud, P. (1978) Local balance and innovations gains, Tech. Rep., IRIA.

[23] Brémaud, P. (1979) Optimal thinning of a point process, SIAM J. Control and Optimization **17**, pp. 222–230.

[24] Brémaud, P. and Jacod, J. (1977) Processus ponctuels et martingales: résultats récents sur la modélisation et le filtrage, Adv. Appl. Probab. **9**, pp. 362–416.

[25] Brémaud, P. and Yor, M. (1978) Changes of filtrations and of probability measures, Z. für W. **45**, pp. 269–295.

[26] Brillinger, D. (1978) Comparative aspects of the study of ordinary

time series and of point processes, *Developments in Statistics*, Vol. 1., Academic Press, New York, pp. 33–133.

[28] Brown, T. C. (1978) A martingale approach to the Poisson convergence of simple point processes, Ann. Probab. **6**, 4, pp. 615–628.

[29] Burke, P. J. (1956) The output of a queueing system, Op. Res. **4**, pp. 699–704.

[30] Burke, P. J. (1972) Output processes and tandem queues, *Proc. Symp. on Computer Communications and Teletraffic*, Wiley, New York, pp. 419–428.

[31] Burke, P. J. (1976) Proof of a conjecture on the interarrival distribution in a $M/M/1$ queue with feedback.

[32] Cameron, H. and Martin, W. (1944) Transformation of Wiener integrals under translations, Ann. Math. **45**, pp. 386–396.

[33] Chou, C. S. and Meyer, P. A. (1974) Sur la représentation des martingales comme intégrales stochastiques dans les processus ponctuels, *Sem. Proba. VIII*, Lect. Notes in Math. **381**, Springer, Berlin.

[34] Clark, J. M. (1969) Conditions for the one-to-one correspondence between an observation process and its innovations, Tech. Rept., Dept. of Computing and Control, Imperial College, London.

[35] Courrège, Ph. (1963) Intégrale stochastique par rapport à une martingale de carré intégrable, *Séminaire Brelot-Choquet-Deny, Théorie du Potentiel*, 7ème année.

[36] Courrège, Ph. and Priouret, P. (1965) Temps d'arrêt d'une fonction aléatoire, Publ. Inst. Stat. Univ. Paris, pp. 245–274.

[37] Cox, D. (1955) Some statistical methods connected with series of events, J.R. Stat. Soc. **B17**, pp. 129–164.

[38] Cramer, H. (1940) On the theory of stationary random processes, Ann. of Math. **41**, pp. 215–230.

[39] Cramér, H. and Leadbetter, M. (1967) *Stationary and Related Stochastic Processes*, Wiley, New York.

[40] Davis, M. H. A. (1976) The representation of martingales of jump processes, SIAM J. Control **14**, pp. 623–638.

[41] Davis, M. H. A. (1976) Martingales of Wiener and Poisson processes, J. London Math. Soc. (2) **13**, pp. 336–338.

[42] Davis, M. H. A. (1976) The structure of jump processes and related control problems, Math. Prog. Study **6**, pp. 2–14.

[43] Davis, M. H. A. (1979) Capacity and cutoff rates for Poisson type channels, preprint.

[44] Davis, M. H. A., Kailath, T., and Segall, A. (1975) Non-linear filtering with counting observations, IEEE Trans. **IT-21**, pp. 143–150.

[45] Davis, M. H. A. and Varaiya, P. (1973) Dynamic programming conditions for partially observable stochastic systems, SIAM J. Control **11**, pp. 226–261.

[46] Davis, M. H. A. and Wan, C. (1977) The general point process disorder problem, IEEE Trans. **IT-23**, pp. 538–540.

[47] Dellacherie, C. (1972) *Capacités et Processus Stochastiques*, Springer, Berlin.

[48] Dellacherie, C. (1974) Intégrales stochastiques par rapport aux processus de Wiener et de Poisson, *Séminaire Proba. VIII*, Lect. Notes in Math. **381**, pp. 25–26.

[49] Doléans-Dade, C. (1970) Quelques applications de la formule de changement de variables pour les semi-martingales, Z. für W. **16**, pp. 181–194.

[50] Doléans-Dade, C. and Meyer, P. A. (1970) Intégrales stochastiques par rapport aux martingales locales, in: *Séminaire Proba. IV*, Lect. Notes in Math. **124**, Springer, pp. 77–107.

[51] Dolivo, F. (1974) Counting processes and integrated conditional rates: a martingale approach with application to detection, Ph. D. Dissertation, Univ. of Michigan.

[52] Doob, J. L. (1953) *Stochastic Processes*, Wiley, New York.

[53] Duncan, T. (1967) Probability densities for diffusion processes with application to non-linear filtering and detection, Ph. D. Dissertation, Stanford.

[54] Duncan, T. (1968) Evaluation of likelihood functions, Inf. and Control **13**, pp. 62–74.

[55] Dynkin, E. B. (1965) *Markov Processes*, Vols. I and II, Academic Press, New York.

[56] Ferguson, T. (1967) *Mathematical Statistics: A Decision Theoretic Approach*, Academic Press, New York.

[57] Finch, P. D. (1959) The output process of the queuing system $M/G/1$; J. Roy. Stat. Soc., Series B, **21**, pp. 375–380.

[58] Fishman, P. H. and Snyder, D. L. (1976) How to track a swarm of fireflies by observing their flashes, IEEE Trans. **IT-21**, pp. 692–694.

[59] Fishman, P. H. and Snyder, D. L. (1976) The statistical analysis of space–time point processes, IEEE Trans. **IT-22**, pp. 257–274.

[60] Franken, P., König, D., Arndt, U., and Schmidt, V. (1981) *Queues and Point Processes*, Akademie-Verlag, Berlin, and Wiley, New York.

[61] Friedman, A. (1975) *Stochastic Differential Equations and Applications*, Vols. 1 and 2, Academic Press, New York.

[62] Fujisaki, M., Kallianpur, G., and Kunita, H. (1972) Stochastic differential equations for the non-linear filtering problem, Osaka J. Math. **9**, pp. 19–40.

[63] Galtchuk, L. and Rozovskii, B. (1971) The disorder problem for a Poisson process, Theor. Probab. and Appl. **16**, pp. 729–734.

[64] Girsanov, I. (1960) On transforming a certain class of stochastic processes by absolutely continuous substitution of probability measure, Theor. Probab. and Appl. **5**, pp. 285–301.

[65] Grandell, J. (1976) *Doubly Stochastic Point Processes*, Lect. Notes in Math. **529**, Springer, Berlin.

[66] Grigelionis, B. (1975) Stochastic point processes and martingales, Litovsky Mat. Sbornik **15**, pp. 101–114.

[67] Grigelionis, B. (1973) On the non-linear filtering theory and absolute continuity of measures corresponding to stochastic processes, *Proc. Japan SSSR Symposium*, Lect. Notes in Math. **330**, Springer, pp. 80–94.

[68] Grigelionis, B. (1974) On the stochastic integral representation of square integrable martingales, Litovsky Mat. Sbornik **14**, 4, pp. 53–69.

[69] Hadidi, N. (1972) On the output process of a state dependent queue, Skandinavisk Aktuarietidskrift, pp. 182–186.

[70] Hoversten, E. V., Rhodes, I. B., and Snyder, D. L. (1977) A separation theorem for stochastic control problems with point process observations, Automatica **13**, pp. 85–89.

[71] Isham, V. and Westcott, M. (1979) A self correcting point process, Stochastic Processes and their Applications **8**, pp. 335–347.

[72] Ito, K. (1944) Stochastic integral, Proc. Imp. Acad. Tokyo, **20**, pp. 519–524.

[73] Ito, K. and McKean, H. P. (1965) *Diffusion Processes and Their Sample Paths*, Springer, New York.

[74] Jackson, J. R. (1957) Networks of waiting lines, Op. Res. **5**, pp. 518–521.

[75] Jackson, R. R. P. (1956) Random queueing processes with phase type service, J. Roy. Stat. Soc., Series B, **18**, pp. 129–132.

[76] Jacod, J. (1975) Multivariate point processes: predictable projection, Radon–Nikodym derivatives, representation of martingales, Z. für W. **31**, pp. 235–253.

[77] Jacod, J. (1976) Un théorème de représentation pour les martingales discontinues, Z. für W. **34**, pp. 225–244.

[78] Jacod, J. (1979) *Calcul Stochastique et Problèmes de Martingales*, Lect. Notes in Math. **714**, Springer, Heidelberg.

[79] Kabanov, Y. (1974) The capacity of a channel of the Poisson type, Theory of Probab. and Appl. **23**, pp. 143–147.

[80] Kadota. T., Zakai, M., and Ziv. I. (1971) Mutual information of the white Gaussian channel with and without feedback, IEEE Trans. **IT-17**, 4, pp. 368–371.

[81] Kailath, T. (1970) The innovations approach to detection and estimation theory, IEEE Proc. **58**, pp. 680–695.

[82] Kailath, T. and Segall, A. (1975) The modeling of randomly modulated jump processes, IEEE Trans., **IT-21**, 2, pp. 135–142.

[83] Kailath, T. and Segall, A. (1975) Radon–Nikodym derivatives with respect to measures induced by discontinuous independent increments processes, *Ann. Probab.* **3**, pp. 449–464.

[84] Kalman, R. E. (1960) A new approach to linear filtering and prediction problems, Trans. ASME, J. Basic Eng. **82**, pp. 35–45.

[85] Karlin, S. (1968) *Introduction to Stochastic Processes*, Academic Press, New York.

[86] Kelly, F. (1975) Networks of queues with customers of different types, J. Appl. Probab. **12**, pp. 542–554.

[87] Kelly, F. (1976) Networks of queues, Adv. Appl. Probab. **8**, pp. 416–432.

[88] Kelly, F. (1979) *Reversibility and Stochastic Networks*, Wiley, London.

[89] Kennedy, D. (1976) Some martingales related to cumulative sum tests and single server queues, Stochastic Processes and Their Applications **4**, pp. 261–269.

[90] Kerstan, J., Matthes, K., and Mecke, J. (1974) *Unbegrenzt teilbare Punktprocesse*, Akademie Verlag, Berlin; translated as *Infinitely Divisible Point Processes*, Wiley Interscience, New York, (1978).

[92] Kolmogorov, A. (1935) Zur Theorie der Markoffschen Ketten, Mathematische Annalen **112**, pp. 155–160.

[93] Kolmogorov, A. (1940) Curves in Hilbert space invariant with regard to a one-parameter group of motions, Dokl. Akad. Nauk. SSSR **26**, pp. 6–9.

[94] Komatsu, T. (1973) Markov processes associated with certain integro-differential operators, Osaka J. Math. **10**, pp. 271–303.

[95] Kunita, H.; Watanabe, S. (1967) On square integrable martingales, Nagoya Math. J. **30**, pp. 209–245.

[96] Kushner, H. (1964) On the differential equations satisfied by conditional probability densities of Markov processes, SIAM J. Control **2**, pp. 106–119.

[97] Lazaro, J. (1974) Sur les hélices du flot spécial sous une fonction, Z. für W. **30**, pp. 279–302.

[98] Lipčer, R. and Shiryayev, A. (1978) *Statistics of Random Processes, Vol. II, Applications*, Springer, Heidelberg.

[99] Loeve, M. (1955) *Probability Theory*, Van Nostrand Reinhold, Princeton.

[100] Macchi, O. and Picinbono, B. (1972) Estimation and detection of weak optical signals, IEEE Trans. **IT-18,** pp. 562–573.

[101] Martins-Netto, A. (1975) A martingale approach to queuing processes, Ph. D. Dissertation, Univ. of Calif., Berkeley.
[102] Martins-Netto, A. and Wong, E. (1976) A martingale approach to queues, Math. Prog. Study **6**, pp. 97–110.
[103] McFadden, J. (1965) The entropy of a point process, SIAM J. Appl. Math. **13**, pp. 988–994.
[104] Melamed, B. (1979) On Poisson traffic processes in discrete space Markovian systems with applications to queueing theory, Adv. Appl. Probab. **11**, pp. 218–239.
[105] Melamed, B. (1979) Characterization of Poisson traffic streams in Jackson queueing networks, Adv. Appl. Probab. **11**, pp. 422–438.
[106] Meyer, P. A. (1965) *Probabilités et Potentiel*, Hermann, Paris; also (1966) *Probability and Potential*, Blaisdell, San Francisco.
[107] Meyer, P. A. (1969) Démonstration simplifiée d'un théorème de Knight, Séminaire Proba V, Lect. Notes in Math. **191**, Springer, pp. 191–195.
[108] Mirasol, N. (1963) The output of an $M/G/\infty$ queueing system is Poisson, Op. Res. **11**, pp. 282–284.
[109] Muntz, R. (1972) Poisson departure processes and queueing networks, IBM Res. Lab., Rep. RC 4145, Yorktown Heights, N.Y.
[110] Neveu, J. (1966) *Mathematical Foundations of Probability Theory*, Blaisdell, San Francisco.
[111] Neveu, J. (1969) Intégrales stochastiques, Cours de 3ème Cycle, Univ. Paris VI.
[112] Orey, S. (1974) Radon–Nikodym derivatives of probability measures, Report of the Dept. of the Foundations of Math. Sciences, Tokyo Univ. of Education.
[113] Papangelou, F. (1972) Integrability of expected increments of point processes and a related change of scale, Trans. Amer. Math. Soc. **165**, pp. 483–506.
[114] Pardoux, E. (1979) Filtering of a diffusion process with Poisson-type observations, in Lect. Notes in Control and Information Sci., **16**, Springer, Berlin.
[115] Prabhu, N. (1965) *Queues and Inventories*, Wiley, New York.
[116] Priouret, P. (1974) Equations différentielles stochastiques, *Ecole d'Eté de Probabilités de Saint Flour III*, Lect. Notes in Math., Springer.
[117] Reich, E. (1957) Waiting times when queues are in tandem, Ann. Math. Stat. **28**, pp. 527–530.
[118] Riesz, F. and Sz.-Nagy, B. (1955) *Functional Analysis*, Frederik Ungar, New York.
[119] Rishel, R. (1976) Optimal control of a Poisson source, *Proc. Joint Automatic Control Conference at Purdue*, pp. 531–535.

[120] Ross, S. (1969) Optimal dispatching of a Poisson process, J. Appl. Probab. **6**, pp. 692–699.

[121] Royden, H. L. (1968) *Real Analysis*, Macmillan, New York.

[122] Rubin, I. (1972) Regular point processes and their detection, IEEE Trans. **IT-18**, pp. 547–557.

[123] Rubin, I. (1974) Regular jump processes and their information processing, IEEE Trans. **IT-20**, pp. 617–624.

[124] Rudemo, M. (1972) Doubly stochastic Poisson processes and process control, Adv. Appl. Probab. **4**, pp. 318–338.

[125] Rudemo, M. (1973) Point processes generated by transitions of Markov chains, Adv. Appl. Probab. **5**, pp. 262–286.

[126] Rudemo, M. (1973) State estimation for partially observed Markov chains, J. Math. Anal. Appl. **44**, pp. 581–611.

[127] Rudemo, M. (1975) Prediction and smoothing for partially observed Markov chains, J. Math. Anal. Appl. **49**, pp. 1–23.

[128] Segall, A. (1973) A martingale approach to modelling, estimation and detection of jump processes, Ph. D. Dissertation, Stanford Univ.

[129] Segall, A. (1976) Dynamic file assignment in a computer network, IEEE Trans. **AC-21**, pp. 161–173.

[130] Segall, A. (1977) The modeling of adaptive routing in data communications networks, IEEE Trans. **COM-25**, pp. 85–95.

[131] Segall, A. (1977) Optimal control of noisy finite state Markov processes, IEEE Trans. **AC-22**, pp. 179–186.

[132] Serfozo, R. (1972) Conditional Poisson processes, J. Appl. Probab. **9**, pp. 288–302.

[133] Serfozo, R. (1972) Processes with conditional independent increments, J. Appl. Probab. **9**, pp. 303–315.

[134] Shannon, C. (1948) Mathematical theory of communication, Bell Syst. Tech. J. **27**, pp. 379–423 (part I), 623–656 (part II).

[135] Shannon, C. (1948) Communications theory of secrecy systems, Bell Syst. Tech. J. **28**, pp. 656–715.

[136] Skorokhod, J. (1965) *Studies in the Theory of Random Processes*, Addison-Wesley, Reading, Mass.

[137] Snyder, D. L. (1972) Filtering and detection for doubly stochastic Poisson processes, IEEE Trans. **IT-18**, pp. 97–102.

[138] Snyder, D. L. (1972) Smoothing for doubly stochastic Poisson processes, IEEE Trans. **IT-18**, pp. 558–562.

[139] Snyder, D. L. (1973) Information processing for observed jump processes, Inf. and Control **22**, pp. 69–78.

[140] Snyder, D. L. (1975) *Random Point Processes*, Wiley, New York.
[141] Srinivasan, S. K. (1974) *Stochastic Point Processes and Their Applications*, Griffin, London.
[142] Stroock, D. and Varadhan, S. (1969) Diffusion processes with continuous coefficients, Comm. Pure and Appl. Math. **22**, pp. 345–400, 479–530.
[143] Takacš, L. (1962) *Introduction to the Theory of Queues*, Oxford Univ. Press, New York.
[144] Van Schuppen, J. (1973) Estimation theory for continuous time processes, a martingale approach, Ph. D. Dissertation, Univ. of Calif., Berkeley.
[145] Van Schuppen, J. (1977) Filtering, prediction and smoothing for counting process observations, a martingale approach, SIAM J. Appl. Math. **32**, pp. 552–570.
[146] Van Schuppen, J. and Wong, E. (1974) Translation of local martingales under a change of law, Ann. Probab. **2**, pp. 879–888.
[147] Varaiya, P. (1975) The martingale theory of jump processes, IEEE Trans. **AC-20**, pp. 34–42.
[148] Varaiya, P. and Walrand, J. (1978) When is a flow in a Jacksonian network Poisson? Report, ERL, U. of Cal., Berkeley.
[149] Varaiya, P. and Walrand, J. (1978) The outputs of Jacksonian networks are Poissonian, Report, ERL, U. of Cal., Berkeley.
[150] Varaiya, P. and Walrand, J. (1980) Interconnection of Markov chains and quasireversible queueing networks, Stoch. Proc. and Their Appl. **10**, pp. 209–219.
[151] Varaiya, P. and Walrand, J. (1981) Flows in queueing networks: a martingale approach, Math. Operat. Res., to appear.
[152] Watanabe, S. (1964) On discontinuous additive functionals and Lévy measures of a Markov process, Japanese J. Math. **34**, pp. 53–70.
[153] Whittle, P. (1968) Equilibrium distributions for an open migration process, J. Appl. Probab. **5**, pp. 567–571.
[154] Wiener, N. (1949) *Time Series*, MIT Press, Cambridge, Mass.
[155] Wold, H. (1954) *A Study in the Analysis of Stationary Time Series*, Almquist and Wiksell, Stockholm.
[156] Wong, E. (1971) *Stochastic Processes in Information and Dynamical Systems*, McGraw-Hill, New York.
[157] Yashin, A. (1970) Filtering of jump processes, Avtomat. i Telemeh. **5**, pp. 52–58.
[158] Yor, M. (1976) Représentation des martingales de carré intégrable relatives aux processus de Wiener et de Poisson à n paramètres, Z. für W. **35**, pp. 121–129.

[159] Zakai, M. (1969) On the optimal filtering of diffusion processes, Z. für W. **11**, pp. 230–243.

Added in Proof

[160] Itmi, M. (1981) Histoire interne des processus ponctuels, Thèse 3ème cycle, U. de Rouen.

[161] Sismaïl, K. (1981) Représentation intégrale des réseaux markoviens de files d'attente, Thèse 3ème cycle, U. de Paris IX.

[162] Aalen, O. (1977) Weak convergence of stochastic integrals related to counting processes, Z. für W. **38**, pp. 261–277.

[163] Rebolledo, R. (1978) Sur les applications de la théorie des martingales à l'etude statistique d'une famille de processus ponctuels, in *Proc. Coll. de Stat. de Grenoble*, Lect. Notes in Math. **636**, pp. 27–70.

[164] Karp, S. and Clark, J. R. (1970). Photon counting: A problem in classical noise theory, IEEE Trans. **IT 16**, pp. 672–680.

Index

Adapted 3, 288

Balance equations 139
  detailed 295
  local, in Markov chains 152
Bayes' formula 171
Bayesian decision 161
Beneš' existence theorem 226
Bernoulli switch 36
Bounded variation 334

Capacity of a point process channel 180
Change of time 41, 300
Characterization
  martingale, of intensity 28
  of doubly stochastic Poisson
    processes 25
  theorem, Levy's 313
  Watanabe's 25
Conditional expectation 275
Control
  admissible 197
  attraction 219
  complete class of 207
  dynamics 197

  impulsive 196, 211
  intensity 196
  local 198
  Markovian 206
  optimal 196
Cost
  final 197
  optimal cost-to-go 205
  per unit of time 197
  structure 197
  switching 201
Countable dependence theorem 264
Counting
  measure 234
  process 19

Dellacherie–Meyer's integration by
    parts formula 382
Detection
  formula 174, 187
  separation of, and filtering 174, 177
Disruption problem 92
Doleans–Dade's exponential
    martingale 165
Dynamic programming 196

Elapsed service 29
Error
  of the first kind 161
  of the second kind 161
Exponential martingale 165
Existence theorem for point
    processes 168

Filtering 84, 240
  formula, Snyder's 111
  recursive 84
  separation of detection and 174, 187
Filtration 287
Flow conservation equations 135

Girsanov
  changes of intensity "à la
      Girsanov" 165
  theorem 327

Hamilton–Jacobi sufficient
      conditions 203, 206
History 2, 287
  completed 309
  global 83
  internal 2, 289
  observed 83
  of a point process, internal 57
  point process 303
  right-continuous 288

Independence 275
Indistinguishable 9, 288
Infinitesimal characteristics of a Markov
    chain 6, 292
Information
  mutual 181
  pattern 3, 197
    complete 198
    partial 198
Innovations 85
  gain 90
  method of filtering 85
  process 90

Input
  exogenous 38
  process 21
  regulation 211
  service facility, stream 39
Insensitivity 169
Integral representation
  of brownian martingales 324
  of point process martingales 64, 74,
      76, 239
Integration theorem 27, 235, 247
Intensity 22, 27
  change of history for 32
  control 196
  kernel 235
  martingale characterization of 28
  predictable 31
  regenerative form of the 61, 239
  stochastic 27
Ito
  differentiation rule 321
  stochastic integral 312

Jackson
  network 131
  product theorem 137
Jensen's inequality 280

Kelly's network 153
  decomposition 147
Kolmogorov equations 293

Last regeneration of service 29
Levy
  characterization theorem 313
  formula 5, 294
Likelihood ratios 7, 159, 162, 226
Local characteristics 237

Mark space 223
Markov
  chain 5, 282
    conservative 293
    homogeneous 292
    stable 293

process 291
Martingale 4
  exponential 72, 165
  local 7
  point process 56, 64
  Poissonian 70
  square integrable 11
    point process 68
  sub- 4
  super- 4
Melamed's independence theorem 152
Modification 288
  of a martingale, right-continuous 75
Monotone class theorem 256

Network 131
  cascade 145
  exogenously supplied 136
  Jackson's 131
  Kelly's 153
    decomposition 147
  open 136
  processor sharing 140
  tandem 142
Neyman–Pearson's lemma 164

Optimal file allocation 219
Optional sampling 7
Orey–Jacod uniqueness theorem 64
Output
  process 21
  stream, service facility 38
  theorem 123, 138, 154
    Burke's 123, 138
    general 138
    Mirasol's 238

Past 3
  at (stopping) time T 3
  strict 3
Point process 18
  cryptographic 44
  integrable 19
  marked 233
  multivariate 20, 234
  non explosive 18

space–time 233
Poisson process 23
  conditional 21
  doubly stochastic 21
    characterization of 24
    driven by another stochastic
      process 23
  homogeneous 22
  marked 237
  standard 22
Predictable 8
  intensity 31
  process 8
    indexed 235
  σ-field 8
Priority
  allocation, optimal 223
  dynamic assignment of 201
Probability of reference 171
  method of the 170
Process
  arrival 21
  birth and death 294
  counting 19
  departure 21
  indistinguishable 9
  predictable 8
  progressive 288
  queueing 20
  stochastic 1, 287
Projection
  dual predictable 245
  theorem 235, 247

Queueing process 20
  feedback representation of a 36
  feedforward 127
  first order equivalent 30
  parameters of a 29
  standard representation of a 36

Radon–Nikodym derivative 6, 166, 187, 271
Random strategies 211
Recursive
  estimation of a Markov chain 92
  filtering 84

Reich's reversibility argument   124
Reversibility
  argument, Reich's   124
  in $M/M/1$ feedback queues   130
  in $M/M/1$ queues   124
  in Markov chains   294
Routing procedure   36

Service
  facility   35
    input stream   39
    output stream   38
  station   35
Snyder's filtering formula   113
State   85, 199
  equation   85
  estimates for Markov chains   106
  estimates for queues   100
  projection of the   87
  semimartingale representation of the   87
Stopping time   2, 289, 382
Stieltjes–Lebesgue
  calculus   332
  integral   332
    exponential formula for   337

  product formula for   336
Stream
  feedback   38
  of arrivals, exogenous   38
System
  $d-$ 258
  $\pi-$ 258

Traffic equations   135, 153

Uniform integrability   286
Usual conditions   75

Value   197
  process   205

Watanabe
  characterization   25
  theorem, multichannel   26
White noise   111, 382
Wiener
  -driven stochastic system   312
  process   313

Springer Series in Statistics

Measures of Association for Cross Classifications
Leo A. Goodman and **William H. Kruskal**
1979 / 146 pp. / cloth
ISBN 0-387-**90443**-3

Statistical Decision Theory: Foundations, Concepts, and Methods
James Berger
1980 / 425 pp. / 20 illus. / cloth
ISBN 0-387-**90471**-9

Simultaneous Statistical Inference, Second Edition
Rupert G. Miller, Jr.
1981 / 299 pp. / 25 illus. / cloth
ISBN 0-387-**90548**-0